Evolution of Communicative Flexibility

The Vienna Series in Theoretical Biology
Gerd B. Müller, Günter P. Wagner, and Werner Callebaut, editors

The Evolution of Cognition edited by Cecilia Heyes and Ludwig Huber, 2000

Origination of Organismal Form: Beyond the Gene in Developmental and Evolutionary Biology edited by Gerd B. Müller and Stuart A. Newman, 2003

Environment, Development, and Evolution: Toward a Synthesis edited by Brian K. Hall, Roy D. Pearson, and Gerd B. Müller, 2004

Evolution of Communication Systems: A Comparative Approach edited by D. Kimbrough Oller and Ulrike Griebel, 2004

Modularity: Understanding the Development and Evolution of Natural Complex Systems edited by Werner Callebaut and Diego Rasskin-Gutman, 2005

Compositional Evolution: The Impact of Sex, Symbiosis, and Modularity on the Gradualist Framework of Evolution Richard A. Watson, 2006

Biological Emergences: Evolution by Natural Experiment Robert G. B. Reid, 2007

Modeling Biology: Structures, Behaviors, Evolution edited by Manfred D. Laublichler and Gerd B. Müller, 2007

Evolution of Communicative Flexibility: Complexity, Creativity, and Adaptability in Human and Animal Communication edited by D. Kimbrough Oller and Ulrike Griebel, 2008

Evolution of Communicative Flexibility
Complexity, Creativity, and Adaptability in Human and Animal Communication

edited by D. Kimbrough Oller and Ulrike Griebel

The MIT Press
Cambridge, Massachusetts
London, England

For information about special quantity discounts, please email special_sales@mitpress.mit.edu

This book was set in Times New Roman on 3B2 by Asco Typesetters, Hong Kong.
Printed and bound in the United States of America.

Library of Congress Cataloging-in-Publication Data

Evolution of communicative flexibility : complexity, creativity, and adaptability in human and animal communication / edited by D. Kimbrough Oller and Ulrike Griebel.
p. cm. — (Vienna series in theoretical biology)
Includes bibliographical references and index.
ISBN 978-0-262-15121-4 (hardcover : alk. paper)
1. Communication. 2. Animal communication. 3. Language and languages—Origin. 4. Human evolution. I. Oller, D. Kimbrough II. Griebel, Ulrike.
P91.E96 2008
302.2—dc22 2008005750

10 9 8 7 6 5 4 3 2 1

Contents

	Series Foreword	vii
	Preface	ix
I	**INTRODUCTION**	1
1	**Signal and Functional Flexibility in the Emergence of Communication Systems: The Editors' Introduction** D. Kimbrough Oller and Ulrike Griebel	3
II	**CROSS-SPECIES PERSPECTIVES ON FORCES AND PATTERNS OF FLEXIBILITY IN COMMUNICATION**	7
2	**Evolutionary Forces Favoring Communicative Flexibility** Ulrike Griebel and D. Kimbrough Oller	9
3	**Vocal Learning in Mammals with Special Emphasis on Pinnipeds** Ronald J. Schusterman	41
4	**Contextually Flexible Communication in Nonhuman Primates** Charles T. Snowdon	71
5	**Constraints in Primate Vocal Production** Kurt Hammerschmidt and Julia Fischer	93
6	**Contextual Sensitivity and Bird Song: A Basis for Social Life** Martine Hausberger, Laurence Henry, Benoît Testé, and Stéphanie Barbu	121

III THE ROLE OF FLEXIBILITY AND COMMUNICATIVE COMPLEXITY IN THE EVOLUTION OF LANGUAGE 139

7 **Contextual Flexibility in Infant Vocal Development and the Earliest Steps in the Evolution of Language** 141
D. Kimbrough Oller and Ulrike Griebel

8 **Scaffolds for Babbling: Innateness and Learning in the Emergence of Contexually Flexible Vocal Production in Human Infants** 169
Michael J. Owren and Michael H. Goldstein

9 **Cognitive Precursors to Language** 193
Brian MacWhinney

10 **Language and Niche Construction** 215
Kim Sterelny

IV UNDERPINNINGS OF COMMUNICATIVE CONTROL: FOUNDATIONS FOR FLEXIBLE COMMUNICATION 233

11 **How Apes Use Gestures: The Issue of Flexibility** 235
Josep Call

12 **The Role of Play in the Evolution and Ontogeny of Contextually Flexible Communication** 253
Stan Kuczaj and Radhika Makecha

V MODELING OF THE EMERGENCE OF COMPLEXITY AND FLEXIBILITY IN COMMUNICATION 279

13 **Detection and Estimation of Complexity and Contextual Flexibility in Nonhuman Animal Communication** 281
Brenda McCowan, Laurance Doyle, Allison B. Kaufman, Sean Hanser, and Curt Burgess

14 **The Evolution of Flexibility in Bird Song** 305
Robert F. Lachlan

15 **Development and Evolution of Speech Sound Categories: Principles and Models** 327
Gert Westermann

Contributors 347
Index 349

Series Foreword

Biology is becoming the leading science in this century. As in all other sciences, progress in biology depends on interactions between empirical research, theory building, and modeling. However, whereas the techniques and methods of descriptive and experimental biology have evolved dramatically in recent years, generating a flood of highly detailed empirical data, the integration of these results into useful theoretical frameworks has lagged behind. Driven largely by pragmatic and technical considerations, research in biology continues to be less guided by theory than seems indicated. By promoting the formulation and discussion of new theoretical concepts in the biosciences, this series intends to help fill the gaps in our understanding of some of the major open questions of biology, such as the origin and organization of organismal form, the relationship between development and evolution, and the biological bases of cognition and mind.

Theoretical biology has important roots in the experimental biology movement of early-twentieth-century Vienna. Paul Weiss and Ludwig von Bertalanffy were among the first to use the term "theoretical biology" in a modern scientific context. In their understanding the subject was not limited to mathematical formalization, as is often the case today, but extended to the conceptual problems and foundations of biology. It is this commitment to a comprehensive, cross-disciplinary integration of theoretical concepts that the present series intends to emphasize. Today theoretical biology has genetic, developmental, and evolutionary components, the central connective themes in modern biology, but also includes relevant aspects of computational biology, semiotics, and cognition research and extends to the naturalistic philosophy of sciences.

The "Vienna Series" grew out of theory-oriented workshops organized by the Konrad Lorenz Institute for Evolution and Cognition Research (KLI), an international center for advanced study closely associated with the University of Vienna (http://kli.ac.at). The KLI fosters research projects, workshops, archives, book projects, and the journal *Biological Theory*, all devoted to aspects of theoretical biology, with an emphasis on integrating the developmental, evolutionary, and cognitive sciences. The series editors welcome suggestions for book projects in these fields.

Gerd B. Müller, University of Vienna, KLI
Günter P. Wagner, Yale University, KLI
Werner Callebaut, Hasselt University, KLI

Preface

Understanding the evolution of human communication presents one of the fundamental challenges of the future of biological science. As far as can be told, humans communicate in vastly more complex ways than any other animal, both in content and in form. The evolutionary roots of this human capability have long been shrouded in mystery, but recent comparative research has begun to shed light on the possible origins of human communication. The key first step in that evolutionary history seems to have been the establishment of basic communicative flexibility—specifically, the ability to vocalize freely along with the capability to coordinate vocalization with communicative intent.

We have been fortunate to have been provided with a magnificent setting and support system for the exploration of the origins of communicative flexibility. The Vienna Series in Theoretical Biology, published by MIT Press and developed through the Konrad Lorenz Institute for Evolution and Cognition Research in Altenberg, Austria is dedicated to fostering growth in understanding of evolution of cognitive capabilities, one of which is communication. The series is based on workshops conducted with small numbers of selected international figures in their respective fields. The workshop on which the present volume is based follows a prior workshop (volume 4 in the series) with the same organizers/editors (Oller and Griebel, 2004, *Evolution of Communication Systems*, Cambridge, MA: MIT Press) that laid groundwork for the present effort. The workshop on which the present volume is based included an extremely lively and informative exchange over two and a half days at the Konrad Lorenz Institute, which is housed in the Lorenz mansion. All of the primary contributors were participants in the workshop and also served as reviewers for manuscripts upon which the chapters are based.

We wish to thank the Konrad Lorenz Institute, both administrators and staff, for the support of the workshop and the volume. The experience of the workshop was a genuine pleasure as was the work that preceded and followed it. We hope this product will provide further inspiration for research and theoretical development to help deepen our understanding of the origins of language.

Reference

Oller DK, Griebel U, eds. 2004. *Evolution of Communication Systems*. Cambridge, MA: MIT Press.

I INTRODUCTION

1 Signal and Functional Flexibility in the Emergence of Communication Systems: The Editors' Introduction

D. Kimbrough Oller and Ulrike Griebel

This volume is founded on the supposition that evolution of complex communication systems, where human language offers the most extraordinary example, requires a foundation of flexibility in both the form and the usage of signals. Without the ability to voluntarily manipulate potential signals, and without the ability to utilize those signals adaptively in an ever-changing physical and social environment, potential communicators are reduced to producing actions limited in communicative power and very unlike language.

Of course some actions may *result* in communication even though the purpose of the actions is not communicative. For example, if an animal leaves a track in the snow, it may thus unintentionally leave the basis for another animal to pursue or avoid it. The track is an indicator (or "cue"; see chapter 2) of a path and so can inform a perceiver, even though walking, and thus leaving tracks in the snow, is not an action that was evolved to communicate. Similarly, the sound of a prey animal's breathing or the sound of its chewing may be audible and may betray its position to a predator, but in neither breathing nor chewing do we see examples of actions that were evolved to be communications.

Further, cues such as tracks in the snow and the sound of chewing yield communicated information because the *perceiver* makes it so—any flexibility in response to such a cue is attributable to flexibility of the perceiver. The producer of a cue, on the other hand, is not an intentional signaler, and consequently the structure of an action that serves as a cue is incidental to other functions (locomotion, ingestion, etc.) and possesses no flexibility for the purpose of communication.

Even in the case of actions that were evolved specifically to communicate, there exist abundant examples lacking the fundamental flexibility of language. For example, the newborn infant cry is communicative in that it inspires caregivers to offer sustenance or comfort. When a house cat hisses in the direction of a conspecific, arches its back, and shows piloerection, the display communicates threat. Both the human infant distress cry and the feline threat were evolved to communicate, but they are inflexible in important ways. Both possess a stereotyped form that is universal within the species, and the function that each can serve is predetermined within the members of the species by natural history. Thus, a newborn cry is a distress signal that cannot be restructured to act, for example, as a greeting, an exultation, or as a name for a household pet. Perhaps even more fundamentally, the cry has a stereotyped form that is itself not fundamentally modifiable by the infant—to serve its purpose, the cry must be produced largely as nature decrees. The same goes for the feline threat display. It *is* what it *is* and cannot be restructured, for example, to offer praise, to announce a resignation, or to say, "The earth is blue when viewed from the moon." And again, the human infant and the house cat do not need to learn how to perform cries and threat displays.

The great bulk of the evolved, specialized signaling that occurs in nonhumans, and some of the signaling that occurs in humans, is fixed in the way that cries and feline threat displays are fixed. Such displays have been termed "fixed signals" by the classical ethologists (Lorenz, 1951; Tinbergen, 1951), and the evidence suggests that at least among primates, communicative evolution has primarily consisted of processes yielding within each species a small class of fixed signals, stereotyped in form and each designed to serve a particular social function: threat, greeting, invitation, exultation, warning, and so on.

Fixed signals are believed to be evolved by a process termed "ritualization." Fixed signals can obviously be important to survival and reproduction, because most animal species, as far as we know, possess fixed signals of this sort, each signal stereotyped in form and each tied to a particular function.

This volume asks, How did some species, and particularly how did ancient hominins, break free of the stereotypy and fixedness that are so widespread in communicative systems? How did it come to be that birds in three broadly different taxa (oscines, parrots, and hummingbirds) can learn to produce a wide variety of vocalizations (see Lachlan, this volume), and how did it come to be that only humans are able to learn such complexities of vocal communication as to make it possible for them to deliver lectures about communication itself or to edit volumes on the topic? Fundamentally, the authors of the chapters in this volume ask, What *is* communicative flexibility and what evolutionary conditions can produce it? Of course, human language is a primary target of explanation, but the questions addressed by the authors are basic and draw upon interest in communicative flexibility in species ranging from humans to fireflies of the genus *Photuris* (see Griebel and Oller, this volume). The theoretical perspectives offered are intended to illuminate these questions through reviews and interpretations of a rapidly growing body of research in diverse disciplines.

It is notable that the perspectives offered in the chapters that follow this introduction contrast in important ways with a variety of previous efforts dedicated to the study of language evolution, specifically by attempting to take the questions of language origins back farther in evolutionary time than has occurred in much previous work; many of the chapters attempt to address the very earliest communicative break of the hominin line from the primate background, and others address the evolutionary origins of flexibility in birds and cetaceans. Previous writings, in contrast, have often focused on how relatively complex features of language were evolved: (1) syntax (Bickerton, 1990; Pinker and Bloom, 1990), (2) symbolism (Deacon, 1997), (3) articulations of the wide range of vowel sounds in natural languages (Lieberman, 1984), or (4) the articulatory complexities evidenced in canonical babbling (MacNeilage, 1998). Such features of language seem likely to have been relatively recent innovations in hominin communicative capability. The present work does not ignore such topics but focuses primarily on answering questions that can be deemed even more fundamental, because the emergence of basic flexibility in communication through primitive vocalizations and gestures must have preceded the evolution of any of the more complex types of capabilities (1)–(4).

The emphasis on the earliest steps of evolution toward language has led in recent research to emphasis on parallels between human communicative flexibility and flexibility found in a wide variety of additional species (see articles in our previous volume in the Vienna Series in Theoretical Biology, Oller and Griebel, 2004). Further, the recent approach has led to special interest in the earliest phases of human development and in the developmental history of humans (Locke and Bogin, 2006). Importantly, the roots of early changes in hominin communication are also being sought in development and evolution of gesture (Tomasello, 1996; Tomasello and Call, 2007) and in the origins of play (Špinka et al., 2001). Finally, there is an exciting growth occurring in mathematical modeling, where scholars are seeking to offer theoretical perspectives on the evolution of language and to simulate scenarios of communicative evolution, an approach that incorporates many new tools that are themselves rapidly evolving (Elman et al., 1996; Niyogi, 2006).

The present volume is organized around themes that are at the cutting edge of these efforts to illuminate the origins of communicative flexibility. The approach entails interest in communication and cognition in hominins as well as in other species.[1]

Note

1. As this volume was going to press, a report appeared in *Science* (Aronov et al., 2008) regarding the neural basis for flexible vocal learning in the zebra finch. The work indicates that the subsong, or "babbling," of the juvenile bird is controlled by a special "forebrain nucleus involved in learning but not in adult singing" (p. 630). The result contributes to the growing excitement in research seeking to reveal the concrete mechanisms of flexible vocal learning in a variety of species.

Acknowledgments

In addition to receiving support for the workshop from the Konrad Lorenz Institute for Evolution and Cognition Research, this work has been supported by a grant from the National Institutes of Deafness and other Communication Disorders (R01DC006099-01 to D. K. Oller, Principal Investigator, and Eugene Buder, Co-principal Investigator) and by the Plough Foundation.

References

Aronov D, Andalman AS, Fee MS. 2008. A specialized forebrain circuit for vocal babbling in the juvenile songbird. *Science, 320*, 630–634.

Bickerton D. 1990. *Language and Species*. Chicago: University of Chicago Press.

Deacon TW. 1997. *The Symbolic Species*. New York: Norton.

Elman JL, Bates EA, Johnson MH, Karmiloff-Smith A, Parisi D, Plunkett K. 1996. *Rethinking Innateness: A Connectionist Perspective on Development*. Cambridge, MA: MIT Press.

Lieberman P. 1984. *The Biology and Evolution of Language*. Cambridge, MA: Harvard University Press.

Locke J, Bogin B. 2006. Language and life history: A new perspective on the evolution and development of linguistic communication. *Behavioral and Brain Sciences* 29: 259–325.

Lorenz K. 1951. Ausdrucksbewegungen höherer Tiere. *Naturwissenschaften* 38: 113–6.

MacNeilage PF. 1998. The frame/content theory of evolution of speech production. *Behavioral and Brain Sciences* 21: 499–546.

Niyogi P. 2006. *The Computational Nature of Language Learning and Evolution*. Cambridge, MA: MIT Press.

Oller DK, Griebel U, eds. 2004. *Evolution of Communication Systems*. Cambridge, MA: MIT Press.

Pinker S, Bloom P. 1990. Natural language and natural selection. *Behavioral and Brain Sciences* 13: 707–84.

Špinka M, Newberry RC, Bekoff M. 2001. Mammalian play: Training for the unexpected. *The Quarterly Review of Biology* 76: 141–68.

Tinbergen N. 1951. *The Study of Instinct*. Oxford: Oxford University Press

Tomasello M. 1996. *The gestural communication of chimpanzees and human children*, Waseda University International Conference Center, Tokyo.

Tomasello M, Call J. 2007. *The Gestural Communication of Monkeys and Apes*. Mahwah, NJ: Lawrence Erlbaum Associates.

II CROSS-SPECIES PERSPECTIVES ON FORCES AND PATTERNS OF FLEXIBILITY IN COMMUNICATION

Part II illustrates ways that comparative research can contribute to the search for origins of communicative flexibility. Clearly, some sorts of communicative flexibility have arisen in multiple taxa in evolution, and evaluation of examples is instructive in the search for perspective on the human case. There is special importance attached to comparative research in other mammals and especially in other primates.

Part II begins with a chapter by the editors, Griebel and Oller, providing definitional background on terms related to topics raised in all the chapters. In particular, the chapter defines "signal flexibility" and "functional flexibility" as two kinds of contextual flexibility in communication. The chapter proceeds to provide examples of both types from widely ranging species and to offer perspectives based on the ecology of each species, shedding light on the possible evolutionary forces that may have led to flexibility. The chapter concludes with speculations about the evolutionary conditions that may have led to vocal flexibility in the hominin line.

Schusterman, in the second chapter, greatly enhances the cross-species perspective given by Griebel and Oller, as he provides remarkable evidence of vocal flexibility in pinnipeds (seals, sea lions, and walruses), species whose learning capabilities are surprising indeed. The research reviewed in the chapter, much of it by Schusterman and his colleagues, makes it clear that pinnipeds are far more capable of vocal production learning than any of the nonhuman primates and that much remains to be learned about flexible vocal capabilities in aquatic species. The work is especially intriguing because it provides support for the idea that lifestyles requiring voluntary breath control (and aquatic species are in particular subject to such a requirement) may provide particularly important foundations for vocal flexibility.

Snowdon's chapter reviews in much greater detail the critically important literature on primate vocalization. The work, much of it based on his own research in New World monkeys, provides a broad perspective emphasizing that while primates are limited in flexibility of vocal production (the ability to produce sounds themselves), they are more flexible in usage (contexts of application) of vocalizations and far more flexible in comprehension. Nonhuman primates show particular limitation in the ability to learn to produce vocalizations that depart from the species-specific repertoire but are quite capable of applying each vocalization type to a wide variety of circumstances that fit the broad functional role of the particular vocalization, and they are further capable of learning to understand vocalizations in flexible ways, even vocalizations from other species. Further, there are important ways in which even the production of vocalizations shows some limited degree of flexibility in many nonhuman primates, as Snowdon illustrates.

Hammerschmidt and Fischer provide additional perspectives on vocalization capabilities in primates, supplementing the viewpoints of Snowdon as well as those of Griebel and Oller. The Hammerschmidt and Fischer chapter includes developmental data from both humans and nonhumans, as well as a state-of-the-art review of the physiology of vocal control in primates, including humans. The studies covered by Hammerschmidt and Fischer include recent data from their own laboratories on what happens to vocal systems when an animal (or a human) grows up without hearing or is reared in acoustically impoverished environments. These studies help to flesh out details regarding the differences in vocal flexibility in humans and nonhumans.

In the final chapter of part II, Hausberger, Henry, Testé, and Barbu seek evidence from songbirds of evolutionary convergence with certain features of the vocal flexibility found in humans. They evaluate social environments and vocal patterns in songbirds, for many years a primary research focus of Hausberger and her colleagues. They introduce evidence that a social contract similar to that postulated for human communication by the social psychologist Ghiglione (1986) may also play a role in bird song. Phenomena such as turn taking and automatization may not, according to their interpretation, require high-level cognition but may represent types of communicative flexibility that can emerge in relatively simple communicative systems. The chapter again greatly enhances the cross-species perspectives offered in other chapters in part II and lays groundwork for subsequent chapters, especially by Sterelny, who addresses social circumstances and language, as well by Lachlan, who presents additional views on bird song flexibility.

Reference

Ghiglione R. 1986. *L'Homme Communiquant*. Paris: Colin.

2 Evolutionary Forces Favoring Communicative Flexibility

Ulrike Griebel and D. Kimbrough Oller

Introduction

There is a huge gap between the complexity of communication in humans and nonhumans. In the nonhuman case, signals and functions are few and show limited flexibility in both function and signal characteristics, whereas in humans, signals are of indefinitely large number, signals have a multitude of functions, and mapping from signal to function and vice versa shows massive flexibility. What were the evolutionary conditions that led to such a discrepancy between humans and nonhumans?

To understand these circumstances in general and to shed light on the specific circumstances that were relevant to the explosion of communication in the human case, we turn to observation of other species that show some degree of communicative flexibility. The evidence suggests that specific circumstances and a limited number of selectional pressures favor the evolution of communicative flexibility in animals. In this chapter we assess environmental/social conditions and communication types that appear to favor selection for variability or complexity in communication systems and that may lead to signal and functional flexibility. The goal is to catalogue circumstances or evolutionary scenarios that have led to communicative flexibility in the past and thus might help explain what happened in the remarkable human case.

What Are Signals and How Do They Evolve?

The exposition we plan requires clarity about several concepts and related terms. We begin with an important recent definition of the notion "signal," as an action or feature coevolved between sender and receiver where both benefit on average from the exchange of signals (Maynard Smith and Harper, 2003). The definition is founded on the idea that without benefit to both participants in interaction, signals would not be naturally selected and could not stabilize. For example, a systematically identifiable alarm call produced by a bird fulfills the requirements of a signal. In Hockett's (Hockett, 1960; Hockett and Altmann, 1968) terminology, a signal (as defined by Maynard Smith and Harper) is thus said to be "specialized" for communication.

In contrast to a signal, a "cue," in the terminology of Maynard Smith and Harper, is a state or action that is perceived by other organisms and used as a guide for action but has not been evolved (has not been specialized) for that purpose. For example, body size can be used as a cue for strength, or substantial concentrations of CO_2 can be used (e.g., by mosquitoes) as a cue to the presence of large animals in the nearby environment. A cue can be interpreted, but the result is not viewed in this terminology as communication—we shall use the term "cue interpretation" in this case.

Signals are believed to evolve from animal traits or actions that begin as cues for certain states of the animal. In the evolution of an alarm call, for example, an animal may at one stage of evolution produce an involuntary vocal cue associated with a bodily reaction when it perceives a predator from afar. Such a vocal cue could result, for example, from a sudden change in breathing pattern in anticipation of flight. If this sound benefits the producer's kin because they hear the sound, associate it with danger, and develop a retreat response, the sound can be systematically selected to function as an alarm call. In this way, it can become specialized for communication through ritualization (Huxley, 1914; Lorenz, 1951; Tinbergen, 1951), a process through which the signal evolves to be easily recognizable. To be easily recognizable, it must be high in contrast, conspicuous, unambiguous, and stereotyped in form. Such ritualized signals usually also develop a typical intensity, because gradations of a signal can only be understood in the context of a typical intensity.

It has been proposed that some signals may not have started out as cues. Owren and Rendall propose a scenario for the origin of some primate vocal signals (Owren and Rendall, 2001) suggesting manipulative use of vocalizations learned associatively. Such learned vocalizations could capitalize on attention-getting features of certain sounds in situations like predator danger or aggression, and their usage could be enhanced by further associative learning on the part of listeners.

Animal signals such as alarm calls evolve to serve specific functions. For each such signal, the function is fixed, that is, there is a coupling of each signal with a particular function. The function associated with a particular signal cannot be changed within an individual animal. For example, a call evolved as an alarm cannot be reassigned as an aggressive or a courtship signal. Signal and function are thus coupled in the majority of animal signals in a one-to-one mapping (Oller, 2004; Oller and Griebel, 2005) that justifies the classical ethologists' term "fixed signal" (Lorenz, 1951; Tinbergen, 1951).

Fixed Signals in Animal Communication Systems

The typical repertoire size of fixed signals in social species, such as the primates, appears to include about five to seven function types, not counting gradations of intensity associated with signals for each function type. Much larger numbers can be derived if one categorizes individual gradations as types (Sutton, 1979). The literature generally suggests that signal categories can be readily identified and also that the occurrence of each signal can be largely predicted from social context. The functions tend to be mostly drawn from the following social expression types: (1) aggression or threat (with signals often used in intra- as well as interspecific communication), (2) courtship or sexual solicitation, (3) appeasement or submission, (4) distress or complaint, (5) greeting, contact calling, or affiliation (often in the context of parental care), (6) feeding announcement, (7) warning or alarm, (8) comfort or pleasure, and (9) exultation or positive excitement. Of course, not all species have all categories; for example, apparently only gregarious species have alarm and feeding announcement calls. Often one or more of the above categories are served by a single signal type (e.g., exultation and feeding announcement might be served by a single positive arousal signal).

It is important to note that fixed signals are often (perhaps usually) not *entirely* stereotyped in signal or function (see Snowdon, as well as Hammerschmidt and Fischer, this volume). As mentioned, signals tend to show gradations of intensity, corresponding to gradations of arousal or urgency in the service of the functions that correspond to each signal. In addition, contextual flexibility occurs to a limited extent in the usage of signals, such that, for example, some primates can inhibit the production of an alarm signal in the absence of kin in the vicinity. Further, maturation and learning appear to play important roles such that primate signals become clearer and their functional usage more appropriate with development. Two factors, however, remain fixed: (1) the destiny of each signal type to form a one-to-one mapping with a functional type, broadly defined by a class of contexts appropriate to the function, and (2) the *relatively* stereotyped form of the signal that serves each function type, making it easily identifiable within the species even with gradations of intensity.

Refinements of the Terms Based on Austinian Distinctions

The distinction between "illocutionary force" and "perlocutionary force" (Austin, 1962) can be adapted usefully here. Illocutionary forces are the functions (aggression, appeasement, greeting, etc.) served by signals (or potential signals) *in the act of signaling*. Receivers interpret illocutionary forces, but since animal signals tend to be stereotyped, there is relatively little danger of misinterpretation at the illocutionary level. Perlocutionary forces are *effects that take place as a result of illocutionary interpretation* by the receiver, and these effects are inherently flexible in intelligent creatures such as primates. Thus, in response to an aggressive signal, a receiver may respond in a variety of ways: with counteraggressive signaling, with attack, with retreat, with appeasement, or with indifference, to name a few possibilities that may be dependent on how dangerous the original signaler is believed to be, how seriously the signal is taken, and whether the signal was directed at the receiver or another animal.

Intelligent animals appear to be selected to produce a small number of signal types that transmit illocutionary forces, but they are also selected to interpret these illocutionary forces with considerable flexibility and range.

When we use the terms "function" or "functional flexibility," we intend the terms to apply to the immediate social functions of communication, not subsequent results of the communications. Consequently, the term "function" refers to illocutionary forces specifically, rather than to perlocutionary forces. Illocutionary forces in animal communication and in very early human development are limited to a small class of social interaction types, as indicated above—aggression, appeasement, and so on. We propose that signals can be naturally selected to transmit illocutionary forces because each illocutionary force is a unified type of social interaction. Perlocutionary effects, on the other hand, include everything that can happen as a result of a communication, and as a consequence, perlocutionary effects provide no unified targets for natural selection in communication among intelligent animals. An alarm call does not directly cause an animal to run up a tree but causes an animal to be alarmed (or at least to *recognize* the alarm call for what it is)—several different reactions can occur (freezing, running up a tree, attacking, merely looking at the caller, etc.). Comprehension of calls usually exceeds production capabilities (e.g., in nonhuman primates see Snowdon, this volume), a fact that corresponds to the observation that individual fixed signals within primate species show perlocutionary flexibility but illocutionary fixedness. In our terminology, then, since fixed signals are illocutionarily fixed, they are not functionally flexible.

In the case of deceptive uses of signals, however, an additional function, not of the illocutionary type, is pursued by the signaler in order to exploit an existing illocutionary function. For example, a warning call can be used to serve an additional function of competition avoidance (e.g., food or mate competition), or a mating signal can be used to function in addition as a lure for prey. Consequently we use the terms "function" and "functional flexibility" to encompass both illocutionary functions and the additional functions that deception may invoke. Deceptive functions are always "parasitic" on an existing communication system including both signals and illocutionary functions (see discussion of deception below).

Illocutionary forces are also distinct from "meanings" following Austin's terminology. Illocutionary forces are interactive events within a social dyad, a sender and a receiver. Illocutionary forces do not require that the signal make reference to any entity external to the dyad. Meanings, on the other hand, involve a triad in our Austinian scheme, the social dyad plus some entity of reference that can be external to the dyad—"me and you and that thing over there (or me and you and the idea we are talking about)." If one points to a leopard and says "leopard," one not only draws attention to the leopard but also invokes a name for the class of all entities that are leopards. Naming the leopard is much more specific and includes more information than simply pointing to the leopard and making a sound such as a grunt or a scream, either of which, of course, can also draw attention to the leopard. A meaning *invokes* a referential entity in this terminology, but an illocutionary force *constitutes* a social action such as threatening, warning, appeasing, and so forth. When one points to a leopard and names it, one produces *both* an illocutionary force (e.g., an action of, for example, naming or warning) and a meaning (by invoking the class of leopards).

Importantly, fixed signals cannot have meaning in this Austinian sense because they possess no capacity for free reference, that is, reference independent of the illocutionary force of the signal in question. A predator-specific alarm call (see Snowdon, this volume, for a review of such calls) is always an alarm and cannot make reference to a predator independent of the act of alarm calling. The capacity for reference to some external source of alarm such as a predator is bound to the circumstance of alarm and depends on the receiver's active interpretation of the situation (aided by looking in the direction the alarm caller is looking, by actively seeking to locate a predator, etc.). These facts about alarm calling indicate that their referential effects are circumstantially bound and may be largely or entirely perlocutionary aftereffects of the combination of the receiver's interpretation of illocutionary force and the context of signal production. In free reference, on the other hand, as occurs in language, we can use a term referring to a predator with or without creating alarm and importantly, we can use the term without the *intention* of creating alarm. The term for the predator is a word and has a meaning in Austin's sense but can be used on differing occasions for differing illocutionary effects. In general, it is useful to clarify that animal signals, including primate vocal signals as described in the literature, transmit illocutionary forces, not meanings. Illocutionary forces in these cases constitute social functions coupled with individual signals.

Functional Decoupling in the Human Infant

Humans on the other hand show quite free decoupling of signals and functions, and this pattern of vocal communication begins very early in life. Research from our laboratories (Kwon et al., 2006; Oller et al., 2003; Oller et al., 2007) has shown that even in prespeech sounds of the first six months of life, decoupling of signal and function occurs. For example, a sound identified as pertaining to a "squeal" category can be used to express positive affect (exultation) on one occasion and negative affect (complaint) on another, and on another occasion it can even be used in a neutral state, as in vocal play when the baby vocalizes alone. The same reversal of emotional valence and consequent illocutionary force occurs with other sound categories in the human infant, although some sounds diversify in function more than others. Squealing, growling, and vowel-like sounds show great diversity of function. Crying, on the other hand, which appears to begin as a fixed signal of physical distress at birth, maintains a negative connotation and cannot acquire a positive one in the infant, even though the range of negative expressions appears to grow as the infant matures; by four months, infants cry instrumentally to get attention as well as to express pain or signal hunger (Green et al., 1987). Only the relatively rare cases of "crying for joy" in adults may present examples of a change to a positive connotation for crying.

Later, when infants start to speak, they not only learn to name a vast number of objects and states but also learn to use these names to serve a wide variety of social functions (or illocutionary forces). For example, humans can use the word "pig" (which always has the effect of invoking the "meaning," which is to say it invokes a reference to the class of animals that are pigs) with a variety of illocutionary forces: as a simple statement of fact ("this is a pig"), as a warning ("watch out, a pig!"), as a question ("is this a pig?"), as an insult ("you pig!"), as an example (consider the word's usage throughout this paragraph), or to serve a variety of other illocutionary functions. In language learning, the decoupling of functions and signals multiplies in early infancy like an explosion until, by the end of the second year of life, essentially any word and/or sentence can be used to serve multiple functions, and every function can be served by multiple words and sentences. The signaling system of the human child is essentially open in this sense from a very early stage, surpassing the functional flexibility seen in nonhuman primate vocal systems at any point in life (Oller, 2000).

This functional flexibility of communication in humans extends to other domains beyond the vocal one. Sign languages have the same sort of functional flexibility as vocal language does (Lyons, 1991; Stokoe, 1960). Individual signs serve multiple illocutionary forces by very early in life in sign learners. Further, human "body language" and other natural gestures are not entirely stereotyped to serve fixed functions. Most body language signals can be employed for many functions, just as in the case of vocalization. Information in natural human gesture and body language seems mostly encoded in timing and rhythm of movements, where flexibility of intent is a key feature in the expressive system (Grammer, 1995; Grammer et al., 1999; Grammer et al., 2000; Grammer et al., 1998).

Apes have been taught to use some human sign language, and in that context they have shown limited functional flexibility of learned signs (Fouts, 1987; Gardner and Gardner, 1969; Savage-Rumbaugh, 1988). Further, various animals have learned to *understand* a considerable number of human words and to use that understanding with some functional flexibility (Pepperberg, 2004; Kaminski et al., 2004). However, this sort of flexibility is clearly limited (Terrace, 1979; Terrace et al., 1979) to a small number of illocutionary functions and has not been shown to extend in any significant way to *vocal* communications produced by nonhuman primates (Gardner and Gardner, 1969). For this reason it remains a crucial puzzle to determine how humans came to have such vast vocal flexibility, a capability that seems to have been critical in the evolution of language.

Special Cases Related to Signaling and Communication

This chapter is intended to review forces that have acted to influence communicative flexibility across a variety of species and in a variety of modalities of communication in order to provide perspective on the hominin case. The goal leads us to take into account not only clear cases of evolved communication but also actions such as camouflage and deception, where our notions of signal and communication are pressed to the limit. Developmental patterns also present challenges to the definitions of signal and communication. The following clarifications are in order.

Camouflage is utilized by organisms not to communicate but to prevent any kind of cuing in order to avoid detection by predators or prey. Camouflage also does not obey the Maynard Smith and Harper definition of signals, since it is not evolved to confer benefit on both sender and receiver. Coevolution between sender and receiver can occur nevertheless in an arms race of camouflage and detection. Yet the mechanisms of camouflage in some species include flexible control of actions that could be used to communicative advantage within or across species. Consequently, although camouflage itself is not a signal system, we review certain camouflage phenomena because they may play roles in establishing foundations for flexible signaling.

Deception is a special kind of communication that, as indicated above, is "parasitic" on a primary signaling system or cue interpretation system. Any deceptive signaling act depends upon the existence of a more frequently occurring honest signal or cue against the background of which the deception can be implemented. Like camouflage, deception violates the Maynard Smith and Harper definition of signal because deception does not occur to benefit both sender and receiver. Like camouflage, deceptive systems can coevolve across senders and receivers in an arms race of deception and detection. Consequently, evolutionary pressure on deception can create signaling flexibility since the action of deception must become increasingly adaptive in order to prevent receivers from recognizing deceptions for what they are. Thus, some examples of animal deception research are reviewed below, again because we seek to provide special perspective on the origins of communicative flexibility in general.

Finally, in the very young of some species, flexible actions are often observed that appear to be precursors to the mature signaling system, but where the developmental "signal" events may have different status for senders and receivers from the standpoint of the notion of illocutionary force. In many songbirds and human infants, for example, vocalizations occur that may be purely a form of practice or exploration of sounds ("play") for the producer, in which case they have no illuctionary function from the standpoint of the producer. On the other hand, these sounds may constitute fitness indicators for caregivers, who may be able to attend to the young more effectively on the basis of that fitness information. Our interest in the origins of flexible signaling leads us to treat these developmental actions as a type of signaling even though their functions may be different for sender and receiver in the course of development.

Selection for Variety of Signaling in Hominins and Other Species

In hominin evolution of an open vocal communicative system, with decoupled vocal signal and function, we propose that a particular change in pressures on selection for vocal capability must have occurred: At some point, there must have existed a selection pressure favoring variable sounds, rather than the stereotyped sounds that are found in fixed vocal signals. At the initial point of selection for variability, we propose that variably produced signals served a specific, single function, presumably fitness indication. In this scenario, the advantages of variability had to outweigh the advantages of stereotypy (speed of production, clarity . . .).

We envision two types of enhanced capabilities that could have been selected to produce variability in vocal signaling. The first type of new capability would have constituted an ability to produce sounds more freely and would have enabled an increase in endogenously produced variable sounds, which were presumably very infrequent in an earlier stage of hominin evolution (as they are in modern nonhuman primates). This capability implies spontaneous vocalization and exploration of vocal capability (see Oller and Griebel, this volume). The second new capability would have been based on learning to produce sounds heard in the external environment. The latter mechanism implies mimicry or imitation. These new capabilities, we propose, are foundations required for the emergence of significant functional flexibility.

We do not, for example, find it plausible that an evolutionary process that diversified a particular signal class into a small number of fixed subtypes (e.g., predator-specific alarm calls or alarm calls that encode locations of potential danger; see Struhsaker, 1967) could have blazed the path to an explosion of functional flexibility such as that seen in humans, because diversified subtypes of alarm calls are still stereotyped and fixed in function as alarms; they are not (and presumably cannot be) used to serve other functions, such as greeting, appeasement, or courtship. Human linguistic signals, however, and even prelinguistic infant vocalizations show precisely this sort of flexibility. According to our reasoning, variable production of signals had to evolve first, and these signals had to later be freed from their original functions and applied in variable contexts/functions.

In the comparative enterprise, we seek parallels to the human case: diverse signals that map onto a single function and/or individual signals that map onto several functions. For language, both are required. Our comparative enterprise also seeks to determine the evolutionary conditions for selection for variety. We ask, in what cases does variability or functional flexibility in communicative signaling occur in nature?

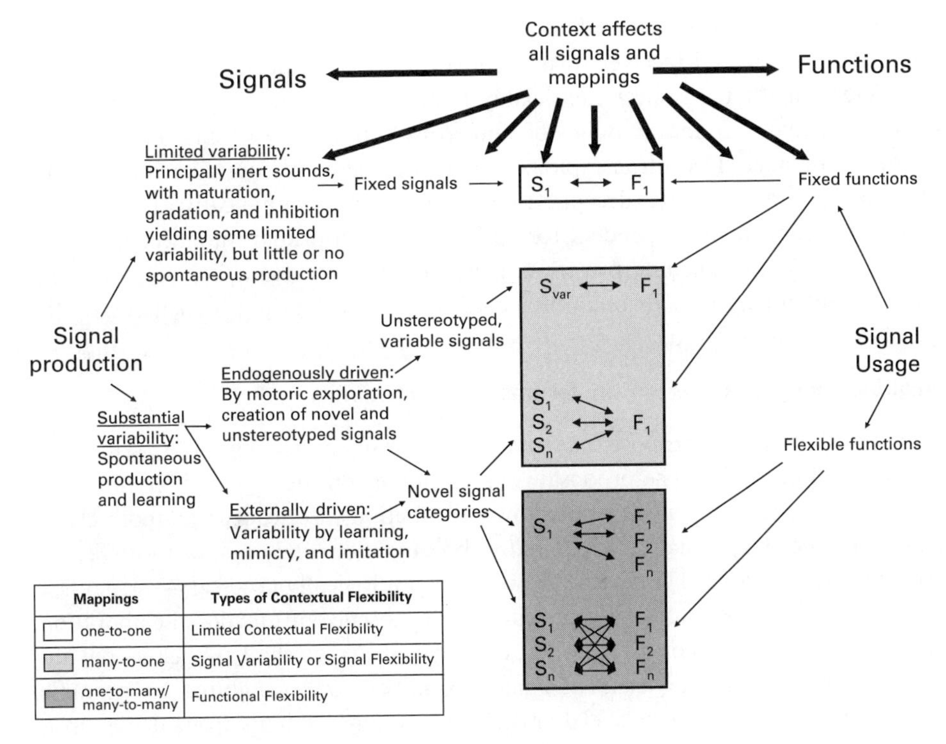

Mappings	Types of Contextual Flexibility
one-to-one	Limited Contextual Flexibility
many-to-one	Signal Variability or Signal Flexibility
one-to-many/ many-to-many	Functional Flexibility

Figure 2.1
Types of contextual flexibility in communication. S, signal; F, function.

Terminological Overview on Flexibility of Communication

Figure 2.1 diagrams the proposed terminology for types of flexibility in communication. Because individual fixed signals in social animals such as primates typically show variation in intensity, maturational change, and inhibitability (e.g., in audience effects), we propose that the term "contextual flexibility" should apply (at least minimally) to all the circumstances of signal mapping that we will be discussing. Fixed signals, however, show limited contextual flexibility and one-to-one mapping. Many-to-one mappings can occur when signals are produced in a variable, nonstereotyped way to serve a single function, or when novel signal categories are developed by individual organisms through motoric exploration and practice or are learned by mimicry or imitation from the environment and then mapped as a group of signal categories to single functions. In many-to-one mappings we will say there is "signal flexibility." Only when one-to-many or many-to-many mappings occur do we apply the term "functional flexibility." Camouflage and deception present special cases not directly represented in the figure.

Camouflage itself is, by our definition, not a signal because it does not evolve to benefit both sender and receiver. It is the inverse of communication, an attempt to avoid detection. At the same time, flexibly implemented camouflage can be argued to involve a many-to-one mapping of numerous actions to the single function of detection avoidance. Deception exploits an existing system of communication or cue interpretation, and although deception can influence a communicative system, it is always secondary, being dependent for its function on a primary nondeceptive base of communication or cue interpretation. Deception can involve either signal flexibility or functional flexibility or both, in accord with our definitions, depending on the type of deception involved.

Signal Flexibility as a Prerequisite for Functional Flexibility

As suggested above, we propose that some degree of signal flexibility is a prerequisite for functional flexibility. The proposal is based in part on the logic that communication requires signals to transmit functions—if functions are to become more elaborate or more flexible, signals are required to become more elaborate or more flexible to serve those functions. The reverse, however, is not logically necessary—signal flexibility can be developed, logically speaking, in the absence of functional elaboration.

Empirical evidence is consistent with our proposal. As we shall see below, the animal kingdom displays many cases of signal flexibility—some songbirds, for example, can learn to produce extremely elaborate songs or other sounds from the environment, but these songs usually serve only one or two functions (see the section on sexual selection below). In contrast to the evidence from bird song, we know of no clear examples where functional flexibility is elaborate but signal flexibility is quite limited.

Deceptions may present at least minor exceptions to this general trend. If an animal uses an alarm call for deception, it uses a fixed signal to manipulate communicative effects without changing anything in the signal itself. In essence, the animal uses the regular signal for warning to serve an additional function such as competition avoidance. But as noted above, deceptive actions are by nature infrequent in occurrence and so do not, we think, fundamentally alter the tendency for signal flexibility to have precedence over functional flexibility in evolution.

Consistent with our reasoning, some degree of functional flexibility is typically seen in species where signal flexibility occurs. Importantly, the numbers of illocutionary functions that can be served in animal communication by flexible signals appear to be quite small (five or less as far as the literature indicates—see below for examples). The only case we know of where extremely elaborate functional flexibility occurs is in humans, and in that case signal flexibility is so highly developed that it allows expression of a much larger number of possible specific illocutionary functions than in any other animal as far as we know—humans control, in addition to the repertoire indicated above for nonhumans, functions such as explanation, criticism, denial, affirmation, acknowledgment, and many others. For a taxonomy of illocutionary functions in humans, see Fraser (1975). Human signal flexibility also allows expression of indefinitely large numbers of words and sentences.

The comparative evidence reviewed here is consistent with our expectation that the evolution of language would have included increases in signal flexibility prior to increases in functional flexibility. This ordering is also consistent with modern human infant development where signal flexibility leads the way in the emergence of communicative skills (see Oller and Griebel, this volume). In what follows we explore the occurrence of the various mapping options for signals and functions in nonhumans from invertebrates to primates. The organization of the exposition is based on presumed forces and circumstances of selection that encourage evolution of flexibility in communication or that encourage evolution of flexibility in capabilities that can serve as foundations for flexible communication as suggested by such special cases as camouflage, deception, and developmental patterns.

Selection Factors Favoring Communicative Flexibility

Sexual Selection

It appears that the most effective force yielding signal variety for variety's sake across many species has been sexual selection. Variety of signaling can be used either to impress the opposite sex in courtship, resulting in intersexual selection, or to impress the same sex in competition for territory or mates, resulting in intrasexual selection. Sexual selection is assumed to have produced the well-known mating songs and territorial songs that are found everywhere from invertebrates to mammals (Alexander, 1961; Payne and McVay, 1971; Hausberger, 1997; Kroodsma, 1999; Nooteboom, 1999). Sexually selected signals are usually based on traits that reflect the sender's resource holding potential (RHP; Parker, 1974). Such traits constitute fitness indicators and thus can contribute to both inter- and intrasexual selection at the same time. Nevertheless, it seems that intersexual selection is somewhat more unpredictable in relation to RHP, since factors such as a specific sensory bias of the female can guide selection toward a seemingly random enhancement of specific traits.

Humpback Song and Bird Song as Indicators of Sexual Selection

The most elaborate songs influenced by sexual selection have been reported to occur in many species of songbirds and, interestingly, in a completely different group, which includes a few of the large whales, specifically the humpback (Payne and McVay, 1971; Payne and Payne, 1997; Tyack and Sayigh, 1997). The songs of the humpback are so beautiful to the human ear that they became a commercial best-seller.

If one speeds up humpback songs to match the frequencies of the acoustic elements of bird song, a noteworthy resemblance emerges. If we compare, for example, the humpback's song to that of the long-billed wren, we find many similarities of structure (Tyack, 1999). Analysis of internal rhythmic structure in both whale songs and bird songs also yields parallels with the rhythm of human vocal communication. Rhythmic groupings of elements are called "songs" in birds and "phrases" in the humpbacks. The minimal rhythmic units within these songs and phrases are sometimes called "notes," but they are also often called "syllables" (Marler and Slabbekorn, 2004). These repeatable minimal units often have comparable durations across many unit types, a pattern that is similar to syllables in speech, which also occur in repeatable units of limited duration. The number of types of these minimal units is one indicator of complexity. Another indicator is the number of transitions between the themes within the songs or phrases, which tend to cycle in a particular order but show some flexibility of cycling. These transitions seem to be more complex in songbirds than in whales; the latter use a smaller number of transitions among fewer than ten themes (Tyack, 1999).

Learning is clearly involved in bird song and humpback song. In many bird species new song elements are acquired by mimicry or imitation. Many species actually show increasing repertoire over the years (Baptista and Petrinovich, 1986; Nottebohm, 1981; Payne and Payne, 1997; West et al., 1997), but in a few species, including the humpback, increase in repertoire size is not seen. Instead all individuals within a population display songs that are very similar at any one time, but the animals slowly change the entire song more or less in synchrony over weeks and months, without increasing repertoire size. New themes appear and old ones disappear many times, so that in a lifetime the total number of elements or themes is essentially open-ended. Some bird song repertoires are known to change seasonally with peaks of repertoire in the mating season.

Clues That Indicate the Operation of Intrasexual and Intersexual Selection

Although mating songs can be used entirely in courtship, they are sometimes shaped by intrasexual and intersexual selection in the same species. In many birds both kinds of shaping seem undeniable. Both selection pressures may also be at work in the humpback whale. Some birds, like the European starling, appear to differentiate their songs into two subtypes, one that they use for male–male competition and the other one for courtship (Hausberger, 1997).

It is complicated to attribute distinct acoustic features to only one of these two selective forces, because the same song elements often appear in both cases. Further, a male sometimes appears to play both roles at once—for example, by singing in a potential mating territory that may attract a female to the territory while simultaneously repelling other males from it. In such cases the general "function" of the variable song may be that of fitness display, serving both courtship and territoriality as subfunctions and thus applying both intra- and intersexually simultaneously. In lekking circumstances where males display for females, the primary force shaping variable song can often be deemed to be specifically intersexual, while in countersinging between males, for example, specifically intrasexual selection may influence the pattern. Intersexual selection appears to produce signal variety in some cases while favoring convergence of songs in others.

Since inter- and intrasexual selection forces are often both at work in variable song development, encouraging spacing between males and providing a means for males to attract females, it is clear that complex bird song (and perhaps humpback whale song) provides evidence of a several-to-several mapping of signal to function; a multitude of signals correspond to *two* functions (courtship and male–male competition/territoriality). Thus, sexual selection forces, depending on social circumstances, can favor both variety of nonstereotyped sounds (signal flexibility) and flexibility in vocal usage (functional flexibility). However, it is important to note that very advanced signal flexibility is seen in many songbirds with *only* two function types being served by that signal flexibility. The very elaborate signal systems of many species of birds (and perhaps the humpback whale) appear not to have resulted from a need for elaborate functional flexibility but to have resulted instead from an arms race in advertisement/fitness signaling and in detection of the characteristics of the elaborated signals by both potential mates and competitors.

A possible third function for the elaborate songs of some birds is based on the "Beau Geste" hypothesis that some birds imitate other species to pretend that the territory is too full for further habitation (Dawkins and Krebs, 1978). If this hypothesis is correct, then these birds may use their elaborate songs in three different ways—in intra- and intersexual selection as well as in deception of other bird species. Birds that imitate are also known to use calls of predators to get rid of food or mate competition—another deceptive use of the general capability to imitate sound.

Social Cohesion as a Force for Communicative Flexibility

Evolution also sometimes appears to produce variety of signaling in cases where social glue is needed between individuals, either between group members or between parent and offspring. In the first case this social glue is needed to keep a group or two individuals together, to help minimize conflicts, to coordinate activities, and to establish boundaries with other groups. Sounds functioning in this way can be referred to as "social cohesion calls." Good examples of this type come from cetaceans and other social mammals.

Pod Repertoires of Killer Whales as Social Cohesion Calls

Killer whales (orcas) live in stable groups called pods, which change only by birth and death (Bigg et al., 1987; Miller and Bain, 2000). The whales produce a variety of sounds, echolocation clicks, tonal whistles, and pulsed calls; some of these are repeated discrete calls, and some are highly variable (Ford, 1991). Discrete calls dominate when killer whales travel or forage. The discrete calls are easy to categorize. Whistles and more variable pulsed calls are common in groups engaged in social interactions; these are difficult to categorize.

Each pod of killer whales has a group-specific repertoire of discrete calls that is stable for many years. Each whale is able to produce the entire repertoire of the pod. Some calls are more common in resting groups, others in more active groups. However, each discrete call in the pod's repertoire can be heard regardless of what the pod is doing. Different pods may share some discrete calls, but no two have the same entire call repertoire. Different pods also have ranges of overlap and also associate together for hours or days before separating. Individual pods also have clearly defined subpods and matrilinear groups that seldom split up but may separate and converge. These group-specific call repertoires are thought to indicate pod affiliation, maintain pod cohesion, and coordinate activities of pod members. The frequencies of the different types could be indications for certain kinds of moods and activities they are associated with, but more research is needed on this topic.

Dolphin Societies and Social Cohesion

At first glance dolphin societies seem similar to killer whale societies. According to McCowan and Reiss (1995, 1997), dolphins also have a group repertoire that varies in composition with context, with presumably similar functions as in orcas. There is also the possibility that these calls can be used by conspecifics to infer the direction of movement of the signaling animal, which would help to coordinate activities in many situations (Lammers and Au, 2003).

Social cohesion could also be supported by a rather unique type of vocalization that has been reported in dolphins. The bottlenose dolphin, for example, has been claimed to show something called a signature whistle unique to each animal, developing within a period of a few months to a few years and remaining stable for a very long time (Caldwell and Caldwell, 1965; Caldwell et al., 1990; Sayigh et al., 1990; Tyack, 1999, Tyack, 2000; Weiß et al., 2006). On the other hand, these so-called signature whistles are also produced by other individuals in the group. Advocates of the signature whistle hypothesis call such occurrences "imitations." Advocates also claim that these imitations are used to "call" or at least initiate interaction with the signature individual, but experimental evidence for this claim is unavailable. We do not know of a parallel or analog case of a "signature whistle" or "signature vocalization" in other animal groups.

The whole idea of the signature whistle is odd in the context of other signal systems. Possessing a signature whistle seems equivalent to swimming around and calling one's name out repeatedly. In an overwhelming variety of species, animals that know each other individually recognize each others' voices whenever they vocalize, so that no special signature sound is required. Therefore, there is reason to question the interpretation of dolphin whistles as signature whistles on plausibility grounds. However, as advocates argue their case, it could be, of course, that because of the acoustic properties of water where sound is easily distorted, it is hard to recognize individual voices and thus necessary to produce individually specific signature whistles in some marine species.

Dolphin societies differ from those of killer whales in that they are not as stable but are more like the fission–fusion societies of chimpanzees. As in chimpanzees, dolphins bond with certain individuals for long periods of time, mostly in male–male and female–female coalitions, and these coalitions may play a role in the maintenance or establishment of signature whistles. It has been suggested by advocates of the signature whistle idea that the coalition partners may imitate each others' signature whistles to a certain extent in order to call each other when they need to coordinate activities. It has also been argued that the signature whistle could be helpful in locating and recognizing a coalition partner in the murky waters of the inshore environment where vision is limited.

Further Remarks on Cetaceans, Social Cohesion, and a Possible Role for Grooming

Whatever the truth may be about the signature whistle hypothesis, one thing is clear in cetaceans: They use a wide variety of different and acquired whistles, many learned by imitation, and they use these signals to achieve cohesion between the members of matrilinear family groups or between coalition partners.

It is interesting also to consider a social, that is, "vocal grooming" function (Dunbar, 1996; Morris, 1967) for vocalizations in cetacean societies. Bodily grooming is limited in cetaceans, so vocal grooming could have important advantages. There is some rubbing and touching that occurs, but usually it is limited to individuals who are good friends or close kin; since the number of familiar and individually recognizable dolphins is quite large, vocal grooming could help to stabilize relations in the group.

It is particularly difficult to determine function and signal relations in the case of cetacean societies because it is difficult to monitor the vocalizations of individuals in their water environments. From what we do know, it appears the communicative systems consist of a several-to-several mapping of signals and functions. We do not know much about the details yet, but the cetacean circumstance, with its high priority on social cohesion, seems to favor selection for variety of communicative signals and for both signal flexibility and functional flexibility.

Interestingly, the existence in cetaceans of "fixed calls" for specific functions like courtship or aggression has only recently been reported (Blomqvist et al., 2005; Connor and Smolker, 1996; Herzing, 1996; McCowan and Reiss, 1995). Blomqvist et al. (2005) even suggest a signal equivalent in function to human "laughter" in dolphins, which is used in the context of play behavior (see Kuczaj and Makecha, this volume). Nevertheless, it is not quite clear yet to what extent these calls are actually "fixed" in the sense of being innate and inflexible in usage. It seems possible that the apparent complexity of vocal communication in cetaceans (and especially dolphins, where most cetacean research has been concentrated) makes it difficult to discern functional unity of the many types of whistles and other sounds.

Recent research suggests that social cohesion signals may be found in primates as well—for example, in the Campbell's monkey where individuals can have variants of a single call type that they share with other individuals in the group and that can change across time (Lemasson et al., 2003; Lemasson and Hausberger, 2004; Lemasson et al., 2005). Also, male chimpanzees, while not producing a group repertoire of different calls, appear to modify a single call category, their pant hoots, through learning, and thereby geographically adjacent groups can be distinguished (see the review in Hammerschmidt and Fischer, this volume).

Pleasure Sounds as Possible Social Cohesion Devices in Cetaceans and Primates

One functional category of signals that has been very little studied is comfort and pleasure sounds (see, e.g., Jürgens and Ploog, 1976; Panksepp, 2000). Pleasure sounds may occur vegetatively in solitary species (and thus could only serve as cues rather than signals in such cases). According to the ritualization hypothesis of classical ethology, pleasure sounds may emerge as signals through shaping by natural selection from such vegetative cues in order to reinforce the source of pleasure, for example, a grooming partner. Purring in cats provides an example of a stereotyped fixed signal of pleasure that is even used in interactions with other species (Case, 2003; Morris, 1986). Since purring seems to enhance the cat's positive state of affect, it may also be used to quell fear, in a way humans sometimes use laughter in circumstances of threat.

In humans, pleasure signaling is very complex, since many elements from the vocal language, vocal nonlanguage, and body language repertoires can be recruited to signal pleasure. For example, sighing is often used to indicate satisfaction with an event but can also be used to indicate weariness. Pleasure sounds in humans can inform the listener of the speaker's positive reaction and, like the cat's purring, can encourage continuation of it. In intimate human relationships the importance of such signals may be fundamental. Research on face-to-face interactions between human infants in the first months of life and their caretakers documents the widespread use by parents across cultures and languages of universal intonation types (superimposed on any sentence spoken to the infant regardless of language) including rising intonations as encouragement to the infant to continue vocal interaction and sighing intonations as soothing devices (see Fernald, 1989; Fernald and Simon, 1984; Fernald et al., 1989; Papoušek et al., 1990; Papoušek et al., 1991). The effect of encouragement and soothing on continuing interaction is consistent with our impression that in many social species pleasure sounds have come to reinforce caretaking, grooming, or other pleasant activities and intimate interactions, including copulation.

Even among chimpanzees, who appear to be particularly limited in vocal flexibility within the primates, females produce apparent passion sounds when copulating with the alpha male but appear to show the flexibility to suppress these sounds when copulating with a low-ranking male (de Waal, 1982). Further, the so-called close calls of many primate species seem especially diverse (Becker et al., 2000; Snowdon, 2004). Such sounds are produced primarily in circumstances of social interaction that may reflect comfort and pleasure.

Ford reports that orcas in Johnstone Strait regularly visit a special beach covered with round pebbles where they rub their bodies along these pebbles. While they do this, they produce a wide variety of (presumably) pleasure sounds, which can be either discrete calls from their own pod, discrete calls from other pods, or completely novel calls and whistles that were never recorded before (Ford, 1991; Riesch et al., 2006). In this special case at the pebble beaches, orcas groom themselves—we do not know whether they produce the same kinds of sounds as pleasure indicators during rubbing against each other when at sea or during copulation. This is a topic that deserves study, because the variability of the sounds produced provides further evidence that expression of comfort or pleasure may be particularly conducive to variability.

We also suspect that there may be an audience effect for pleasure sounds in social species. Such an effect clearly occurs in humans—we do not sigh with pleasure to a stranger the way we might to someone with whom we share intimacy. It has already been mentioned above that chimpanzees show audience effects with passion sounds.

Dunbar suggests that increases in group size in primates can make it impossible for bodily grooming to serve the function of maintaining group cohesion, and consequently that a pressure favoring vocal grooming can thus develop (Dunbar, 2004). In particular, he thinks the shift to vocal grooming occurred in the hominin case. Since comfort-inducing sounds are so variable in humans, it would appear that vocal grooming was another circumstance that encouraged variability in vocalization.

We know of relatively little data on this topic, so we are mostly speculating, but it seems that both in signal and in function, pleasure or comfort sounds may be diverse in social species. The pressure for stereotypy seen in other cases (as with alarm or threat calls, e.g.) appears to be weak here, and instead there may be selection for variety; it could be, for example, that the richness of vocal production by the individual being groomed serves as a reinforcer to the grooming individual. Thus, we suspect that a several-to-several mapping of signal and function may be naturally selected in comfort and pleasure sounds. Given the empirical evidence available, it does appear that pleasure sounds are particularly likely to show signal flexibility.

Parent–Offspring Bonding

Another case of social cohesion occurs in the context of parental care where there is the necessity to establish a strong bond between parents and offspring. In some circumstances, signals between parent and offspring can be quite stereotyped—for example, in pinniped rookeries where mother–pup recognition calls are usually stereotyped and need to be recognizable over a distance (Schusterman et al., 1992). Human parents, however, use at least two different intonation types that can be superimposed on any sentence for purposes of soothing and encouragement of vocal interaction with infants (see the citations above).

Further, there are cases where variability in signal production by the offspring seems to pay off. Human infants provide one such example. Variable sounds interpreted by parents and researchers as expressing pleasure have long been recognized as signals to parents of infant well-being and as bonding devices (Locke, 1993; Papoušek, 1994; Stark, 1978; Stark, 1980). Another case of variability in infant vocalization is seen in the socially breeding marmosets where variable "babbling" (production of sequences of sounds that appear to consist primarily of infant versions of several different adult fixed signals in systematic repetition and alternation) seems to elicit attention from parents and other caregivers (Elowson et al., 1998; Snowdon, 2004; Snowdon et al., 1997; and see Snowdon, this volume). These juvenile sounds are interpreted as babbling because during babbling episodes they do not serve the functions that the same sounds serve in mature marmosets. Such babbling appears to occur in some other New World monkeys as well, but observational evidence is less clear in these cases.

Producing variable vocalizations may reinforce the bond between offspring and parent, elicit parental care, and at the same time give the parent clues concerning the offspring's fitness. Locke (2006) reviews extensive data indicating that human parents use vocalizations of infants as fitness indicators, and both he and Oller and Griebel (2005) argue for parental selection of infants in part based on assessment of infant fitness through vocalization. The evidence is especially suggestive in child abuse and neglect, where infants and children with communication disorders have been shown to be especially vulnerable. In evolution of the hominin line, differential parental (and other caregiver) investment in infants could have played a major role in selection of infants whose vocalizations were particularly indicative of good health and viability. This fitness-indicating "babbling" would present a several-to-one mapping of signal to function and would thus constitute a case of signal flexibility. It could be the case that in some species these babbling sounds are largely drawn from the pleasure and comfort sounds described in the prior section.

Deception

Deceptive Behavior as a Factor in Communicative Flexibility

Another powerful source of selection for variability of signal as well as function in communication is deception. In deceptive behavior, signals that give a false or exaggerated portrayal of the environment or of the state or identity of the sender are used. Deceptive signaling is "parasitic" on existing nondeceptive signaling systems or on cue interpretation as indicated above.

Deceptive signaling can occur in several forms (Mitchell and Thompson, 1986) including at least the following: (1) genetically preprogrammed (innate) deception, in, for example, Batesian mimicry, where, for instance, a nonpoisonous species evolves a coloration corresponding to a poisonous species and thus avoids being preyed upon, (2) instinctive behavioral programs for deception, as has been assumed in the case of injury feigning in birds, (3) conditioned behaviors such as learned injury feigning reported anecdotally for dogs and other species, and (4) intentionally planned deceptions, for which real evidence is rare in the animal kingdom (Byrne and Whiten, 1985, 1988, 1990; Gibson, 1990; Mitchell and Thompson, 1986; Whiten and Byrne, 1988) but plentiful in humans. Even though anecdotes of deceptive signaling are numerous in birds and mammals, it is difficult to prove that learned deceptive signaling occurs in animals, and the evidence is not as persuasive as we would wish it to be. The reason for the lack of convincing data in this field is, of course, that in order for deception to work it must be rare, and thus the burden of proof is heavy on those who wish to assert the existence of learned deception in animals.

In most cases of deceptive signaling, it appears that both deceiver and victim are capable of learning, and an arms race of deception and detection ensues. However, in the case of Batesian mimicry only the victim of the deception appears to learn, for example, to avoid the danger of poison in prey with aposematic (warning) coloration, a fact that can be exploited in deception by a nonpoisonous species that evolves the same coloration.

One requirement for deceptive behavior to form foundations for flexibility in communication is that the behavior has to be learned. In genetically preprogrammed deception such as Batesian mimicry the deceptive signal/cue is fixed within each individual of a species and thus cannot form the foundation for either signal or functional flexibility. The same reasoning seems to apply to instinctive behavioral programs for deceptive behavior that do not allow for learning.

However, functional flexibility can result if some learning occurs in the functions for which the signal can be used. For example, an instinctive alarm call could be used to make other animals scatter and thus get rid of food competition or mate competition or to get out of a precarious aggressive interaction with either a conspecific or a predator. When a deceptive behavior is learned, it can form foundations for signal flexibility as well as functional flexibility. The neurological substrate necessary for these learning processes could be capitalized upon in the evolution or development of communication. Of course, as communicative flexibility increases, communicative acts can increasingly be used for deception.

It is noteworthy that in many cases of deceptive signaling, the signaler changes neither the signal (e.g., the alarm call) nor its illocutionary force (alarm) but exploits the intrinsic illocutionary force to achieve a perlocutionary effect (the other animals scatter after perceiving the alarm), eliminating, for example, competition for food. Another example would be a false infant distress call that causes the infant's mother to attack an animal who is not threatening the infant but is in possession of, for example, a food item the infant wants to obtain.

Surprisingly perhaps, there is actually some evidence of learned deception among invertebrates. Since examples for deception in higher vertebrates are relatively well-known, we review some interesting data for an invertebrate species here.

Fireflies and Communicative Flexibility

Fireflies are nocturnal insects that manufacture their own visual signals with the aid of photochemical equipment at the tip of their abdomens. The lighting signals would seem conspicuous to predators, but fireflies are protected in part by being poisonous. In different species, unique patterns of light pulses have evolved that vary by species in duration, intensity, frequency of occurrence, and color (Lloyd, 1986). This shows that a selection pressure for distinctiveness and stereotypy is at work in the basic species-specific visual signaling system of fireflies. However, the firefly flashes can be shown to be flexibly mapped to a variety of functions.

Fireflies of the genus *Photuris* seem to have the most complex flashing behavior of any firefly studied to date. Females use the flashes in situations other than mating and courtship. They use their light for illumination as they take off or land, when walking on the ground, when ovipositing (i.e., laying eggs), and when climbing through mazes of Spanish moss. They sometimes flash just at the moment they attack a flying flashing firefly in the air, perhaps as a startling device. Such an aerial attack is used in conjunction with aggressive signal mimicry or sometimes against hesitant suitors.

In aggressive signal mimicry, the females of this genus *Photuris* answer the mating flashes from males of certain other species in the genus *Photuris*. Each species of *Photuris* has its own code with a female's answering her male's distinctive flash pattern by giving a flash pattern of her own after a precise time interval. Some females can respond correctly to male signals of three different *Photuris* species. If a *Photuris* female succeeds in luring a male close enough, she will grab and eat him, and she can absorb the poison and transmit it to her eggs, making her own eggs more poisonous and thus better protected. However, the males have developed a counterstrategy. They do not throw themselves onto the females but land at a safe distance and walk cautiously toward them, ready to escape in case of a trap. This may explain why the females sometimes switch to aerial attacks instead.

What makes the case of the fireflies so interesting is that this mimicry is a learned behavior, because the female fireflies can learn novel human-made flash patterns that no real firefly uses. Thus, fireflies present a powerful example of a predisposition for learning to reproduce variable patterns that they encounter in their environment.

According to Lloyd, the story is even more complex. Many *Photuris* males produce not only one type of mating pattern but two or more different ones. They imitate mating patterns of males of other firefly species that live in the same area. Lloyd has speculated that they imitate the flash patterns of males that their own females try to lure with false signaling, in order to try to find one of their hunting females to be able to mate with her. Thus, in this case this male mimicry could have evolved as a mate-seeking tactic.

To understand this seemingly odd system, one has to know that competition for females is fierce in fireflies. A mating takes a female out of the game for some time because she will be busy laying eggs; the average ratio of males to females ready for mating is about 100 to 1. One strategy for males to find a rare mating option is to learn to mimic males of other species that the rare females of their own species may be trying to lure into a trap. Another important strategy evolved by the males seeking a mate is a very cautious approach to a female that gives the right answer. The males usually land at some distance and approach carefully. Typically, a small group of males approaches the same female while exchanging flashes. In this group the males sometimes answer the other males with "transvestite flashes" to lead them down the wrong track and to slow them down. They also emit flash patterns of other species as well as other "extraneous" flash patterns that might be used to deter competition or to test the truthfulness of the female.

Another interesting thing *Photuris* males do is to interrupt the courtship pattern of competitive males with an "injected" flash. Females do not usually answer to patterns that have an injected flash. The effect of injected flashes on flying males is that they sometimes land immediately at the location of the injected flash, presumably to join the competition for a female. Needless to say, some *Photuris* females have learned to inject a flash at the right spot to make males of other species land right next to them (Lloyd, 1986).

Even if only half the story is true, communication in fireflies seems to be a very complicated matter. Both male and female *Photuris* are able to use a variety of signals in a variety of contexts, so these species present a case of several-to-several mappings with both signal flexibility and functional flexibility. An interesting aspect of this system is that the variability in communication in this mating system seems to be driven not by sexual selection alone but mainly by advantages that can be achieved through deception and counterdeception. The complexity arises because signal-tracking predators or competitors tend to eliminate simpler, straightforward signalers from the gene pool and thus feed a spiral of more and more complex signaling and signal detection. The firefly data hint at the possibility that complexity may be a universal countermeasure for dealing with deception in communication in general. This possibility may be relevant to the human case because the extreme elaborateness of human language may also in part be a response to detection and counterdetection pressures (see also Sterelny, this volume).

Camouflage, Protean Behavior, and Complex Signaling in Cephalopods

Even though camouflage does not count as a signal by the definition we are using because successful camouflage prevents communication rather than transmitting it and because in many cases the receiver does not have a chance to coevolve with the sender, we will consider it here as a possible source for a signal production mechanism, especially for mechanisms that may yield flexibility. A prime example of this possibility is found in the cephalopods. Usually communication research is heavily biased toward acoustic communication. In the following section we look at a complex visual communication system and its possibilities in some detail.

Cephalopods are the masters of flexible camouflage, being capable of seemingly disappearing instantaneously into a variety of marine backgrounds. Cephalopods have developed a chromatophore system in their skin with which they can match their body surface to the background (Messenger, 2001). Surprisingly, the variety of patterns they need for this behavior is quite limited; two or three different grain types and general brightness matching to the background usually do the trick in the aquatic environment, which does not favor high visual acuity anyway.

Evolution of Chromatophores as Camouflage and Signaling Devices

As the theory goes, cephalopods originated as shelled creatures with sluggish swimming capability. In order to compete in the open water with fish, they gave up their outer shells for speed and flexibility and became streamlined, fast, and brainy predators. But at the same time they gave up the protection afforded by their shells. Their vulnerability without shells encouraged the evolution of a camouflage system on their skin that would fool predators (Packard et al., 1980). Many believe that chromatophores evolved for camouflage purposes and were later adapted for social communication (Packard et al., 1980). However, this is just speculation; it could have been the other way around. Consider, for example, chameleons, a group commonly believed to use color change for camouflage. In fact, the primary function of color change in chameleons is social communication, and secondary functions are physiological ones to regulate body temperature (LeBerre et al., 2000).

Pelagic (open water) creatures are best camouflaged either by transparency or by reflective surfaces and countershading. Consequently, chromatophores would be of little use in camouflage for creatures that evolved from a shelled life to that of pelagic predators. Therefore, it is plausible to assume that chromatophores might have developed first for reasons other than camouflage, for example, for communication, and indeed the chromatophore system is used extensively for communication. In either case the development of the chromatophore system has led to a very flexible system of displays on the skin in cephalopods that can be brought to the service of communication as well as camouflage.

Of course, the cephalopods do rely partly on background matching and countershading strategies, but additionally they have evolved a whole array of efficient strategies in case they are detected anyway. One of them is deimatic or startling behavior, where they use their chromatophores to produce false eyespots, a very popular startling pattern in the animal kingdom. In some cephalopod species deimatic spots can appear in a variety of body locations.

Protean Behavior in Cephalopods Utilizing the Chromatophore System

Another strategy cephalopods use to avoid predators is called "protean behavior" (Chance and Russell, 1959; Driver and Humphries, 1988), or highly unpredictable behavior. It is universally used in the animal kingdom to confuse either predators or prey. Consider the zigzag flight paths of many animals pursued by predators or the "crazy dance" in weasels or foxes during hunting. In cephalopods the protean display for escaping a predator includes inking and, most importantly, the production of a wide variety of color patterns that present the predator with an ever-changing target (Hanlon and Messenger, 1996; Holmes, 1940). Thus, what a squid or an octopus does when it is detected is not predictable, and the predator is prevented from developing a search image for these targets because they tend to change in appearance with every encounter.

One species, the Caribbean reef squid, *Sepioteuthis sepioidea*, produces a wide variety of body patterns out of several different background colors, along with several pattern types such as horizontal bars, vertical stripes, a mottle, and roundish units (Moynihan and Rodaniche, 1982). All background colors are used as camouflage patterns either alone or combined with patterns such as stripes, bars, or a combination of stripes and bars called "plaid" or combined with mottles and/or spots. However, the social signals for aggression and courtship are composed of the same patterns; thus, several pattern combinations are used for two different functions. One is social communication; for example, the pattern Stripe can be a courtship signal, and the other function is that of hiding when it is used as a camouflage pattern. The aggressive pattern Zebra (which consists of a type of mottle) is used in both intraspecific and interspecific aggression but is also used as a camouflage pattern when the animal is close to the sea bottom. Pale is used as a default pattern during the night, but it is also used as a startling pattern during the day, for camouflage during rapid flight, and to indicate a state of high arousal during courtship. Essentially all body patterns can be used as protean or "startling" patterns.

Deception through Chromatophore Signaling in Cephalopods

We have observed on several occasions that when a male Caribbean reef squid was being chased by another male and challenged for a contest, it sometimes happened that the chased male produced a Saddle, which is a female courtship pattern (Griebel et al., 2004). The effect was that the chasing male stopped the chase immediately, and, probably confused, sometimes even produced the mating pattern that answers Saddle, which is Stripe. At first it appeared that this behavior might be an appeasement signal, but the rarity of occurrence of such cases argues against its being a consistent appeasement signal. More likely, it appears the female courtship signal is produced by the threatened male as a kind of deception—causing the attacker to cease aggression.

We find another case of deception in a different cephalopod species. The giant cuttlefish uses the female "dress" to sneak into position in order to copulate with the female even under the guarding eyes of a dominant male (Hanlon et al., 2005). So in *Sepioteuthis*, and some other cephalopod species, we find a several-to-several mapping of signal to function, with both signal flexibility and functional flexibility.

Multiple Signaling in *Sepioteuthis Sepioidea*

Another specialty feature of the Caribbean reef squid communication system is seen in production of two to three different signals at the same time. The squid either uses the dorsal and ventral side of the body or divides the body in half longitudinally, thus sending different signals to the receiver on the left and on the right. In addition, *Sepioteuthis sepioide*a can combine social signals with startling signals like deimatic spots at the same time if it is advantageous to do so—for example, if a fish plunges into a social interaction among squid (Griebel et al., 2004). We propose that because of the selection pressures for unpredictable or protean behavior, the Caribbean reef squid has evolved a multiple signaling system that allows for combining almost all the signals in the repertoire in a very flexible way. The big question for which we do not have an answer at this point is whether these behaviors are innate or learned in cephalopods. Either way, flexible camouflage capabilities are clearly related to flexible communication capabilities in cephalopods.

Summary on Factors Favoring Contextual Flexibility

Consistent with our reasoning above, we propose that there exist two classes of contradictory selection pressures affecting signaling, one favoring stereotyped fixed signals and one favoring variability of signaling for specific purposes. The relative balance of these contradictory forces can be expected to incline the process of evolution either toward a stereotyped system that stays stereotyped or toward evolution from a primarily stereotyped system to one that also includes variability of production and usage.

The conditions that we have observed in this cross-species review that seem to favor the selection of one-to-several, several-to-one, or several-to several mapping for signals and functions include sexual selection, social cohesion (either between group members or parents and offspring), deception, camouflage, and possibly protean behavior. We do not intend to suggest that this list is complete but merely that these circumstances seem to have favored both signal and functional flexibility in cases that have been described sufficiently to justify their inclusion in the list.

Evolutionary Pressures Pertaining to Human Communication in Particular

Comments on the Possible Role of Gesture in the Origins of Language

Our focus with regard to the origins of human language is on the vocal–auditory channel of communication that is the primary one in all human societies other than those composed primarily of deaf individuals. On the other hand, there has been considerable advocacy for the idea that language originated in gestural acts rather than in vocal ones (see the review in Corballis, 2000). The argument takes stock of the tendency of human infants to use gesture importantly in pointing and requesting early in life and of the fact that languages of trade often include an extensive role for gesture. Further, the argument in favor of gestural origins draws on the fact that nonhuman primates show much more gestural flexibility than vocal flexibility (see Call, this volume) and that apes have been capable of learning the rudiments of a sign language system even though they appear incapable of vocal learning to any significant extent.

It is, of course, possible that gesture played a role in early language evolution, just as gesture plays a role in modern language. However, even if gesture did play a particularly important role at some early stage in hominin evolution, it is still required in the search for evolutionary origins of language to determine how vocal flexibility emerged. A gestural theory does not automatically solve that problem. The reasoning regarding the possible origins of vocal flexibility in humans presented in this chapter would apply even if gesture did play a particularly important role in the early evolution of language.

Given what is presently known about language, development of language, and comparative communication systems, we differ from gestural-origin advocates and favor the view that the earliest origins of language were primarily vocal. Here are a few summarial reasons for our preference: (1) Gestural language is never the primary communicative system in human societies composed of hearing individuals, (2) human infants communicate vocally with considerable elaborateness in the first months of life (see Oller and Griebel, this volume), but their gestural communication (as important as it is) begins most importantly with pointing late in the second half year of life, (3) the manual channel is largely nullified when the hands are occupied as with carrying ojects or children, and importantly, vocal communication does not interfere with use of the hands for these other purposes, (4) the vocal channel is largely free for communication (it has no other primary uses), while the manual channel is complicated by many hand movements that are intended for other purposes, and consequently a vocalization tends to be interpreted as a potential communication to a much greater extent than a hand movement, (5) vocalization is effective in darkness, when the potential audience is out of view, or when the potential audience is not looking at the sender, and (6) acknowledging that some sign language can be learned by certain apes, it is important to remember that no sign language has emerged in the wild in a nonhuman primate and that natural gestural communication in apes is very unlike language (see Call, this volume, for examples of natural ape gestures). Thus, the present chapter seeks to explain the flexibility of communication in humans primarily in the vocal domain, even though the review also takes account of flexible communication and foundations for it in other modalities.

Comments on the Vocal Origin of Language

In the human case, natural selection produced explosive flexibility in vocalization and vocal usage. Our hypotheses about the special circumstances that led to this evolutionary pattern include the following: Since human infants are probably the most altricial offspring in the animal kingdom and dependent on their parents for a period of six to eight years (and ancient hominin infants were already more altricial than other primates and dependent on their parents for long periods), we think a likely scenario for the evolution of signal flexibility was based on a special premium in ancient hominins for parent–offspring bonding. We have proposed (Oller and Griebel, 2005) along with Locke (2006) that modern infants use (and ancestral hominin infants used) early vocal behavior and babbling to elicit attention and care, a contention that is also consistent with reasoning of Snowdon (2004). The more variable and elaborate the sounds produced by the infant, the more effectively the sounds could elicit parental care, and thus, according to the reasoning, there was selection for variability, that is, signal flexibility itself. Details regarding how we propose infant vocal behavior to have provided the basis for parental selection on variable vocalization as well as empirical data on parental selection based on vocalization in modern humans are found in Oller and Griebel (2005) and Locke (2006).

Importantly, we reason that the immediate effect of attention-getting through variable vocalization may have been supplemented by even more important effects in terms of long-term commitments of parents to infants (Locke, 2006). The balance of forces appears to have favored variability especially in the hominin case because the hominin infant was in special need, given its altriciality, to prove its fitness, and it was selected to do so, we reason, by exhibiting a complex vocal capability.

Another likely pressure favoring vocal variability in the hominins at a very early stage in differentiation from the primate background could have emerged in the context of social grooming as suggested above (Dunbar, 1996, 2004). Pleasure and comfort sounds may have been evolved in the hominins to reinforce grooming and, in an extended function, to reinforce social relationships directly, especially as group sizes increased across early hominin evolution.

Even further, once vocalizations were used to reinforce social relations in the bodily grooming circumstance, they could have begun to function as social pacifiers through vocalization alone. In this latter way, vocalizations may have begun to supplant bodily grooming, taking the form of vocal grooming (Morris, 1967) and making it possible for group sizes to increase further. We also note that hominins whose infancies included variable vocalizations developed to elicit parental investment would have been in a particularly strong position to use the same vocalization capability in the vocal grooming circumstance. Thus, we reason, at an early stage of hominin evolution, the special parental bonding needs of the hominin infant may have tipped the balance in favor of variability as opposed to stereotypy in vocalization, and additional social cohesion forces associated with grooming may have added to the advantages of variable vocalization.

The reasons we do not think it is likely that sexual selection played a particularly major role at the earliest phase of special hominin vocal evolution are based on comparative assessments. We do not know of any elaborate and variable mating songs among the primates. Such sexual displays appear to be rare in mammals in general and only occur in specific circumstances such as those pertaining to lekking systems (consider, e.g., marine mammals such as humpback whales or walruses). Sexual selection may, of course, have played a significant role in the evolution of language at a later point in time; our contention is merely that it seems likely to have played little role at the point when hominins began to evolve away from the general primate pattern of vocalization.

Similarly we do not know of any example where deception or protean behavior created major innovations in vocal variability in primates or other mammals. And so we doubt that these factors played important roles at the very beginning of hominin evolution of vocal variability. At the same time, it is clear that deception and protean behavior (razzle-dazzle 'em, and baffle 'em with yer blarney; see Miller, 2000) are major usages of modern human language, and so we reason that these factors could have come into play once a neurological substrate for vocal contextual flexibility had been established. Thus, our proposal is that human communicative flexibility emerged first under pressures of social cohesion, especially related to parent–infant bonding and the needs of grooming.

Acknowledgments

This work has been supported by a grant from the National Institutes of Deafness and other Communication Disorders (R01DC006099-01 to D. K. Oller, Principal Investigator, and Eugene Buder, Co-Principal Investigator), by the Konrad Lorenz Institute for Evolution and Cognition Research, and by the Plough Foundation.

References

Alexander RD. 1961. Behaviour aggressiveness, territoriality and sexual behavior in field crickets (Orthoptera: Gryllidae). *Behaviour* 17: 130–223.

Austin JL. 1962. *How to Do Things with Words*. London: Oxford University Press.

Baptista LF, Petrinovich L. 1986. Song development in the white-crowned sparrow: Social factors and sex differences. *Animal Behaviour* 34: 1359–71.

Becker ML, Buder EH, Ward JP. 2000. Investigating communicative functions of vocalization patterns in *Otolemur garnetti* mothers and infants. *International Journal of Primatology* 35: 415–456.

Bigg MA, Ellis GM, Ford JKB, Balcomb KC. 1987. *Killer Whales—A Study of Their Identification, Genealogy and Natural History in British Columbia and Washington State*. Nanaimo, BC: Phantom Press.

Blomqvist C, Mello I, Amundin M. 2005. An acoustic play-fight signal in bottlenose dolphins (*Tursiops truncatus*) in human care. *Aquatic Mammals* 31: 187–94.

Byrne RW, Whiten A. 1985. Tactical deception of familiar individuals in baboons. *Animal Behaviour* 33: 669–73.

Byrne RW, Whiten A. 1988. *Machiavellian Intelligence: Social Expertise and the Evolution of Intellect in Monkeys*. Oxford: Clarendon Press.

Byrne RW, Whiten A. 1990. Tactical deception in primates: The 1990 database. *Primate Report* 27: 1–101.

Caldwell MC, Caldwell DK. 1965. Individualized whistle contours in bottlenose dolphins (*Tursiops truncatus*). *Science* 207: 434–5.

Caldwell MC, Caldwell DK, Tyack PL. 1990. A review of the signature whistle hypothesis for the Atlantic bottlenose dolphin, *Tursiops truncatus*. In *The Bottlenose Dolphin: Recent Progress in Research*, ed. S Leatherwood, R Reeves, pp. 199–243. San Diego, CA: Academic Press.

Case LP. 2003. *The Cat: Its Behavior, Nutrition, and Health*. Ames, IA: Iowa State University Press.

Chance MRA, Russell WMS. 1959. Protean displays: A form of allaesthetic behavior. *Proceedings of the Zoological Society of London* 132: 65–70.

Connor RC, Smolker RA. 1996. ‘Pop’ goes the dolphin: A vocalization male bottlenose dolphins produce during consortships. *Behaviour* 133: 643–62.

Corballis MC. 2002. *From Hand to Mouth: The Origins of Language*. Princeton, NJ: Princeton University Press.

de Waal F. 1982. *Chimpanzee Politics*. London: Jonathan Cape.

Dawkins R, Krebs JR. 1978. Animal signals: Information or manipulation? In *Behavioural Ecology: An Evolutionary Approach*, ed. JR Krebs, NB Davies, pp. 282–309. Sunderland, UK: Sinauer Associates.

Driver PM, Humphries N. 1988. *Protean Behavior: The Biology of Unpredictability*. Oxford: Oxford University Press.

Dunbar R. 1996. *Gossiping, Grooming and the Evolution of Language*. Cambridge, MA: Harvard University Press.

Dunbar RIM. 2004. Language, music and laughter in evolutionary perspective. In *The Evolution of Communication Systems: A Comparative Approach*, ed. DK Oller, U Griebel, pp. 257–74. Cambridge, MA: MIT Press.

Elowson AM, Snowdon CT, Lazaro-Perea C. 1998. ‘Babbling’ and social context in infant monkeys: Parallels to human infants. *Trends in Cognitive Sciences* 2: 31–7.

Fernald A. 1989. Intonation and communicative intent in mothers’ speech to infants: Is the melody the message? *Child Development* 60: 1497–510.

Fernald A, Simon T. 1984. Expanded intonation contours in mothers’ speech to newborns. *Developmental Psychology* 20: 104–13.

Fernald A, Taeschner T, Dunn J, Papoušek M, de Boysson-Bardies B, Fukui I. 1989. A cross-language study of prosodic modifications in mothers’ and fathers’ speech to preverbal infants. *Journal of Child Language* 16: 477–501.

Ford JKB. 1991. Vocal traditions among resident killer whales (*Orcinus orca*) in coastal waters of British Columbia. *Canadian Journal of Zoology* 69: 1454–83.

Fouts RS. 1987. Chimpanzee signing and emergent levels. In *Cognition, Language and Consciousness: Integrative Levels*, ed. G Greenberg, E Tobach, pp. 57–84. Hillsdale, NJ: Lawrence Erlbaum Associates.

Fraser B. 1975. Hedged performatives. In *Syntax and Semantics: Speech Acts*, ed. P Cole, JL Morgan, pp. 187–210. New York: Academic Press.

Gardner RA, Gardner BT. 1969. Teaching sign language to a chimpanzee. *Science* 165: 664–72.

Gibson KR. 1990. Tool use, imitation, and deception in cebus. In *"Language" and Intelligence in Monkeys and Apes: Comparative Developmental Perspectives*, ed. ST Parker, KR Gibson, pp. 205–18. New York: Cambridge University Press.

Grammer K. 1995. *Signale der Liebe, 3: Neu Überarbeitete Auflage*. Munich, Germany: dtv-Wissenschaft.

Grammer K, Honda R, Schmitt A, Juette A. 1999. Fuzziness of nonverbal courtship communication: Unblurred by motion energy detection. *Journal of Personality and Social Psychology* 77: 509–24.

Grammer K, Kruck K, Juette A, Fink B. 2000. Non-verbal behavior as courtship signals: The role of control and choice in selecting partners. *Evolution and Human Behavior* 21: 371–90.

Grammer K, Kruck K, Magnusson MS. 1998. The courtship dance: Patterns of nonverbal synchronization in opposite-sex encounters. *Journal of Nonverbal Behavior* 22: 3–29.

Green JA, Jones LE, Gustafson GE. 1987. Perception of cries by parents and nonparents: Relation to cry acoustics. *Developmental Psychology* 23: 370–82.

Griebel U, Mather JA, Oller DK. 2004. *Double signaling in the Caribbean reef squid Sepioteuthis sepioidea*, 65th Annual Conference of the Association of Southeastern Biologists, Memphis, TN.

Hanlon RT, Messenger JB. 1996. *Cephalopod Behaviour*. Cambridge, England: Cambridge University Press.

Hanlon RT, Naud M-J, Shaw PW, Havenhand JN. 2005. Transient sexual mimicry leads to fertilization. *Nature* 433: 212.

Hausberger M. 1997. Song acquisition and sharing in the starling. In *Social Influences on Vocal Development*, ed. CT Snowdon, M Hausberger, pp. 57–84. Cambridge, England: Cambridge University Press.

Herzing DL. 1996. Vocalizations and associated underwater behaviour of free-ranging Atlantic spotted dolphins, *Stenella frontalis* and bottlenose dolphins, *Tursiops truncatus. Aquatic Mammals* 22: 61–79.

Hockett C. 1960. Logical considerations in the study of animal communication. In *Animal Sounds and Communication*, ed. WE Lanyon, WN Tavolga. pp. 392–430. Washington, DC: American Institute of Biological Sciences.

Hockett CF, Altmann SA. 1968. A note on design features. In *Animal Communication: Techniques of Study and Results of Research*, ed. TA Sebeok. pp. 61–82. Bloomington: Indiana University Press.

Holmes W. 1940. The color changes and color patterns of *Sepia officinalis L. Proceedings of the Zoological Society of London* A 110: 17–35.

Huxley JS. 1914. The courtship-habits of the great crested grebe (*Podiceps cristatus*); with an addition to the theory of sexual selection. *Proeedings of the Zoological Society of London* 35: 491–562.

Jürgens U, Ploog DW. 1976. Zur Evolution der Stimme. *Archiv für Psychiatrie und Nervenkrankheiten* 222: 117–37.

Kaminski J, Call J, Fischer J. 2004. Word learning in a domestic dog: Evidence for "fast mapping." *Science* 304: 1682–3.

Kroodsma DE. 1999. Making ecological sense of song development. In *The Design of Animal Communication*, ed. MD Hauser, M Konishi, pp. 319–42. Cambridge, MA: MIT Press.

Kwon K, Buder EH, Oller DK. 2006. *Contextual flexibility in precanonical infant vocalizations: Its role as a foundation for speech*, International Child Phonology Conference, Edmonton, AB, Canada.

Lammers MO, Au WL. 2003. Directionality in the whistles of Hawaiian spinner dolphins (Stenella longirostris): A signal feature to cue direction of movement? *Marine Mammal Science* 19: 249–64.

LeBerre F, Bartlett RD, Bartlett P. 2000. *The Chameleon Handbook*. New York: Barron's.

Lemasson A, Gautier JP, Hausberger M. 2003. Vocal similarities and social bonds in Campbell's monkey. *Comptes Rendues Biologiques* 326: 1185–93.

Lemasson A, Hausberger M. 2004. Patterns of vocal sharing and social dynamics in a captive group of Campbell's monkeys (*Cercopithecus campbelli campbelli*). *Journal of Comparative Psychology* 118: 347–59.

Lemasson A, Hausberger M, Zuberbuhler K. 2005. Socially meaningful vocal plasticity in adult Campbell's monkeys (*Cercopithecus campbelli*). *Journal of Comparative Psychology* 119: 220–9.

Lloyd JE. 1986. Oh, what a tangled web. In *Deception: Perspectives on Human and Nonhuman Deceit*, ed. RW Mitchell, NS Thompson, pp. 113–28. Albany, NY: State University of New York Press.

Locke JL. 1993. *The Child's Path to Spoken Language*. Cambridge, MA: Harvard University Press.

Locke JL. 2006. Parental selection of vocal behavior: Crying, cooing, babbling, and the evolution of language. *Human Nature* 17: 155–68.

Lorenz K. 1951. Ausdrucksbewegungen höherer Tiere. *Naturwissenschaften* 38: 113–6.

Lyons J. 1991. *Natural Language and Universal Grammar*. Cambridge, England: Cambridge University Press.

Marler P, Slabbekorn H. 2004. *Nature's Music: The Science of Birdsong*. Amsterdam: Elsevier Academic Press.

Maynard Smith J, Harper D. 2003. *Animal Signals*. Oxford: Oxford University Press.

McCowan B, Reiss D. 1995. Quantitative comparison of whistle repertoires from captive adult bottlenose dolphins (*Delphinidae, Tursiops truncatus*): A re-evalutation of the signature whistle hypothesis. *Ethology* 100: 193–209.

McCowan B, Reiss D. 1997. Vocal learning in captive bottlenose dolphins: A comparison with humans and nonhuman animals. In *Social Influences on Vocal Development*, ed. CT Snowdon, M Hausberger, pp. 178–207. Cambridge, England: Cambridge University Press.

Messenger JB. 2001. Cephalopod chromatophores: Neurobiology and natural history. *Biological Review* 76: 473–528.

Miller G. 2000. *The Mating Mind: How Sexual Choice Shaped the Evolution of Human Nature*. New York: Doubleday.

Miller PJO, Bain DE. 2000. Within-pod variation in the sound production of a pod of killer whales, *Orcinus orca*. *Animal Behaviour* 60: 617–62.

Mitchell RW, Thompson NS. 1986. *Deception: Perspectives on Human and Nonhuman Deceit*. Albany, NY: State University of New York Press

Morris D. 1967. *The Naked Ape*. New York: Dell.

Morris D. 1986. *Catwatching: Why Cats Purr and Everything Else You Ever Wanted to Know*. New York: Random House.

Moynihan MH, Rodaniche AF. 1982. The behaviour and natural history of the Caribbean reef squid *Sepioteuthis sepioidea* with a consideration of social, signal and defensive patterns for difficult and dangerous environments. *Advances in Ethology* 125: 1–150.

Nooteboom SG. 1999. Anatomy and timing of vocal learning in birds. In *The Design of Animal Communication*, ed. MD Hauser, M Konishi, pp. 63–110. Cambridge, MA: MIT Press.

Nottebohm F. 1981. A brain for all seasons: Cyclical anatomical changes in song control nuclei of the canary brain. *Science* 214: 1368–70.

Oller DK. 2000. *The Emergence of the Speech Capacity*. Mahwah, NJ: Lawrence Erlbaum.

Oller DK. 2004. Underpinnings for a theory of communicative evolution. In *The Evolution of Communication Systems: A Comparative Approach*, ed. DK Oller, U Griebel, pp. 49–65. Cambridge, MA: MIT Press.

Oller DK, Buder EH, Nathani S. 2003. *Origins of speech: How infant vocalizations provide a foundation.* Miniseminar for the American Speech-Language Hearing Association Convention, Chicago.

Oller DK, Griebel U. 2005. Contextual freedom in human infant vocalization and the evolution of language. In *Evolutionary Perspectives on Human Development*, ed. R Burgess, K MacDonald, pp. 135–66. Thousand Oaks, CA: Sage Publications.

Oller DK, Nathani Iyer S, Buder EH, Kwon K, Chorna L, Conway K. 2007. *Diversity and contrastivity in prosodic and syllabic development*. Proceedings of the International Congress of Phonetic Sciences, ed. J Trouvain, W. Barry, pp. 303–308. Saarbrucken, Germany: International Phonetics Society.

Owren MJ, Rendall D. 2001. Sound on the rebound: Bringing form and function back to the forefront in understanding nonhuman primate vocal signaling. *Evolutionary Anthropology* 10: 58–71.

Packard NH, Crutchfield JP, Farmer JD, Shaw RS. 1980. Geometry from a time series. *The American Physical Society* 45: 712–6.

Panksepp J. 2000. The riddle of laughter: Neuronal and psychoevolutionary underpinnings of joy. *Current Directions in Psychological Science* 9: 183–6.

Papoušek M. 1994. *Vom Ersten Schrei zum Ersten Wort: Anfänge der Sprachentwickelung in der Vorsprachlichen Kommunikation*. Bern: Verlag Hans Huber.

Papoušek M, Bornstein MH, Nuzzo C, Papoušek H, Symmes D. 1990. Infant responses to prototypical melodic contours in parental speech. *Infant Behavior and Development* 13: 539–45.

Papoušek M, Papoušek H, Symmes D. 1991. The meanings of melodies in motherese in tone and stress languages. *Infant Behavior and Development* 14: 414–40.

Parker GA. 1974. Assessment strategy and the evolution of animal conflicts. *Journal of Theoretical Biology* 47: 223–43.

Payne RB, Payne LL. 1997. Field observations, experimental design, and the time and place of learning bird songs. In *Social Influences on Vocal Development*, ed. CT Snowdon, M Hausberger, pp. 57–84. Cambridge, England: Cambridge University Press.

Payne RS, McVay S. 1971. Songs of the humpback whale. *Science* 173: 583–97.

Pepperberg IM. 2004. The evolution of communication from an avian perspective. In *The Evolution of Communication Systems: A Comparative Approach*, ed. DK Oller, U Griebel, pp. 171–92. Cambridge, MA: MIT Press.

Riesch R, Ford JKB, Thomsen F. 2006. Stability and group specificity of stereotyped whistles in resident killer whales, *Orcinus orca*, off British Columbia. *Bioacoustics* 71: 79–91.

Savage-Rumbaugh ES. 1988. A new look at ape language: Comprehension of vocal speech and syntax. In *Comparative Perspective in Modern Psychology, Nebraska Symposium on Motivation*, ed. D Leger, pp. 201–56.

Sayigh LS, Tyack RS, Wells RS, Scott MD. 1990. Signature whistle of free-ranging bottlenose dolphins, *Tursiops truncatus*: Stability and mother–offspring comparisons. *Behavioral Ecology and Sociobiology* 26: 247–60.

Schusterman RJ, Hanggi EB, Gisiner R. 1992. Acoustic signalling in mother–pup reunions, interspecies bonding, and affiliation by kinship in California sea lions (*Zalophus califonianus*). In *Marine Mammal Sensory Systems*, ed. J Thomas, RA Kastelein, AY Supin, pp. 533–51. New York: Plenum Press.

Snowdon C. 2004. Social processes in the evolution of complex cognition and communication. In *Evolution of Communication Systems: A Comparative Approach*, ed. DK Oller, U Griebel, pp. 131–50. Cambridge, MA: MIT Press.

Snowdon CT, Elowson AM, Rousch RS. 1997. Social influences on vocal development in New World primates. In *Social Influences on Vocal Development*, ed. CT Snowdon, M Hausberger, pp. 234–48. New York: Cambridge University Press.

Stark RE. 1978. Features of infant sounds: The emergence of cooing. *Journal of Child Language* 5: 379–90.

Stark RE. 1980. Stages of speech development in the first year of life. In *Child Phonology, Vol. 1*, ed. GY Komshian, J Kavanagh, C Ferguson, pp. 73–90. New York: Academic Press.

Stokoe WC. 1960. Sign language structure: An outline of the visual communication systems of the American Deaf. *Studies in Linguistics, Department of Anthropology, University of Buffalo* Occasional papers 8: 1–78.

Struhsaker TT. 1967. Auditory communication among vervet monkeys (*Cercopithecus aethiops*). In *Social Communication among Primates*, ed. SA Altmann, pp. 281–324. Chicago: Chicago University Press.

Sutton D. 1979. Mechanisms underlying learned vocal control in primates. In *Neurobiology of Social Communication in Primates: An Evolutionary Perspective*, ed. HD Steklis, MJ Raleigh, pp. 45–67. New York: Academic Press.

Terrace HS. 1979. *Nim: A Chimpanzee Who Learned Sign Language*. New York: Columbia University Press.

Terrace HS, Petitto LA, Sanders RJ, Bever TG. 1979. Can an ape create a sentence? *Science* 206: 891–902.

Tinbergen N. 1951. *The Study of Instinct*. Oxford: Oxford University Press.

Tyack PL. 1999. Communication and cognition. In *Biology of Marine Mammals*, ed. JE Reynolds, SA Rommel, pp. 287–323.

Tyack PL. 2000. Dolphins whistle a signature tune. *Science*: 1310–1.

Tyack PL, Sayigh L. 1997. Vocal learning in cetaceans. In *Social Influences on Vocal Development*, ed. CT Snowdon, M Hausberger, pp. 208–33. New York: Cambridge University Press.

Weiß BM, Ladich F, Spong P, Symonds H. 2006. Vocal behavior of resident killer whale matrilines with newborn calves: The role of family signatures. *Journal of the Acoustical Society of America* 119: 627–35.

West MJ, King AP, Freeberg TM. 1997. Building a social agenda for the study of bird song. In *Social Influences on Vocal Development*, ed. CT Snowdon, M Hausberger, pp. 41–56. Cambridge, England: Cambridge University Press.

Whiten A, Byrne RW. 1988. The manipulation of attention in primate tactical deception. In *Machiavellian Intelligence*, ed. RW Byrne, A Whiten, pp. 211–23. Oxford: Clarendon Press.

3 Vocal Learning in Mammals with Special Emphasis on Pinnipeds

Ronald J. Schusterman

Introduction

At the most general level, questions about the relative rigidity or plasticity of animal communication systems are relevant to a wide variety of subject areas, including behavioral ecology, sociobiology, comparative cognition, and evolutionary and developmental linguistics. As a result, scientists from diverse fields have attempted to identify the form and function of communicative signals produced by a variety of different species, particularly those occurring in the acoustic domain. In the late 1950s, the prevailing view among prominent classical ethologists and behavioral psychologists was that the acoustic signals emitted by animals were ritualized and not modifiable as they are in humans. Behavioral scientists such as B. F. Skinner (1957) typically believed that behavioral changes (as expressed by various nonvocal responses like bar pressing in rats) could occur as a result of environmental experiences but that the vocal responses of nonhuman animals were far less flexible than those of humans. Instead, emotional/motivational constraints, anatomical limitations, and genetic predispositions were believed to control animal vocalizations while environmental or behavioral consequences controlled human spoken language. Despite scattered accounts and observations that some songbirds and parrots appeared capable of mimicking sounds, behavioral scientists of the time generally remained convinced that animal vocalizations were involuntary responses that were exclusively used to express emotions signaling the probability of attack, flight, or courtship. As a result, the potential influence of learning on animal vocal responses was largely discounted and poorly investigated.

This traditional viewpoint was later updated for songbirds, as a result of experimental evidence that revealed a propensity for vocal learning that can be characterized as a sensorimotor skill (see Wilbrecht and Nottebohm, 2003). This skill is acquired when a young bird, during a sensitive period of ontogeny, perceives sound produced by adult conspecifics, remembers what it has heard, and at some later stage of development vocally imitates the adult model. This definition of vocal learning as the modification of vocal output through reference to extrinsic auditory information has at its core an ethological focus and leans heavily on such constructs as "learning instinct," "critical period," and "sensory gating mechanisms."

This general model of vocal learning in songbirds was derived in large part from a series of field and laboratory observations on the song dialects of white-crowned sparrows carried out by Peter Marler and his colleagues in the San Francisco Bay area of California (e.g., Marler and Tamura, 1962). In the field, these sparrows were observed to exhibit small-scale regional differences in song structure that could not easily be explained by genetic differences between overlapping populations. In order to further investigate whether experiential factors were contributing to these differences in vocal production, Marler's team moved to the laboratory and applied research methodologies that depended on passively tutoring young birds by exposing them to tape recordings of conspecific and extraspecific song stimuli. In doing so, they experimentally demonstrated that while vocal learning does occur in songbirds following passive exposure to auditory information, the process is instinctive, restricted to a sensitive period, and constrained by a species-specific song template. Marler's pioneering studies of vocal learning in songbirds relied on tape-recorded songs and thus eschewed live tutors capable of interacting socially with young birds. A more behaviorally interactive approach developed by Baptista and Petrinovich (1984) later showed that vocal learning may occur beyond the recognized sensitive period and can include anomalous or allospecific vocalizations when avian social tutors replace recorded songs.

The degree to which the vocalizations of some birds may be modified by learning occurring through social feedback continues to be clarified. For example, West and King (1985) found that female cowbirds perform a subtle wing stroke display during a male's serenade in response to acoustic structures in the male's performance. Mature males respond to the wing stroke by altering their songs to include more of those vocal patterns that are more effective at eliciting the female's copulatory posture. In summary, females react more positively to some song variants than others, and such social reinforcement modifies male vocal behavior. Further, other investigators have shown that some young male songbirds with a history of socially interactive experiences can acquire a repertoire of song syllables by means of operant conditioning (Adret, 1993a) or by a process that Marler (1990) has termed "action-based learning." One example of such learning comes from Adret (1993b), who concluded from a study of young zebra finches that operant conditioning, with conspecific songs serving as reinforcers, increases the attention of listeners toward the detail of song phrases, thereby increasing the motivation of the bird to imitate the tape-recorded song. Pepperberg (1986) has also shown that social interactions between a human tutor and an African gray parrot greatly facilitate the imitation of human speech by the parrot.

Studies such as those just described, which show that the call learning by some adult passerines and psittacines are influenced by social experiences, have been expanded to show that operant conditioning of vocalizations can be accomplished with not only social reinforcers but food rewards as well. Innovative operant procedures devised by Manabe and his colleagues (Manabe, Kawahima, and Staddon, 1995; Manabe, Staddon, and Cleaveland, 1997) have used computer-based systems and food reinforcement to train budgerigars to alter frequency characteristics of their calls and to emit these calls in novel contexts.

Collectively, the evidence reviewed here suggests that the production or acoustic structure and contextual usage of natural vocalizations emitted by mature birds of certain species can be modified by the selective distribution of social and food reinforcement. It is likely that the behavioral scientists of just a generation ago would consider such findings to be remarkable. While human utterances have traditionally been viewed and continue to be viewed as subject to powerful learning effects, there remains little evidence for evolutionary continuity in vocal learning processes between human and nonhuman animal species. The introduction of contingency learning—the modification of behavior through selective reinforcement—as a field of special interest within both human and nonhuman animal vocal learning may provide significant clues to the common bases of communication systems. For example, recent research on human speech acquisition indicates that contingency learning and social shaping may serve as general processes that underlie the ontogeny of speech in humans as well as song development in birds (Goldstein, King, and West, 2003). This research has shown that the babbling of eight-month-old infants sounds progressively more speech-like when mothers socially reinforce recognizable syllables emerging during this maturational period than do comparable sounds that are produced by control subjects who are not given such positive feedback. This work has striking parallels to the research previously described on social influences in song learning by male cowbirds.

Thus, the extent to which the vocalizations of nonhuman mammals may be modified through experience is especially interesting from a comparative perspective. While more is currently known about how some mammals learn to respond to different auditory cues, little is understood about if and how readily they learn to adapt their own vocal behavior in response to changing auditory and social environments. In this chapter, I focus on contingency learning as a tool for understanding how the vocalizations that are emitted by mammals may be influenced by reinforcement contingencies. If we are to understand the evolution of spoken language or human speech, then it is necessary to study the vocal behavior of those species that possess adequate vocal complexity. Such an animal group must have "a capacity for vocal learning and a vocal tract with a wide phonetic range" (Fitch, 2000). Moreover, such candidates for studying the evolution of spoken language must vocalize frequently.

This is necessary because vocal behavior is more likely to develop depending on the degree to which it competes with other responses (e.g., limb movement) for access to positive reinforcement (Salzinger, 1973).

The Search for Vocal Learning in Nonhuman Mammals

Support for Classical Perspectives on Flexibility in Vocalizations

In contrast to the research on certain bird species that has identified vocal learning as a significant influence in avian acoustic communication, for the most part, the traditional viewpoint about the relative inflexibility of animal vocal responses has been substantiated to the present day by studies of mammalian vocal production. Results from a good deal of research, especially on nonhuman primates, suggest that the acoustic structure of species-typical mammalian vocalizations develops within a relatively closed genetic program. For example, deafening at birth produces only minor modifications in vocal development in some monkeys (Hammerschmidt, Freudenstein, and Juergens, 2001). Further, nonhuman primates raised either in isolation or with foster mothers of a different species still emit vocalizations that are species typical (Owren, Dieter, Seyfarth, and Cheney, 1993). In a recent review of vocal development in nonhuman primates, Seyfarth and Cheney (1997) point out that while there is abundant evidence of flexibility and modification in the response that individuals perform upon hearing calls, the evidence for plasticity in the contextual usage of emitted vocalizations is more limited and there is virtually no evidence for learned (nonmaturational) modifications in call structure. These observations suggest that it is unlikely that nonhuman primates have voluntary control of their vocal production apparatus, which is more likely under the direct control of affectively linked systems.

It is also significant that experimenters attempting to place primate vocalizations under stimulus control in laboratory situations have typically had limited or no success. While failed attempts have remained generally unpublished, the difficulties inherent in such conditioning procedures are well-known among primatologists (M. Owren, personal communication; R. J. Schusterman, personal observation). These include the observations that only certain call types, combined with specific, appropriate reinforcement types, may be useful in conditioning procedures at all (see Pierce, 1985).

These observations with nonhuman primates are consistent with the view that the vocal signals of mammalian species are automatic and a fixed part of instinctive behavior. This view has been partially explained by theories suggesting that mammalian vocal musculature is constrained by direct emotional controls and is therefore unavailable for modification by means of operant conditioning techniques (Salzinger and Waller, 1962). If true, this would predict that mammals should lack voluntary control over the usage of their vocal behavior and therefore be unable to learn when and under what conditions and contexts to use a call. According to this concept, a call type that is associated with a particular eliciting stimulus should not easily be mapped onto novel arbitrary stimuli associated with a different or neutral affective context, and therefore, the call type cannot come to acquire novel functions. Based on this rationale, mammals other than humans have generally been considered incapable of learning to use their vocalizations independent of emotional contexts. Such contextual flexibility in vocal responses has recently been termed "vocal usage learning" by Janik and Slater (2000). Moreover, mammals were also believed to be incapable of what Janik and Slater have termed "vocal production learning," or the ability to arbitrarily alter the physical structure of their sounds or to imitate sounds from social or nonsocial sources.

A final idea with respect to the fixed nature of vocal behavior in mammals can be derived from well-established principles regarding biological constraints on learning. These principles state that there are limits on the extent to which some behaviors can be modified by contingencies that depend on reinforcers that have no natural relationship to the response being emitted. For example, while it is quite easy to train a rat to press a lever to receive food or water reinforcement, it is difficult, if not impossible, to teach the same behavior using shock avoidance. Conversely, the same rat can be easily trained to jump or run using shock avoidance techniques, while these escape behaviors are not readily learned using food reinforcement (Bolles, 1970). One might expect that, in situations where the natural vocalizations of mature animals have no normal functional or causal relationship with food, the modification of vocal behavior using food rewards might be at best difficult or at worst not possible. On the other hand, perhaps vocal conditioning using food rewards can more readily occur in species that emit acoustic signals to detect and locate food (e.g., the use of echolocation signals by dolphins and bats). Of course, any of these possibilities involving vocal conditioning would depend on call production's being under volitional control to some degree.

For all of the above reasons, including the apparent fixedness of mammalian sound production, probable emotional constraints, presumed anatomical inflexibility, and possible biological limitations on vocal learning, it would seem unlikely that comparative studies of mammalian vocalizations would play an important role in the search for precursors of complex communication systems such as human speech. However, there is now accumulating evidence to suggest that signals used in vocal communication by some mammals are indeed accessible to discriminative control and selective shaping by the process of contingency learning using positive reinforcement (see box 3.1).

Box 3.1
Instrumental and operant conditioning in relation to the vocal behavior of mammals

Instrumental conditioning (as first described by Edward Thorndike) and operant conditioning (as articulated by B. F. Skinner) are usually treated as functionally equivalent concepts describing general principles of learning. However, Thorndike, unlike Skinner, emphasized the importance of species-typical or species-specific responses elicited by identifiable stimuli and strengthened by appropriate reinforcement (Thorndike, 1911). Therefore, Thorndike's early ideas about learning actually fit well with our current understanding of biological constraints on learning. Skinner, on the other hand, emphasized that operants were never elicited responses but rather novel or arbitrary responses (Skinner, 1938). Skinner considered these behavioral units to be *emitted* rather than *elicited* responses because these operants have no identifiable eliciting stimulus. Reinforcement increases the future probability of operant responding when its delivery is made contingent on the emitted behavior. Moreover, an arbitrary stimulus can become discriminative for the behavior because it evokes but does not elicit the operant response that it precedes and accompanies.

With respect to vocal learning, it is likely that the general principles shared by both instrumental and operant conditioning paradigms can lead to discriminative control of vocal behavior. Once such control has been established, modification of sound production through selective reinforcement is possible, and thus, vocal learning through the process of contingency learning can occur. The distinction between instrumental and operant conditioning in this regard becomes relevant when we consider both the nature of the response and the reinforcer. The finding that vocal behavior in some species can come to occur reliably in the presence of an arbitrary discriminative cue that is established by using food reinforcement suggests vocal conditioning may more closely follow the instrumental paradigm because the anticipation of food is clearly important. Skinner (1957) himself did not believe that mammalian vocalizations were voluntary, and therefore were not subject to operant control. This is clearly not the case, as the last four decades of research on conditioning of "involuntary" autonomic responses has shown. It is quite possible that Skinnerian conditioning procedures can be used to acquire discriminative control of vocal responses, and it is likely that such procedures, when applied experimentally to the study of sound production in some mammals, will reveal a degree of plasticity and environmental influence heretofore not described. However, there is a strong caveat here because it is well-known that vocalizations like food coos of macaque monkeys as well as baboon grunts and the food calls of chimps are critically dependent on the underlying affective/motivational systems that elicit calling (Owren, personal communication).

Finally, because contingency learning essentially boils down to reinforcement training, it is prudent to carefully evaluate the properties of the reinforcer involved. While the simplest examples involve food (or water) reinforcement, other biological reinforcers, such as access to sexual partners, shelter, or safety can also be substituted into the reinforcement contingency. Further, as classical conditioning experiments have demonstrated, any other cues associated with such biological reinforcement can come to share reinforcement value as well. Thus, arbitrary cues as well as social stimuli can come to enter into situations where behavior is modified through positive reinforcement. As a consequence, it is possible that social reinforcement may be effective at strengthening and influencing sound production in social settings. This can be illustrated by the example of the cowbirds discussed earlier, where a vocal response (song production by a male) can be modified by a discriminative cue (the wing stroke of a female) that has come to be associated with biological reinforcement (copulatory posture).

Evidence in Support of Flexible Vocal Responses

The evidence suggesting inherent inflexibility in the vocal responses of mammals outlined in the previous section is somewhat paradoxical in light of the great malleability in both the structure and function of human utterances, which incidentally are readily accessible to a variety of operant conditioning situations including vocal mimicry. The notion of a total evolutionary disconnect between the vocal communication of humans and other animals was summarized most clearly and succinctly by Lenneberg (1967), who rejected the study of animal vocal communication because he claimed that the processes of human language were "deeply rooted, species-specific, innate properties of man's biological nature" (p. 394) and because "no living animal represents a direct primitive ancestor of our own kind and, therefore, there is no reason to believe that any one of their traits is a primitive form of any one of our traits" (p. 234–235). Today, however, studies of animal vocal learning and communication are considered relevant to larger scientific investigations of human language evolution. This approach is typified by Hauser, Chomsky, and Fitch (2002) and Fitch, Hauser, and Chomsky (2005), who have both recently pointed out that any faculty involved in language (e.g., vocal learning) is worthy of comparative study because it is part of a language faculty in a broad sense. These faculties may be contrasted, they say, with those mechanisms that are both specific to language and uniquely human, such as recursion, which is involved in the faculty of language in the narrower sense.

Consistent with this approach, Janik and Slater (1997) have recently pointed out that investigations within a comparative context "... enable us to compare and contrast birds and mammals and to consider the possible functional significance of vocal learning" (p. 60). In addition to this review, there have been several other reviews of vocal learning in the past thirty years (Salzinger, 1973; Adret, 1993a; Janik and Slater, 2000). These show that the phonations of such ecologically and phylogenetically diverse species as sea lions, dolphins, beluga whales, monkeys, cats, dogs, and rodents, can—at least in part and under the right circumstances—be acquired by operant conditioning techniques using food reinforcement. Furthermore, while several studies demonstrate that animals can learn to produce vocalizations in response to specific arbitrary cues, complementary studies have found that mammals can also learn to inhibit their vocalizations in the presence of different arbitrary cues (Schusterman and Feinstein, 1965; Schusterman, 1967). These findings, which demonstrate volitional control of sound production as well as inhibition of sound production (both of which occur at very different levels of internal control), show that the vocal response involved is not merely a function of an animal's state of arousal. They also provide evidence for call usage learning in the species whose vocalizations have been brought under stimulus control, as vocalizations controlled by particular emotional contexts come to be controlled by qualitatively different cues. Some laboratory experiments similar to those just described also touch on call production learning, where individuals learn to modify certain structural properties of their vocal emissions through experimenter manipulation of reinforcement contingencies. These properties include call rate, pitch, loudness, and duration (e.g., Adret, 1993a). While the acoustic structure of operantly conditioned calls can sometimes be purposefully shifted along these dimensions through selective reinforcement, Janik and Slater (1997) point out that these are all dimensions that are also intimately connected with arousal and affect and that modifications along these dimensions can be accomplished by mechanisms other than direct control of vocal musculature per se (e.g., through respiratory action). This again raises the issue of the influence of emotional constraints on flexibility in vocal responses. There is weaker experimental evidence for modifications of more arbitrary aspects of call structure, such as fundamental frequency and frequency modulation, and it remains to be seen to what extent these call dimensions are broadly available for modification. In addition to experimental data demonstrating that some mammals are able to control and modify their sound production, there is evidence from natural settings showing that some marine mammals, including whales, dolphins, and seals, and perhaps some terrestrial mammals, such as bats, can alter the acoustic structure of their sound emissions on the basis of auditory feedback from other individuals. This observational evidence, which is probably the strongest evidence of vocal learning in nonhuman mammals, is reviewed in detail elsewhere (Tyack and Sayigh, 1997; Janik and Slater, 1997).

While the information available on vocal learning in mammals is relatively sparse and limited to a handful of phylogenetically disparate species, the data that are available support the view that the study of vocal learning in mammals is relevant to the search for the evolutionary and ontogenetic origins of complex communication systems including human speech. Evidence that the vocalizations of several mammals, like humans and some birds, have some degree of emotional (fixed) control in addition to some degree of voluntary (flexible) control is significant. When taken alongside observations suggesting that the calls of these mammals may be modified by learning with respect to both structure and function, these findings suggest that contingency learning may be a general process that supports more complex vocal learning abilities.

More field and laboratory investigations are required to better understand how reinforcement contingencies can modify vocal responses. An example of a relevant field study comes from Seyfarth and Cheney (1986), who have shown that when an infant vervet monkey gives a contextually correct alarm call, it is frequently given positive feedback in the form of a second appropriate alarm call by an adult. These adult calls are consistent with the infant's alarm call and might serve as reinforcers that provide essential feedback on vocal responses. Positive feedback of this nature should facilitate the infant monkey's developing an understanding of an equivalence relation between different alarm calls and their referents (see Schusterman, Reichmuth Kastak, and Kastak, 2003). However, it is essential that experiments exploring such learning under highly controlled conditions, such as those conducted with human infants (Goldstein et al., 2003) and cowbirds (West and King, 1996), be explored with nonhuman mammals. Such experiments may help to disentangle the ecological, phylogenetic, and ontogenetic variables affecting vocal learning. Additionally, it is clear that more research is needed in the area of vocal conditioning using food reinforcement. Some recent examples of this approach comes from Adret (1993a) and Shapiro, Slater, and Janik (2004), who have outlined progressively more complicated examples of operant conditioning in the context of experimental studies. These include an initial experimental phase in which an animal's specific natural call types come to be controlled by different arbitrary stimuli. In the next phase, the animal has to use its generalization and cognitive skills to vocally respond to classes of stimuli. In the final phase, the subject has to match vocally the stimulus that it has just heard (i.e., barking when it hears a bark and growling when it hears a growl). Experiments such as these may be useful for a wide variety of reasons. For example, they would make possible anatomical investigations of vocal musculature in trained animals which may help to identify structural constraints on vocalizations. Further, because such operant conditioning research can be conducted with precision, such experiments may enable us to set procedural and measurement standards by which we can compare the vocal learning abilities of different mammalian species.

In the following sections, I introduce a group of mammals that may be especially well-suited for controlled studies of flexibility in sound production using food reinforcement—the pinnipeds. Following a brief introduction to these animals, I describe published experiments and new observations on their vocal usage and production learning capabilities. In contrast to several contemporary reports on primate vocal communication (see, e.g., Seyfarth and Cheney, 1997) this material emphasizes plasticity in the production of vocalizations rather than in the comprehension of acoustic signals. The research discussed here represents some of the most comprehensive and successful efforts to operantly condition vocal responses in nonhuman animals. The data will show that some pinnipeds can learn to alter the contexts in which they emit natural vocalizations, as well as modify their call structure along several structural dimensions, to an extent not known to occur in most terrestrial mammals.

The Case for Vocal Learning in Pinnipeds

A Brief Introduction to the Pinnipeds

The pinnipeds are semi-aquatic, carnivorous mammals consisting of 33 living species in three major families: phocids (true seals), otariids (sea lions and fur seals), and odobenids (walruses). Their ancestors were terrestrial carnivores that originated in temperate waters of the northern hemisphere about 30 million years ago and spread throughout the world's oceans, colonizing isolated shores and feeding in nearby productive waters. Pinnipeds combine two characteristics which occur together in no other mammals. They give birth to and attend their pups on land or ice and they feed exclusively at sea (Bartholomew, 1970). These amphibious mammals are especially interesting mammals to study regarding vocal learning for numerous reasons, some of which are briefly itemized here, and several of which are further explained in the following sections: (1) All pinnipeds spend a significant amount of time at sea as well as ashore. Because their visual range is limited in both media, most species rely heavily on acoustic modes of communication. (2) Like humans, pinnipeds not only produce many of their vocal signals in the larynx, but they have relatively good control over articulatory movements occurring above the larynx. Thus, the sounds produced by the larynx can be influenced by oral and nasal cavities as well as by movement of the tongue, lips, and mouth. Certain species also have highly specialized structures used in sound production. (3) Directly related to the above, some pinnipeds have specialized oral feeding musculature which makes them good candidates for motor conditioning of movements of the lips, mouth, tongue, teeth, and throat muscles. (4) Pinnipeds are diving mammals that have superior breath control relative to terrestrial mammals. Such breath control may be used during sound production both in air and under water but the mechanism for doing this is not well understood. (5) Since all pinniped pups are born on land or ice and utter airborne distress calls with their mouth open, upon entering the water, they must in some manner modify the way they produce their vocal signals. It is interesting to note that many of the sounds emitted by pinnipeds under water do not involve the release of air. (6) Several species that mate aquatically or semi-aquatically emit acoustically complex vocalizations or songs during the breeding season. (7) In at least seven species, there is evidence of intraspecific geographic variation in vocalizations. In some cases, this variation is suggestive of regional dialects similar to those found in humans, birds, and cetaceans (Tyack and Miller, 2002). (8) Vocal mimicry has been clearly demonstrated in a single male harbor seal. This seal, called "Hoover," was reared in New England by human caretakers before being transferred to an aquarium; as he matured, he spontaneously emitted a repertoire perceived as about twelve English words produced through a variety of vowel-like sounds. Anybody who listens to recordings of Hoover cannot fail to be impressed and amused by the seal's New England dialect and flawless belly laugh (Hiss, 1983; Rawls, Fiorelli, and Gish, 1985). Hoover imitated human speech at least as proficiently as any parrot I have ever heard. In an effort to replicate the Hoover finding, another male harbor seal named "Chimo" was trained using operant conditioning to emit human speech-like sounds but only eventually gave poor approximations of "hello" and "how are you" as well as several vowel sounds (Moore, 1996).

Acoustic Communication in Pinnipeds

Although contemporary pinniped species inhabit unique behavioral and ecological niches, another significant commonality among them is the apparent structural complexity of their phonations and the degree to which they used these vocal signals in communicative contexts (Schusterman and Van Parijs, 2003). Whereas whales and dolphins are known to use many of their sounds for the purpose of detecting prey, pinnipeds do not appear to possess similarly specialized echolocation abilities (Schusterman, Kastak, Levenson, Reichmuth, and Southall, 2000). Indeed, the sounds that are produced by pinnipeds are used almost exclusively for influencing or being influenced by other individuals—in other words, in some form of social communication (see Dawkins, 1995). Their amphibious lifestyle requires that social communication occur both in the atmosphere and the hydrosphere, therefore, most pinnipeds need to communicate acoustically in both media. All pinnipeds exhibit some degree of polygyny in their breeding behavior, and consequently, their vocal behavior is highly influenced by competition for mates among males.

Many pinnipeds, including all of the otariids and some of the terrestrially breeding phocids, congregate seasonally on traditional sites to breed and give birth. These rookeries can be very dense, with up to tens of thousands of individuals inhabiting relatively small stretches of coastline. The vocalizations produced in these settings are often loud and repetitive, and the call types produced include male threat/sexual displays, female threats, mothers calling to their pups, and pups calling for their mothers. Variation in many of these vocalizations tends to be greatest between individuals and more stereotyped within individuals. Playback experiments in the field demonstrate that individuals can be identified by distinctive acoustic cues, and that vocal recognition may occur between females and their dependent pups as well as between males competing for access to females during the reproductive season (see Insley, Phillips, and Charrier, 2003, for a review). In contrast to land breeding pinnipeds, many species of phocid seals interact sexually in the water and are more loosely spaced while ashore. Mature males of these aquatically breeding species tend to have unusual and complex underwater vocal repertoires or songs (for reviews, see Tyack and Miller, 2002; Van Parijs and Schusterman, 2003). Seal pups of most pinniped species rarely swim prior to weaning but, in highly precocial species such as the harbor seal, pups may swim soon after birth and can emit their signature calls in water as well as air (Perry and Renouf, 1988).

In addition to the obvious selective advantage of an acoustic mode of communication imposed by the physical environment, it is likely that crowding and competition for space have also played significant roles in making the vocal-auditory channel such an important part of pinniped communication systems. Vocal signaling by males functions primarily as a mechanism to disperse other males and may potentially influence females' choice of mates. The calls of male pinnipeds appear to serve as "honest signals," that is, they advertise to others the presence of the male and they are associated with a cost. During the breeding season, dominant and territorial males often produce their vocal signals nearly incessantly, and observations in captivity suggest that the energetic cost associated with this behavior is significant (Schusterman and Gentry, 1970). The high density of animals within breeding colonies appears to favor the production of loud, harsh, redundant calls to make signal detection by others more likely.

Crowding has probably functioned as a selective pressure on parturient females as well. The constant threat of mother-pup separation in a rookery leads to the frequent exchange of calls between females and their pups, and rapid learning of unique call characteristics by females, and, in some species, by pups. Later on mothers and their adolescent offspring probably use these calls to locate one another following extended periods of separation (see Insley et al., 2003). Competition between males and interactions between mothers and their pups comprise the most significant social behaviors in pinniped groups, and it is noteworthy that these social relationships are maintained by vocal signaling that can be easily viewed in the framework of contingency learning: Responses to affiliative calls are associated with positive consequences, and responses to aggressive calls are associated with the avoidance of negative consequences.

The mechanisms involved in pinniped vocalizations are relatively unstudied but very intriguing (see Tyack and Miller, 2002, for a review). The structures that are common to those used by terrestrial mammals include the larynx and the soft tissue of the nasal membranes, lips, mouth, and tongue; however, these structures are modified to various extents in pinnipeds. Some species also appear to have tracheal mechanisms for sound generation that are probably related to adaptations for diving. These include compressible airways comprised of flexible cartilage and membranes in the respiratory tract that produce sounds by vibration when air is passed over and between these and associated structures, all without necessity of inhalation and exhalation. Additionally, some species have highly derived structures that are involved in sound production, for example, the resonating chambers of the elephant seal proboscis, the inflatable nasal hood and septum of the hooded seal, and the unique pharyngeal pouches of the walrus, each of which can be used to generate or modify sounds. Different sound generating mechanisms can be used alone or in combination to produce distinctive call types, and some species are actually capable of emitting two structurally different calls simultaneously.

Modes of sound production may also change as an individual moves from land to water. For example, male California sea lions display vocally by barking along the boundaries of their shoreline breeding territories. At high tide, a male's territory may become partially or wholly submerged; however, the attending male continues to patrol the same boundaries with vocal displays comprised of underwater barks. While these barks sound similar in air and under water, it is clear that the motor patterns involved are different: Males barking in air do so with their mouths open, while males barking underwater do so with their mouths closed, without releasing any air (Schusterman and Balliet, 1969). Obviously, the various mechanisms and motor behaviors involved in sound production by pinnipeds may be quite complex, therefore, it seems likely that individuals must exercise some voluntary control over their sound emissions.

There are good examples from the field demonstrating that California sea lions can voluntarily control the usage of vocal signals and learn the context in which to use a call (see Schusterman et al., 2003). Nonterritorial subadult males, who frequently vocalize on other parts of the rookery, will sometimes "sneak" through a bull's territory, moving low to the ground and inhibiting their vocalizations during their entire period on the territory. If a territorial bull discovers one of these young individuals, he emits a highly directional, loud series of barks aimed directly at him (see Schusterman, 1977, 1978) and the intruder flees without emitting any vocalizations (Schusterman et al., 2003). Indeed, such evidence gleaned from the field is supported by observations made in captivity of male California sea lions that inhibit their barking in the presence of a territorial male but do not inhibit their barking in the absence of a territorial male (Schusterman and Dawson, 1968).

What Is the Experimental Evidence for Vocal Learning in Pinnipeds?

The most recent experimental work to show that individuals of a pinniped species have voluntary motor control over their vocalizations and can learn the context in which to use a call involves grey seals, which are known to naturally emit growls, clicks, and moans in air and underwater (Schusterman, Balliet, and St. John, 1970). In this recent study, Shapiro et al. (2004) used the withholding of anticipated fish rewards (a frustration technique) to initiate constant aerial calling from two young captive grey seals. To obtain discriminative control of each of the seal's vocal responses, the trainer presented a light and then reinforced the first vocal response given by the seal by pressing a clicker (a conditioned reinforcer) that was followed by the presentation of a fish reward. Therefore, the fish reward was contingent upon vocalizing only in the presence of the light, which served as a discriminative stimulus. The investigators report that under these conditions, both seals rapidly learned to control their vocal musculature and emit sounds only when appropriate. Thus, in the first phase of this study, call usage learning was demonstrated because the young grey seals signaled vocally given one arbitrary discriminative cue and refrained from emitting vocal signals given another.

I have described this initial phase of the experiments by Shapiro et al. (2004) on grey seals in detail because previous research on another pinniped (California sea lions) done forty years earlier contained nearly the same procedural ingredients and resulted in the same outcome (Schusterman and Feinstein, 1965). In a later study, three more sea lions were added to the sample (Schusterman, 1978) and vocalizations from the three were readily elicited by what I then called the "frustration technique" (i.e., by withholding fish rewards while a sea lion worked at an underwater target-pressing task, a response that hitherto had resulted in reinforcement). When a sea lion was shifted to the vocal conditioning task, it was promptly reinforced with fish for vocalizing when the target-pressing response did not "pay off." Next, the vocalization was brought under the control of the size of circular and triangular stimuli. A click burst emitted in the presence of a large or small stimulus was reinforced, and silence in the presence of the opposite sized stimulus was also reinforced. Thus, vocalization or silence in the presence of the appropriate stimuli defined the correct response. For two subjects, discriminative control of sound emissions was complete after one hundred randomized presentations of large and small stimuli. The third subject required nearly 600 trials before reliably learning to control its vocalizations in this context. As expected, for all of the sea lions, most errors made during the early period of learning were "vocal" and only as the experiment continued did "silent" errors occur. In general, these early studies with California sea lions show that call usage learning occurs more rapidly in these pinnipeds than has been shown in monkeys (e.g., Myers, Horel, and Pennypacker, 1965), cats (Molliver, 1963), and dogs (Salzinger and Waller, 1962). Indeed, several investigators have found some monkeys incapable of call usage learning (e.g., Yamaguchi and Myers, 1972). In a relatively recent study on vocal learning in cats, only twenty-two of thirty-two cats were found to be capable of the simplest type of call usage learning; all subjects in this experiment received extensive training before any of them learned a straightforward vocal conditioning task (Farley, Barlow, Netsell, and Chmelka, 1992). These laboratory results comparing terrestrial and marine mammals are consistent with what many marine mammal trainers have reported to me over the years: Namely, vocalizations are as easy to condition in sea lions and dolphins as are gross motor movements of the head, torso, and limbs. In contrast, when it comes to training nonhuman terrestrial mammals, usually vocalizations are much harder to condition than gross motor movements.

In a follow-up to the sea lion study just described, I was interested in determining whether a sea lion could transfer its knowledge about when to vocalize and when to remain silent to many pairs of visually presented stimulus configurations that differed simultaneously in both shape and size. A single sea lion was presented with forty-five such problems, each of which repeated for 160 trials. The stimuli from each problem were randomly ordered in successive trials, and then the next problem was presented. For each new problem, differential fish reinforcement was arranged for vocalizing to one stimulus and for remaining silent to the other stimulus. Following 160 trials, a new pairing was presented and differential reinforcement was arranged for vocalizing or remaining silent. The results showed that during the first five problems, correct responses by the subject averaged 59%. However, on the next ten problems, correct responses jumped to 75%, and on the last five problems, correct responses averaged 88% (Schusterman, 1978). Apparently, using conditioned vocalization and silence as indicator responses, the sea lion had solved each new problem more rapidly than the earlier one. In other words, the sea lion had learned the rule: If vocalizing to one stimulus on the earlier trials of a new problem pays off, then on succeeding trials, keep vocalizing in the presence of that stimulus and remain silent in the presence of the other stimulus. It is the behavior reflecting this rule that established the "learning set" or "learning to learn" phenomenon (Harlow, 1949). In later experiments my colleagues and I used the tightly controlled, operantly conditioned vocalizations of our sea lion subjects as indicator responses for determining their aerial and underwater visual acuity (Schusterman and Balliet, 1970) as well as their aerial and underwater auditory sensitivity (Schusterman, Balliet, and Nixon, 1972). The learning set study, the psychophysical studies, and another study on conditioning rate of barking underwater to different sounds (Schusterman, 1978) clearly show that California sea lions use their generalization skills and are capable of forming broad mental concepts to respond vocally to particular classes of stimuli across sensory modalities.

The grey seal study introduced at the start of this section continued after the calls of two subjects were placed under stimulus control using fish reinforcement. In an interesting attempt to investigate vocal production learning using auditory feedback, Shapiro et al. (2004) trained the same seals to emit two different vocalizations (moan and growl) in response to two different auditory stimulus types, which consisted of a small set of tape-recorded moans and growls. While the training of appropriate (or "type" matched) responses proceeded in an encouraging fashion, when the experimenters introduced novel moans and growls to the seals, both subjects biased their response and produced only growls. These findings led the investigators to conclude that earlier, the seals had learned specific associations between specific exemplars of the two classes of calls but were unable, with the training they were given, to generalize novel auditory stimuli belonging to each class by matching their vocal response to the perceived call type. This study, although demonstrating stimulus control over grey seal vocalizations in several different contexts, failed to show that grey seals are capable of one type of production learning, that is, producing sounds that match the sounds they hear, a capability that in contrast seems to have been demonstrated with "Hoover," the harbor seal (Rawls et al., 1985).

In a discussion of these results, Shapiro et al. (2004) raised what I consider to be an important issue in the topic of vocal learning in pinnipeds. This is the issue of examining the natural vocal development of these animals in order to determine how their vocal repertoire emerges with maturation. To various degrees, such research has already been done in some nonhuman primate species as well as in dolphins (for several reviews, see Snowdon and Hausberger, 1997).

New Experimental Evidence Supporting Vocal Learning in Pinnipeds

Most of my own research on pinniped vocal learning was completed over three decades ago. In those experiments, my colleagues and I worked mostly with individual animals representing a single species, the California sea lion, and much of what was learned was incidental to other research goals. During my career, I was fortunate to learn about vocal behavior in sea lions and some other pinnipeds through my own active experimentation and personal observations. These experiences included opportunities to learn about natural vocal signaling, amphibious sound production, social and environmental factors influencing vocalizations, sound production over annual and developmental scales, and discriminative control of sound emissions using visual and acoustic cues, all under the umbrella of research on perception, cognition, and communication in these animals. Throughout my career, I have remained intrigued and impressed by the changes in vocal behavior that occur as a result of maturational and experiential processes. I often wish that I had made time for more formal studies specifically dealing with vocal learning in the pinnipeds that I worked with; while the captive studies discussed earlier yielded a wealth of useful information, I am certain that focused efforts using similar reinforcement training techniques to systematically modify vocal responses would have been worthwhile.

Recently, I had the opportunity to participate in some new investigations of vocal learning (along with several close colleagues, who are listed in the acknowledgments) in two species of pinniped that are particularly interesting from the standpoint of sound production and vocal learning. The general goal of this effort was to explore vocal learning using food reinforcement with trained animals with known behavioral and vocal histories. These two case studies are ongoing efforts, but the findings thus far are illustrative of many of the concepts described in this chapter and are useful in suggesting what might be gained from applying such conditioning approaches to the study of vocal learning in nonhuman animals.

Case Study 1: Harbor Seal Vocal Conditioning

At first glance, harbor seals would not appear to be particularly interesting subjects for studies of vocal communication. In air, these seals are relatively silent with acoustic signaling among adults consisting of short-range growls and grunts to maintain social spacing and access to haul-out spaces. Pups are born in the spring and are highly precocial. The period of maternal dependency is quite short, and pups are weaned and abandoned within a month from parturition, leaving relatively little opportunity for meaningful nonauditory maternal feedback that might influence vocal development. Harbor seal pups produce stereotyped calls that can be emitted in air and underwater, with structural variation sufficient for individual recognition (Perry and Renouf, 1988). Once pups are weaned, however, they cease to produce these signature vocalizations. In contrast to many other pinniped species, harbor seal mothers do not emit stereotyped calls while attending their pups and they do not exchange calls with their pups. The lack of complex sound production by mature harbor seals in air belies the observation of complex underwater vocal signaling among adult males during the breeding season (Hanggi and Schusterman, 1994). These acoustic displays appear to play a significant role in the breeding behavior of this species. The production of underwater acoustic displays by seals is common to all species that mate in the water rather than on land (for a general review, see Van Parijs and Schusterman, 2003).

The subject of the current study was "Sprouts," a 17-year-old male harbor seal (*Phoca vitulina*) housed at Long Marine Laboratory in Santa Cruz, California. Sprouts was born into a captive colony of harbor seals at Sea World, San Diego, where he lived with his mother and other harbor seals for the first nine months of his life. At Long Marine Lab, he lived with other pinnipeds but no conspecifics. Sprouts was relatively silent until the age of seven, when he spontaneously began producing underwater vocal displays (see figure 3.1A) that were similar in general structure to the typical calls of wild harbor seal males. He has continued to produce these stereotyped vocalizations during the spring of each year, during a time period that coincides with the harbor seal breeding season along the central California coast. These calls appear to have changed little over time.

At the age of five, Sprouts was conditioned to produce an airborne growl-like sound by selectively shaping a low-level guttural sound occasionally made while he was having his teeth brushed. This growl was placed under the control of the discriminative signal (S_D) "speak," and it has been part of his training repertoire since that time. Sprouts rarely if ever produced that or any other airborne sound at any time other than when presented with the discriminative cue, and it seemed to have no relationship to the spontaneous underwater breeding vocalizations that appeared at the onset of sexual maturity.

In the spring of 2005, experimental sessions were conducted with Sprouts to explore how his conditioned airborne vocal response might be modified using fish reinforcement. Prior to this study, little selective shaping of his conditioned vocal response had been conducted and his sound production in response to the S_D

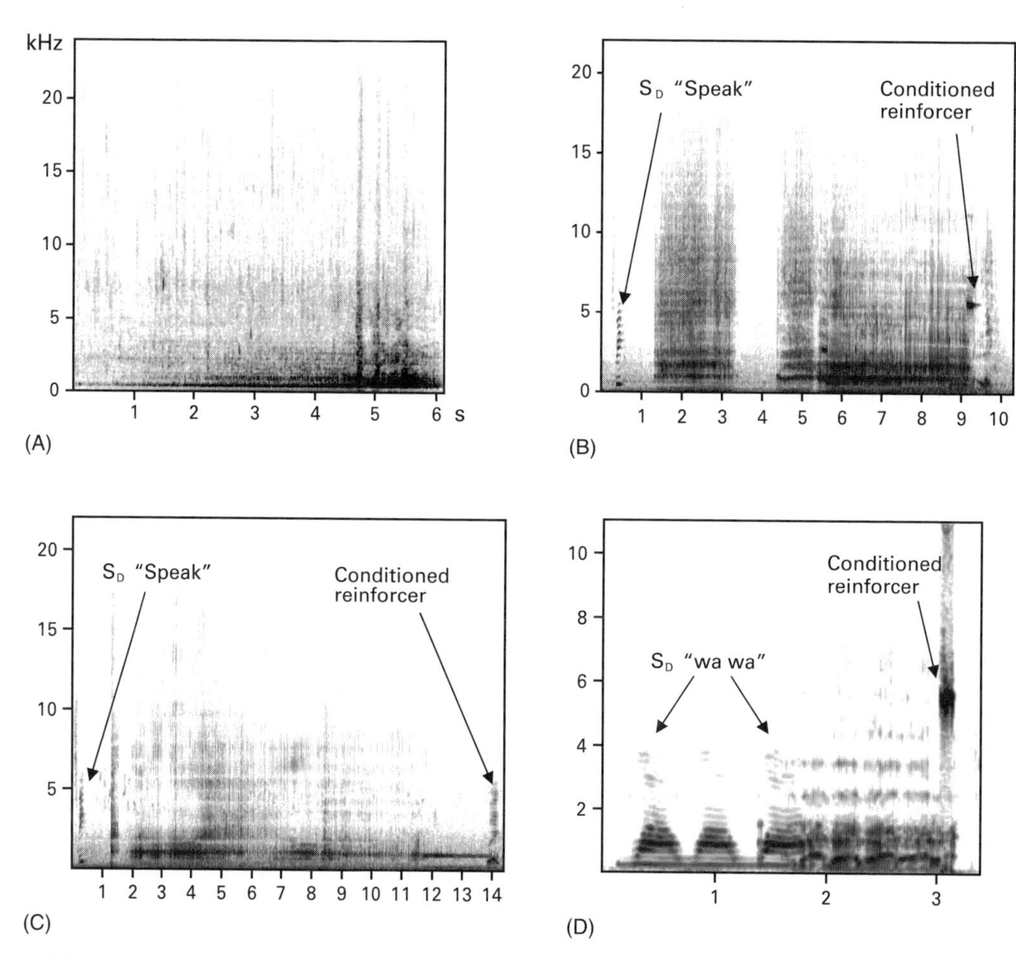

Figure 3.1
Spectrograms of sounds produced by a 17-year-old male harbor seal named Sprouts. Spectrograms show frequency in kilohertz on the vertical axis and time in seconds on the horizontal axis. (A) Underwater roar produced spontaneously during the spring breeding season; the sound is a guttural growl that ramps up in amplitude over 6 seconds to end in three broadband pulses. (B) Example of the growl-like conditioned vocalization that was produced in air when the seal was given the S_D "speak" prior to selective shaping of the call structure. Note the presence of a discriminative stimulus (a verbal cue) and the conditioned reinforcer (a whistle) given by a human trainer, which are indicated in (C) and (D). (C) Example of an airborne vocalization that was relatively novel compared to previously reinforced vocal responses; this sound was relatively pulsed and tonal and sounded similar to a motorboat. (D) Example of the selectively shaped airborne vocalization that had vowel-like qualities; the S_D was the verbal cue "wa-wa-wa" given by the trainer, and the seal's response was a vocalization that sounded like "wa-wa-wa-wa-wa."

"speak" was relatively stereotyped (see figure 3.1B). During the initial experimental sessions, we recorded these "typical" or baseline vocal responses onto videotape as well as with a directional microphone connected to a digital audiotape recorder. Once we had established his baseline sound production in response to the S_D, we began reinforcing only relatively novel variants of his conditioned vocal response following presentation of the discriminative cue. This contingency led to a very rapid expansion of the sounds produced in response to the S_D and a persistence in vocal behavior until reinforcement was provided. We occasionally encouraged movement of the mouth by gently prompting him with a light touch to move his mouth while vocalizing. As he explored which sound types would result in food reinforcement, the sounds Sprouts produced varied gradually and continuously between growls, snores, sneezes, sputters, moans, and pulses (see, e.g., figure 3.1C). The sounds also became progressively more varied along the dimensions of amplitude, frequency, modulation, and rate. He continued to change his vocalizations on each trial until he was signaled with a whistle that a fish reward would be provided. During his second session of this type, he emitted a sound that had some vowel-like syllabic qualities (a more tonal "wa" that was emitted while the mouth was moving) and this sound was reinforced. At that point, we opted to begin selective shaping of this particular sound, and we continued to reinforce only variants of this sound. Sprouts rapidly learned to produce the "wa" sound in response to the S_D "speak," and this sound was progressively shaped to become more tonal and more repetitive "wa-wa-wa-wa" (see figure 3.1D).

Once a vocal response such as Sprouts' "wa-wa" is reliably produced, it can be placed under the control of a particular S_D. This S_D can be arbitrary (a cue that has no structural similarity to the sound being produced) or it can match certain qualities of the desired vocal response (e.g., the S_D could be a tape recording of a particular sound or a vocal cue that sounds similar to the desired response). In this fashion, different acoustic responses can be placed under the control of different discriminative cues. In this example, Sprouts' "wa-wa" sound was eventually controlled by a similar verbal cue given by the trainer.

Case Study 2: Walrus Vocal Conditioning

Walruses are perhaps the most interesting species among the pinnipeds from the standpoint of sound production, having multiple well-developed modes of sound production, which include manipulation of the larynx, lips, mouth, nose, tongue, and specialized pharyngeal pouches. The soft tissue structures of the mouth are especially muscular and mobile, which is likely a by-product of adaptations for suction feeding. Walruses produce a great diversity of sounds in air and underwater using various combinations of these structures. The sounds produced in the larynx and shaped by the mouth include highly variable barks, coughs, grunts, guttural sounds, and roars. Highly differentiated whistles are produced by blowing air in and out over the tongue and the lips while a variety of sucking and smacking sounds are produced by movement of the entire mouth. Distinctive taps, knocks, pulses, and gong-like sounds (which are often emitted underwater or at the water's surface) are produced by controlled air movement associated with the pharyngeal sacs (see Tyack and Miller, 2002). At times the gong-like sound is augmented by a flipper striking the throat.

The breeding system of walruses can be classified as polygynous, and adult walruses exhibit a pronounced degree of sexual dimorphism. Males produce diverse and elaborate acoustic displays above and below the water's surface near ice edges where females congregate (Stirling, Calver, and Spencer, 1987). Reproductive behavior has been described but not well studied. Walruses are not easy animals to observe in the wild, especially during the spring breeding season because they gather in groups on the pack (free-floating) sea ice to give birth and mate. Of all pinniped species, they have the longest period of maternal dependency, with mothers nursing their pups for up to three years. Such an extended period of maternal dependency might provide an opportunity for elaboration of a sound-producing ability by the maturing walrus that with development could be considerably more flexible and spontaneous than in other pinniped species.

We had the opportunity to observe the vocal development of four captive walruses at Six Flags Marine World in Vallejo, California. The walruses were orphaned in 1994 and recovered from the open pack ice off the coast of Saint Lawrence Island, Alaska. The four animals (one male and three females) were approximately two weeks of age when they were brought into captivity. The pups were hand-raised on formula provided by human caretakers until they were weaned at approximately one year of age. The pups produced short barks as their primary spontaneous vocalizations for the first three to five years of their lives. As the walruses matured beyond that point, their sound-producing repertoires in air and underwater became more varied and began to reflect more of the sound types found in wild populations, including buzzes, growls, gulps, moans, whistles, barks, pulses, and sputters. At the time of these observations, the walruses were eleven years old. The male had begun to seasonally produce a variant of the knocking noises produced by rutting mature males using the pharyngeal pouches; the metallic gong-like sound used in adult acoustic displays had yet to emerge.

All four of these captive walruses were conditioned to produce sounds in air using food reinforcement. This process occurred in the context of an ongoing husbandry training program. Vocal conditioning began when the walruses were approximately two years old through selective reinforcement of vocalizations that occurred in the context of feeding situations. Delay of food reward increased vocal behavior, and the walruses quickly learned the relationship between sound production and feeding.

The male walrus, "Sivuqaq," was particularly engaged in this training process, but the training of the females followed the same general process.

Once Sivuqaq learned to reliably produce sounds in response to a trainer's prompts, four initial call types were placed under the control of four different discriminative stimuli over a period of about ten months. These call types were coded as "oooh," "growl," "talk," and "burp." All of these calls were shaped from undifferentiated guttural sounds produced in the larynx and emitted through the mouth. The calls were selectively shaped along particular dimensions—for example, "oooh" was shaped for acoustic elements including longer duration, louder intensity, and lack of movement of the mouth and lips during sound emission (see figure 3.2A). The trainer also cued the animal to position itself in an upright posture with its head oriented vertically and his mouth open when giving this vocal response. The call type "burp" was similarly shaped using selective reinforcement to be a very brief, loud, guttural sound (see figure 3.2B). The vocalizations "growl" and "talk" were both conditioned with Sivuqaq in a head-lowered posture. These were both shaped to be lower intensity sounds. "Growl" was shaped to include more moan-like qualities and was emitted with the mouth mainly closed (see figure 3.2C). "Talk" was a similar call that was selectively shaped to include modulation by rapid movement of the mouth and lips with the mouth partially open (see figure 3.2D).

Over the next several years, Sivuqaq was conditioned to produce at least four more distinctive acoustic emissions in response to different discriminative stimuli (see figures 3.2E, 3.2F, 3.2G, and 3.2H). One of these sounds, "whistle," was shaped by selectively shaping mouth position while simultaneously reinforcing exhalation (see figure 3.2E). Another call, coded as "ting," was also shaped as Sivuqaq became more mature. This was a metallic sound produced by the pharyngeal pouches, and it was shaped by prompting Sivuqaq to assume a chest-out posture with his mouth closed and then opportunistically reinforcing small movements and sound emissions originating from near the throat. The selective shaping of this sound resulted in a rapid series of pulsed "tings" correlated with visible movement of the chest and neck (see figure 3.2F). Two other calls, "sniffle" and "raspberry," were shaped by physical manipulation of the nose and mouth during exhalations combined with differential reinforcement of sound emissions that occurred either through nose only ("sniffle"; see figure 3.2G) or through the mouth and lips with the tongue extended ("raspberry"; see figure 3.2H).

Sivuqaq's accuracy at emitting the correct sound (which, of course, involved using the correct anatomical structures) in response to a given discriminative stimulus was quite high. He emitted the correct response at least 80% of the time even with very limited rehearsal. The conditioned vocal responses of the females followed a similar pattern of acquisition and performance. Interestingly, the female walruses were trained to produce at least two distinct sounds types, whistles and tings, not reported

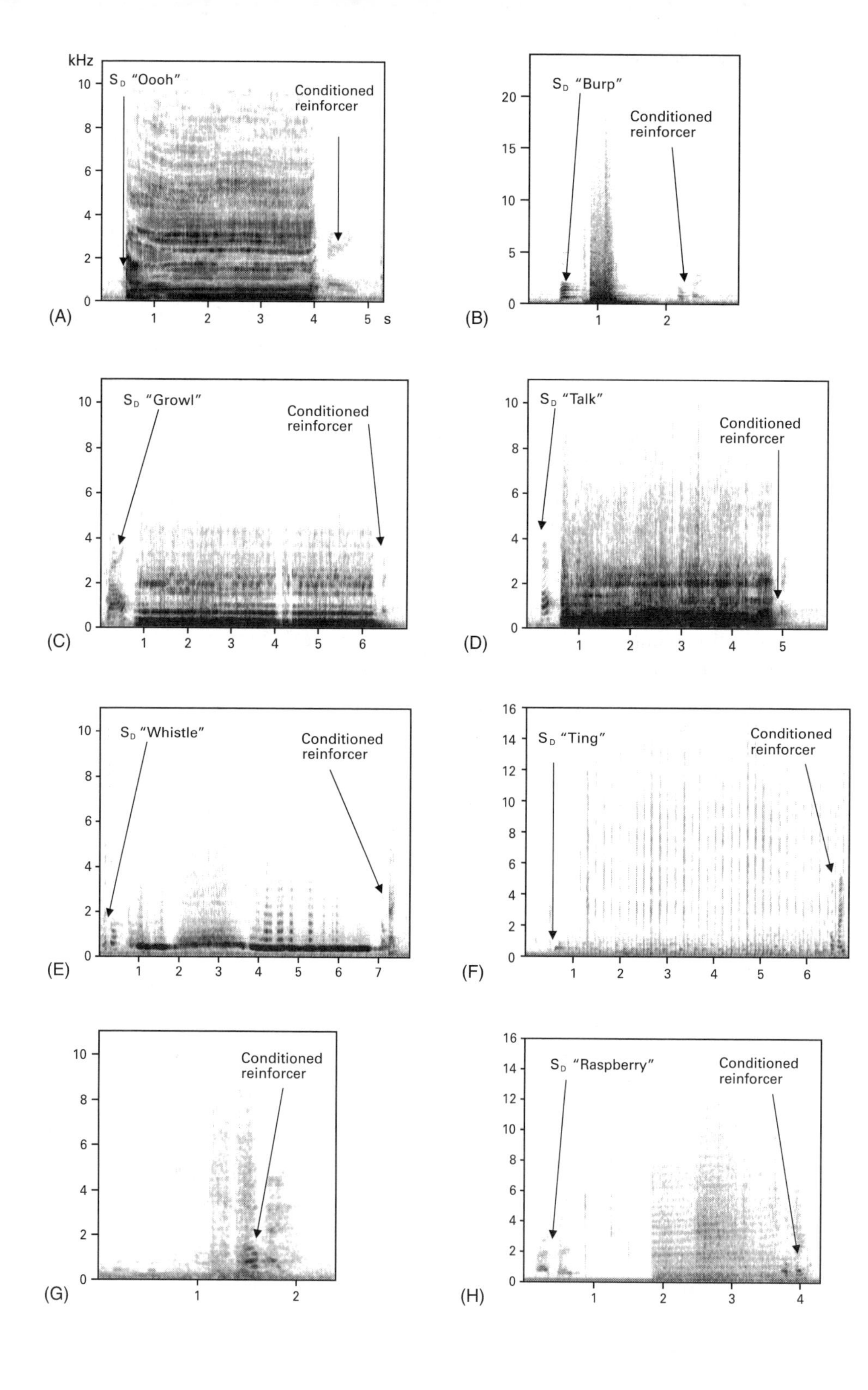

kHz
S_D "Oooh"
Conditioned reinforcer
s
(A)
S_D "Burp"
Conditioned reinforcer
(B)
S_D "Growl"
Conditioned reinforcer
(C)
S_D "Talk"
Conditioned reinforcer
(D)
S_D "Whistle"
Conditioned reinforcer
(E)
S_D "Ting"
Conditioned reinforcer
(F)
Conditioned reinforcer
(G)
S_D "Raspberry"
Conditioned reinforcer
(H)

to be emitted by wild female walruses, and not in the natural repertoire of these individuals prior to conditioning.

Summary and Conclusions of Case Studies

The preliminary observations gleaned from these two ongoing studies reveal the process by which vocal responses can be modified when reinforcement contingencies are related to specific call characteristics. They demonstrate the ease with which vocal responses can come to be produced in response to arbitrary experimental contexts, and they highlight the potential for selective shaping to influence aspects of vocal production including vocalizations produced at the levels of the respiratory tract, the larynx, and the supralaryngeal structures of the oral and nasal cavities that may serve to filter emitted sounds. The results add to the growing body of evidence suggesting that pinnipeds do have some degree of voluntary control over their sound emissions. The example described for the harbor seal is an interesting case. Here is a subject raised for most of his life without conspecifics who still displays apparently species-typical seasonal vocal behavior patterns in response to intrinsic (hormonal) or nonsocial environmental (photoperiod) cues. Despite this apparently innate pattern of sound production, the findings described here indicate that his vocal behavior is still readily accessible to modification by reinforcement contingencies, showing that both fixed and flexible vocal responses can occur concomitantly within an individual under the right set of circumstances. The walrus observations reveal plasticity in sound production in a particularly vocal species with a strong potential for complex communicative interactions during development. This leads to the question of whether reinforcement contingencies may come into play in natural social settings and influence vocal emissions.

The Ontogeny of Vocal Conditioning

If vocal conditioning is relevant in natural settings, then it is reasonable to wonder when this process might begin. Consider the following examples drawn from observations of pinniped behavior during the period of pup dependency. As noted earlier, there is a constant threat of mother–pup separation on a crowded sea lion rookery, resulting in frequent calling between the individuals involved. In most cases, the

Figure 3.2
Spectrograms of eight different types of conditioned vocalizations produced in air by an 11-year-old male Pacific walrus named Sivuqaq. All spectrograms show frequency in kilohertz on the vertical axis and time in seconds (s) on the horizontal axis. Also included are the discriminative stimuli (S_D) and conditioned reinforcers that were verbal cues given by a human trainer. The sounds were identified by their discriminative stimuli as follows: (A) "oooh"; (B) "burp"; (C) "growl"; (D) "talk"; (E) "whistle"; (F) "ting"; (G) "sniffle"; and (H) "raspberry" (note that no discriminative stimulus was recorded for G).

vocal exchange between a mother and her pup is terminated when the pup's calling is positively reinforced with nursing and contact comfort. For example, in northern elephant seals, a close physical association between the mother and her pup is formed soon after parturition and is maintained throughout lactation. The female and pup sleep and rest together, and she responds to its nuzzling and suckling attempts by assuming a nursing position, by rolling on her side and exposing her ventrum toward the pup's face. Given these cues, the pup begins to suckle in a reflexive manner. From the nursing female's viewpoint, these first attempts by the pup to suckle may act as a tactile cue eliciting the milk let-down reflex. This glandular response moves the milk toward her nipple, where it becomes available to the suckling pup. Once successful suckling has been established, a female that has been resting or sleeping while lying on its ventrum frequently waits until the pup emits a high-pitched distress cry before rolling on her side and exposing her teats. I speculate that once vocal utterances precede nursing, the milk let-down reflex of a lactating female may become gradually conditioned as a respondent (see Skinner, 1938) to the cries of the pup. At any rate, the dependent pup is likely to learn that its sound emissions can exert control over the female's orientation postures, giving it ready access to the teat and subsequent nursing. This type of relationship between vocalizing and nursing that I have described in northern elephant seals is ubiquitous and found not only in pinnipeds but in nearly all mammals.

The nursing calls of mammals are similar in many ways to the nursing cries of human infants in that all such calls signal need and their purpose is to elicit attention. Because they are correlated with access to milk, these highly stereotyped and undifferentiated nursing calls may become operantly conditioned. Presently, little is known about the role of learning in the vocal ontogeny of most mammalian species (Boughman and Moss, 2003). However, it is possible that the early relationship established between calling and nursing is relevant to changes in vocal behavior that may occur with maturation.

For example, in some songbirds where there is ontogenetic continuity between food begging and subsong, the origin of song learning in birds can be traced back to the earliest sounds they produce, namely, food-begging calls (Nottebohm, 1972). If this is true for some birds, then perhaps analogous ontogenetic processes in mammals may help us to trace how nursing calls in neonatal mammals play a role in the more complex development of vocalizations by mature individuals.

When we account for such early experiences, it becomes clear that the findings discussed earlier showing discriminative control of vocal behavior established with food reinforcement may not be inconsistent with theories on biological constraints on learning; rather, these findings suggest that one of the original functions of vocal behavior is related to nursing. Thus, infant calls may modify and be modified by nursing and social interactions. This synergistic interaction may be the earliest known social mechanism for discriminative vocal behavior.

General Discussion

Animal vocal communication as a form of social learning has been divided into three kinds of phenomena: receiver comprehension of calls, call usage, and call production (Seyfarth and Cheney, 1997; Janik and Slater, 2000). The first deals with behavioral changes that occur as a result of learning about the function of certain calls that are perceived, the second and third deal with changes that occur as a result of learning that affects the context and structure of calls that are produced. There has been general agreement based on evidence for a variety of mammals that learning and memory play a very important role in responses to the calls of others, that is, in the comprehension of vocalizations. In contrast, vocal usage and vocal production show much less flexibility in nonhuman mammals as compared to birds and humans.

The traditional ideas used to explain this lack of vocal plasticity have been that the production and usage of vocalizations by mammals is innate, that is, relatively fixed with little potential for any modification by learning. The only changes in vocal responses believed to occur were dependent on maturational factors and not experiential factors. These ideas were promoted by investigators who primarily researched vocal learning in terrestrial species of mammals, particularly nonhuman primates, and who did not take a broad comparative approach that included vocal learning in marine mammals. Having begun my career working with primates, I probably would have persisted in this thinking had I not had the opportunity to shift my efforts to the study of marine mammals. It was the research of John Lilly that first brought to my attention the ease with which the operant conditioning paradigm could be used to study the significance of vocal usage and production learning in marine mammals such as dolphins (Lilly and Miller, 1962). Today, the important role of learning and experience in sound production and usage by cetaceans such as dolphins has been well documented (McGowan and Reiss, 1997; Tyack and Sayigh, 1997), and these findings have been viewed as significant exceptions to traditional ideas about mammalian vocal communication. Until now, however, aside from the unusual case of Hoover, the harbor seal, much less has been made of experimental evidence for call production and usage learning of another marine mammal group—the pinnipeds (see Schusterman, 1978; Rawls et al., 1985). In this chapter there has been an attempt to make up for this deficiency by demonstrating that several different pinniped species show remarkable flexibility in the ways that they can use and modify their vocal emissions. In general, the material reviewed here has shown that changes in vocal responses may occur as a result of contingency learning using operant conditioning paradigms with food reward as a positive reinforcer.

Pierce (1985) argues that food reward may not be an appropriate reinforcer for modifying the ways that calls are used and produced, particularly in nonhuman primates. I counter this argument with the hypothesis that perhaps all mammalian infants acquire some experience with call usage in their nursing and contact-comfort interaction with their mothers. This could explain why all mammals that have been studied using operant paradigms show some degree of vocal usage learning, even though most nonhuman terrestrial mammals do not appear to learn how to use their calls as efficiently as some diving mammals like whales, dolphins, seals, sea lions, and walruses.

Why do marine mammals appear more likely to learn to use and produce vocalizations compared to nonhuman terrestrial mammals? Hypotheses related to this question are fairly straightforward but clearly require further comparative research. This needs to be done not only on the pinnipeds but also on those terrestrial mammals like elephants and bats that depend on vocal communication to a much greater extent than some other terrestrial mammals. Without knowing the detailed structural mechanisms that produce pinniped vocalizations, we can still point to at least two important adaptations that are probably strongly involved in voluntary control and modification of vocal emissions both in air and underwater. These are adaptations related to (1) breath control and therefore laryngeal control and (2) control over articulatory movements of tongue, lips, mouth, and teeth occurring in the oral cavity above the level of the larynx. We look forward in the near future to collaborative studies in which we tease out variables related to the mechanics of laryngeal control, supralaryngeal filtering, and the brain structures mediating voluntary vocal behavior. These studies should clarify the degree to which pinnipeds, rather than nonhuman primates, have a capacity for vocal production learning, which more closely resembles that of humans. Such convergence in the precise control of vocal emissions, although probably related—at least in part—to diving and feeding activities in pinnipeds, should provide clues to the type of morphological, neural, and behavioral mechanisms designed for human spoken language.

Acknowledgments

First, I would like to thank Colleen Reichmuth from Long Marine Laboratory in Santa Cruz, California, and Debbie Quihuis from Six Flag Marine World in Vallejo, California, for being involved with most of the training of the harbor seal and the four walruses who served as subjects of this investigation. The research reported here could not have been accomplished without the participation of these people as well as the animals. Next, I want to thank Marla Holt for working with me on the spectrographic analysis of the sounds and for helping, along with Colleen Reichmuth, with the preparation of this chapter. Colleen's reorganization of some of my earlier material was particularly important for the structure and meaning of the final chapter. In addition to these members of the team, I would also like to thank Kristy Lindemann for her assistance at both Long Marine Laboratory and Marine World. Kathy Streeter from the New England Aquarium provided me with information about harbor seal vocal conditioning that was extremely helpful. Tecumseh Fitch and Vincent Janik sent me several preprints and reprints as well as messages of encouragement that helped to guide me in all stages of writing this chapter. All of the reviewers who commented on the original version provided invaluable feedback. I especially want to thank Michael Owren for his extremely detailed and thorough critique. Last, but not least, I want to thank Uli Griebel and Kim Oller for inviting me to participate in the wonderful workshop on "The Evolution of Communicative Creativity" held at the Konrad Lorenz Institute in Altenburg, Austria, hosted by Gerd Müller. The comradeship of the participants was splendid, and the organizers provided a very entertaining program including tours of a castle and a monastery on the Danube and dining on fine food and wine in and around Vienna, Austria. This research, writing, and chapter preparation were partially supported by Office of Naval Research Grant N00014-04-1-0284.

References

Adret P. 1993a. Vocal learning induced with operant techniques: An overview. *Netherlands Journal of Zoology* 43: 125–42.

Adret P. 1993b. Operant conditioning, song learning and imprinting to taped song in the zebra finch. *Animal Behaviour* 46: 149–59.

Baptista LF, Petrinovich L. 1984. Social interaction, sensitive phases and the song template hypothesis in the white-crowned sparrow. *Animal Behaviour* 32: 172–81.

Bartholomew GA. 1970. A model for the evolution of pinniped polygyny. *Evolution* 24: 546–59.

Bolles RC. 1970. Species-specific defense reactions and avoidance learning. *Psychological Review* 77: 32–48.

Boughman JW, Moss CF. 2003. Social sounds: Vocal learning and development of mammal and bird calls. In *Acoustic Communication*, ed. J Simmons, AN Popper, RR Fay, pp. 138–224. New York: Springer-Verlag.

Dawkins MS. 1995. *Unraveling Animal Behaviour*, 2nd edition. Harlow: Longman.

Farley GR, Barlow SM, Netsell R, Chmelka JV. 1992. Vocalizations in the cat: Behavioral methodology and spectrographic analysis. *Experimental Brain Research* 89: 333–40.

Fitch WT. 2000. The evolution of speech: A comparative review. *Trends in Cognitive Sciences* 4: 258–67.

Fitch WT, Hauser MD, Chomsky N. 2005. The evolution of the language faculty: Clarifications and implications. *Cognition* 97: 179–210.

Goldstein MH, King AP, West MJ. 2003. Social interaction shapes babbling: Testing parallels between birdsong and speech. *Proceedings of the National Academy of Sciences* 100: 8030–5.

Hammerschmidt K, Freudenstein T, Juergens U. 2001. Vocal development in squirrel monkeys. *Behaviour* 138: 1179–1204.

Hanggi EB, Schusterman RJ. 1994. Underwater acoustic displays and individual variation in male harbour seals, *Phoca vitulina*. *Animal Behaviour* 48: 1275–83.

Harlow HF. 1949. The formation of learning sets. *Psychological Review* 56: 51–65.

Hauser MD, Chomsky N, Fitch WT. 2002. The faculty of language: What is it, who has it, and how did it evolve? *Science* 298: 1569–79.

Hiss A. 1983. Hoover. *New Yorker* 3 Jan: 25–7.

Insley SJ, Phillips AV, Charrier I. 2003. A review of social recognition in pinnipeds. *Aquatic Mammals* 29: 181–201.

Janik VM, Slater PJB. 1997. Vocal learning in mammals. In *Advances in the Study of Behavior*, ed. PJB Slater, JS Rosenblatt, CT Snowdon, M Milinski, pp. 59–99. San Diego, CA: Academic Press.

Janik VM, Slater PJB. 2000. The different roles of social learning in vocal communication. *Animal Behaviour* 60: 1–11.

Lenneberg EH. 1967. *Biological Foundations of Language*. Oxford: Wiley.

Lilly JC, Miller AM. 1962. Operant conditioning of the bottlenose dolphin with electrical stimulation of the brain. *Journal of Comparative and Physiological Psychology* 55: 73–9.

Manabe K, Kawahima T, Staddon JER. 1995. Differential vocalization in budgerigars: Towards an experimental analysis of naming. *Journal of the Experimental Analysis of Behavior* 63: 111–26.

Manabe K, Staddon JER, Cleveland JM. 1997. Control of vocal repertoire by reward in budgerigars (*Melopsittacus undulates*). *Journal of Comparative Psychology* 111: 50–62.

Marler P. 1990. Song learning: The interface between behavior and neuroethology. *Philosophical Transactions of the Royal Society of London B* 329: 109–14.

Marler P, Tamura A. 1962. Song "dialects" in three populations of white-crowned sparrows. *Science* 146: 1483–6.

McGowan D, Reiss D. 1997. Vocal learning in captive bottlenose dolphins: A comparison with humans and nonhuman animals. In *Social Influences on Vocal Development*, ed. CT Snowdon, M Hausberger, pp. 178–207. Cambridge, England: Cambridge University Press.

Molliver ME. 1963. Operant control of vocal behavior in the cat. *Journal of the Experimental Analysis of Behavior* 6: 197–202.

Moore BR. 1996. The evolution of imitative learning. In *Social Learning in Animals: The Roots of Culture*, ed. CM Heyes, BG Galef, pp. 245–65. San Diego, CA: Academic Press.

Myers SA, Horel JA, Pennypacker HS. 1965. Operant control of vocal behavior in the monkey. *Psychonomic Science* 3: 389–90.

Nottebohm F. 1972. The origins of vocal learning. *American Naturalist* 106: 116–40.

Owren MJ, Dieter JA, Seyfarth RM, Cheney DL. 1993. Vocalizations of rhesus (*Macaca mulatta*) and Japanese (*M. fuscata*) macaques cross-fostered between species show evidence of only limited modification. *Developmental Psychology* 26: 389–406.

Pepperberg I. 1986. Acquistion of anomalous communicating systems: Implications for studies on interspecies communication. In *Dolphin Cognition: A Comparative Approach*, ed. RJ Schusterman, JA Thomas, FG Wood, pp. 289–302. London: Lawrence Erlbaum Associates.

Perry EA, Renouf D. 1988. Further studies of the role of harbour seal (*Phoca vitulina*) pup vocalizations in preventing separation of mother–pup pairs. *Canadian Journal of Zoology* 66: 934–8.

Pierce JD. 1985. A review of attempts to condition operantly alloprimate vocalizations. *Primates* 26: 202–13.

Rawls K, Fiorelli P, Gish S. 1985. Vocalizations and vocal mimicry in captive harbor seals, *Phoca vitulina. Canadian Journal of Zoology* 63: 1050–6.

Salzinger K. 1973. Animal communication. In *Comparative Psychology: A Modern Survey*, ed. DA Dewsbury, DA Rethlingshafer, pp. 161–93. New York: McGraw-Hill.

Salzinger K, Waller BW. 1962. The operant control of vocalization in the dog. *Journal of the Experimental Analysis of Behavior* 5: 383–9.

Schusterman RJ. 1967. Perception and determinants of underwater vocalization in the California sea lion. In *Les Systemes Sonars Amimaux, Biologie et Bionique*, ed. RG Busnel, pp. 535–76. Jouy-en-Josas, France: Laboratoir de Physiologic Acoustique.

Schusterman RJ. 1977. Temporal patterning in sea lion barking (*Zalophus californianus*). *Behav Biol* 20: 404–8.

Schusterman RJ. 1978. Vocal communication in pinnipeds. In *Studies of Captive Wild Animals*, ed. H Markowitz, V Stevens, pp. 247–309. Chicago: Nelson Hall.

Schusterman RJ, Balliet RF. 1969. Underwater barking by male sea lions (*Zalophus californianus*). *Nature* 222: 1179–81.

Schusterman RJ, Balliet RF. 1970. Conditioned vocalization as a technique for determining visual acuity thresholds in the seal lion. *Science* 169: 498–501.

Schusterman RJ, Dawson RG. 1968. Barking, dominance, and territoriality in male sea lions. *Science* 160: 434–6.

Schusterman RJ, Feinstein SH. 1965. Shaping and discriminative control of underwater click vocalizaitons in a California sea lion. *Science* 150: 1743–4.

Schusterman RJ, Gentry R. 1970. Development of a fatted male phenomenon in California sea lions. *Developmental Psychobiology* 4: 333–8.

Schusterman RJ, Balliet RF, Nixon J. 1972. Underwater audiogram of the California sea lion by the conditioned vocalization technique. *Journal of the Experimental Analysis of Behavior* 17: 339–50.

Schusterman RJ, Balliet RF, St. John S. 1970. Vocal displays underwater by the gray seal, harbor seal and Steller sea lion. *Psychonomic Science* 18: 303–5.

Schusterman RJ, Kastak D, Levenson DH, Reichmuth CJ, Southall BL. 2000. Why pinnipeds don't echolocate. *Journal of the Acoustical Society of America* 107: 2256–64.

Schusterman RJ, Reichmuth Kastak C, Kastak D. 2003. Equivalence classification as an approach to social knowledge: From sea lions to simians. In *Animal Social Complexity: Intelligence, Culture, and Individualized Societies*, ed. FBM DeWaal, PL Tyack, pp. 179–206. Cambridge, MA: Harvard University Press.

Schusterman RJ, Van Parijs SM. 2003. Pinniped vocal communication: An introduction. *Aquatic Mammals* 29: 177–80.

Seyfarth RM, Cheney DL. 1986. Vocal development in vervet monkeys. *Animal Behaviour* 34: 1640–58.

Seyfarth RM, Cheney DL. 1997. Some general features of vocal development in nonhuman primates. In *Social Influences on Vocal Development*, ed. CT Snowdon, M Hausberger, pp. 249–73. Cambridge, England: Cambridge University Press.

Shapiro AD, Slater PJB, Janik VM. 2004. Call usage learning in gray seals (*Halichoerus grypus*). *Journal of Comparative Psychology* 118: 447–54.

Skinner BF. 1938. *The Behavior of Organisms*. New York: Appleton-Century-Crofts.

Skinner BF. 1957. *Verbal Behavior*. East Norwalk, CT: Appleton-Century-Crofts.

Snowdon CT, Hausberger M. 1997. *Social Influences on Vocal Development*. Cambridge, England: Cambridge University Press.

Stirling I, Calver W, Spencer C. 1987. Evidence of stereotyped underwater vocalizations of male Atlantic walruses (*Odobenus romarus romarus*). *Canadian Journal of Zoology* 65: 2311–21.

Thorndike EL. 1911. *Animal Intelligence*. New York: Macmillan.

Tyack PL, Miller EH. 2002. Vocal anatomy, acoustic communication and echolocation. In *Marine Mammal Biology: An Evolutionary Approach*, ed. AR Hoezel, pp. 142–84. Oxford: Blackwell Science.

Tyack PL, Sayigh LS. 1997. Vocal learning in cetaceans. In *Social Influences on Vocal Development*, ed. CT Snowdon, M Hausberger, pp. 208–33. Cambridge, England: Cambridge University Press.

Van Parijs SM, Schusterman RJ. 2003. Aquatic Mammals: Special Issue on Animal Vocal Communication, ed. SM Van Parijs, RJ Schusterman, *Aquatic Mammals* 29: 1–319.

West MJ, King AP. 1985. Social guidance of vocal learning by female cowbirds: Validating its functional significance. *Ethology* 70: 225–35.

West MJ, King AP. 1996. Social learning: Synergy and songbirds. In *Social Learning in Animals: The Roots of Culture*, ed. CM Heyes, BG Galef, pp. 155–78. San Diego, CA: Academic Press.

Wilbrecht L, Nottebohn F. 2003. Vocal learning in birds and humans. *Mental Retardation and Developmental Disabilities Research Review* 9: 135–48.

Yamaguchi S, Myers RE. 1972. Failure of discriminative vocal conditioning in rhesus monkey. *Brain Research* 37: 109–14.

4 Contextually Flexible Communication in Nonhuman Primates

Charles T. Snowdon

Historical Introduction

Konrad Lorenz, Niko Tinbergen, and Karl von Frisch shared the Nobel Prize for Medicine in 1973. Not only were they the first (and still only) animal behavior researchers to receive a Nobel Prize, they each were noted for work on communication. Through their work we have the concepts of "innate releasing stimulus" and "fixed-action patterns" with the implication that in animals, at least, communication signals have evolved with a clear one-to-one mapping. Organisms respond reflexively to a highly specific signal, such as a male three spined stickleback reacting with aggression toward another male in breeding condition as indicated by a red belly. Lorenz demonstrated that individual animals reared in isolation would still respond appropriately to a releasing stimulus when presented with it for the first time. Only von Frisch, in his classic studies of communication in honeybees, described any flexibility in communication. The work of these famous ethologists developed the conceptual framework for a generation of animal communication researchers: Identify the releasing stimuli, the precise situations in which they appear, and the innate responses produced.

A few years before the Nobel Prizes, W. John Smith published a brief but seminal paper on "Message, Meaning and Context in Ethology" (Smith, 1965). He distinguished "message" (what state the communicator coveys) from "meaning" (what a recipient infers from the stimulus) and "context" (defined in terms of both "immediate context," those events accompanying a signal that modify its effect, and "historical context," the role of prior experience and memory on responding to a signal). For an example of these concepts, let's consider a bird singing early on a spring morning. In many temperate zone species, increasing day length leads to increased levels of testosterone and development of testes. Singing is driven by testosterone (castrated males do not sing), and testosterone is associated both with male sexual behavior and with increased aggression. Thus, the message of song is likely to convey both higher aggression and a state of reproductive readiness. From the recipient's perspective, there are a variety of meanings. To a conspecific male, song represents a territorial male whose strength needs to be evaluated before any potential challenge. If the song is from a familiar male with whom the recipient has had prior encounters, the decision as to whether to attack or not is conditional upon these past experiences as well as the recipient's own aggressive abilities. For a conspecific female, the meaning of song is of a territorial male who may or may not be a good mate. A female may use the complexity, duration, or vigor of a song to evaluate the male. If she is already paired, she can use contextual features in the song to distinguish her mate from another male and, if it is from another male, to evaluate that male with respect to extrapair copulations. To another species of songbird, the male's song is simply a source of noise to be overcome with one's own song. To a predator, song provides an easy way to find a meal. To a poet, the bird's song conveys the meaning of a musical tribute to the advent of spring. Thus, there are multiple messages and multiple meanings to song, all affected by context and the social relationships between sender and receiver.

Smith's depiction of a triadic relationship between caller, signal, and receiver is a common one in the pragmatics of human communication. Although Smith's terminology is quite different, the concepts share some things in common with those of Austin (1962), who defined illocutionary force as social functions of a signal and perlocutionary force as actions taken by the recipient after recognition of the social function.

In Smith's own research he found several examples of vocalizations that appeared to be explained only by context. Thus, the Eastern kingbird has a vocalization that was described most parsimoniously as having a message of locomotor hesitancy—whether approaching a perch, a mate, or a rival, the call communicates something about the ambivalence of the communicator and might be construed as a social inquiry: "Will I be welcome here?" Smith also describes the choking displays of gulls (Tinbergen, 1959), which are given during hostile interactions within pairs, between rivals, in nest site relief or nest site selection, and in sexual interactions. The display appears to occur at times of conflict between approach and escape tendencies and might be construed as expressing social ambivalence.

Smith (1969) described a list of twelve basic messages in communication (subsequently broadened to nineteen with several modifiers of intensity, probability, or direction as well as identifiers such as species, sex individuality, reproductive state or condition (Smith, 1978). Smith (1969) argued that across a wide range of species and the different modalities within a species one typically finds a small number of discrete signals (fifteen to forty-five per species considering all modalities). The small number of discrete signals in a repertoire coupled with a small number of distinct message types means that context must carry much of the weight of subtle communication in animals.

With increased sophistication of recording and analytical methods, many more discrete signals were identified and the messages were more precise, making context appear less important for animal communication, especially in nonhuman primates. Green (1975a) described seven variants of coo vocalizations in Japanese macaques, most of which could be identified not by the human ear but only through spectral analyses. These coo variations were highly specific in usage with one form given by a confident infant on its own, another by an infant seeking contact with its mother, another by a dominant toward a subordinate, still another by a subordinate to a dominant, and so on. Green arranged these different coo types along a continuum of hypothesized arousal. Since there was often overlap between structure and context, this broad classification of coo variants along a continuum of arousal was apt. Other studies demonstrated subtle variants in primate calls. Pola and Snowdon (1975) demonstrated several trill variations in the repertoire of pygmy marmosets, three of which were used in an affiliative context of staying in vocal contact with other group members and a fourth that was used in agonistic settings. Then, Cleveland and Snowdon (1982) described eight types of chirp vocalizations in cotton-top tamarins, each relating to a different context—one was associated with mobbing, another with alarm, two with feeding, one with aggression toward a rival group, another with calm ongoing activities. There was no overlap between chirp structure and context, and there was no single social or motivational state that could unite all chirp types in a single construct. Suddenly, communication, at least in primates, was much more complex than thought previously.

Subsequent experimental studies using operant methods or playbacks of calls and observing behavioral responses indicated that the monkeys themselves perceived the differences in these calls (Zoloth et al., 1979; May et al., 1989, with Japanese macaques; Snowdon and Pola, 1978, with pygmy marmosets; Bauers and Snowdon, 1990, with cotton-top tamarins). The subtle variations observed through improved acoustic analysis methods were not artifacts but represented distinctions in perception and production made by the monkeys themselves.

About the same time several studies indicated that some signals might be highly specialized in providing reference to particular objects. Owings and Virginia (1978; Owings and Leger, 1980) found that California ground squirrels had two different types of alarm calls, one apparently directed at aerial predators like hawks and the other directed at terrestrial predators. Shortly thereafter Seyfarth, Cheney, and Marler (1980) showed that vervet monkeys at Amboseli National Park in Kenya used three types of alarm calls, each with high specificity toward a particular type of predator—one toward an aerial predator, one toward mammalian predators like leopards, and one toward snakes. Playback studies showed that recipients behaved as if each type of call communicated specific information about each predator. Thus, upon playback of an "eagle" alarm, monkeys would run out of trees and take cover in the brush, and upon playback of a "leopard" alarm, they would climb into trees and scan. Subsequent studies on lemurs (Macedonia, 1990) and on Diana's monkeys (Zuberbühler, 2003) found similar predator-specific calls in other species. The Diana's monkeys have specific calls for leopards and eagles, tailored to the different hunting strategies of these two species. Leopards depend upon surprising their prey, and when a Diana's monkey sees a leopard and gives a leopard alarm call, the leopard moves on to find other prey. In contrast, if chimpanzees hear monkeys, they approach and hunt them. When Diana's monkeys hear chimpanzees, they do not give an alarm call but freeze and remain silent. Clearly, monkeys can recognize different predators and use different strategies with each predator.

These results were exciting because they suggested a potentially much greater cognitive ability in animal communication than previously thought. Not only did primates have subtle vocal variants that communicated about different functions but also they now appeared to have direct reference to some object in the environment. At the same time these highly specific signals could be interpreted as simply a more precise use of the innate releasing signals and fixed-action patterns of Lorenz and Tinbergen, although the response patterns displayed by primates were flexible depending upon the relationship between the location of the listener and the type of predator.

However, in research with ground squirrels, the specificity of reference appeared to be contextually relevant. Closer observation of ground squirrels showed that "aerial" alarm calls were not given to aerial predators alone but also could be given to a terrestrial predator that had approached closely without being detected and that "terrestrial" alarms were given to birds of prey that were not displaying hunting behavior (Owings and Hennessey, 1984). Therefore, the alarm calls appeared to be contextually linked to the degree of urgency or type of response required. With "terrestrial" alarms, squirrels were vigilant and scanned the environment. With "aerial" alarms, the squirrels took cover in their burrows. Thus, the response urgency or context of the predator's behavior, not predator identity, was communicated. Flexibility in communicating about the response required independent of the identity of the predator is a more flexible use of alarm calls that is likely to be more adaptive for ground squirrels.

In primates Macedonia (1990) with ring-tailed lemurs and Digweed et al. (2005) with capuchin monkeys found only a partial differentiation of alarm calls with predators. Digweed et al. reported a type of alarm call that was specific to aerial predators, but a single variant call type was given to both snakes and terrestrial mammalian predators and nonpredators and was on occasion given to other group members. Animals giving the aerial alarm always descended from trees and took cover, whereas a variety of responses including approach and mobbing of a potential predator occurred when the second call type was given. In capuchin monkeys we have one predator-specific alarm call to birds and at the same time an alarm call that is contextually much more variable.

In the first description of vervet monkey alarm calls (from a different population) Struhsaker (1967) reported that the same type of alarm call might be given to one of several predators. Thus, he describes one alarm call that was given to martial eagles, crowned hawk-eagles, as well as to lions, leopards, and other cats. A call given to snakes was also very similar to a call given toward an observer and toward other vervet groups. Thus, even with vervet monkey alarm calls, there was not a consistent one-to-one mapping of signal to object, suggesting alarm calls are not simply sophisticated innate releasers but are subject to other contextual factors. Context matters and affects the meaning extracted by recipients, even if the predominant production and usage of calls is segregated by predator type.

A second area where referential communication has been argued to exist is with food calls. Species as diverse as chickens, cotton-top and golden lion tamarins, spider monkeys, toque and rhesus macaques, and chimpanzees give distinct calls in association with feeding. One difficulty in determining whether a call associated with food is referential or not is the need to separate motivation from reference. That is, is a call truly associated with food, or does it instead communicate excitement at discovering something positive? In a field study of toque macaques Dittus (1984) described food calls that were directed toward rare and highly preferred foods 97% of the time, but the same calls were also given in response to the first sunshine after the monsoon season and to the first clouds at the end of the dry season. Chickens give calls to food (Marler et al., 1986a, b) with higher rates of calling to more preferred foods. Gyger and Marler (1988) found food calls were given by males and seemed to attract females toward them. In nearly 50% of the cases the male had no food, and the interpretation was that males were being deceptive. However, chickens use food transfers as part of courtship, and, with a contextual view of communication, the same vocalization could serve in both feeding and courtship contexts. Cotton-top tamarin food-associated calls were specific to food, and the rate of calls correlated with each individual's food preferences (Elowson et al., 1991). Thus, food calls can both be referential and communicate an affective state, perhaps of social invitation.

Significance of Contextual Flexibility

The notion of referential signals as used in animal studies is seductive in suggesting a one-to-one correspondence between signal and meaning that is superficially suggestive of words. Elegantly designed playback studies provide strong evidence that in a few primate species, different alarm calls appear to evoke differential representations of predators and therefore may be analogous to words.

However, human words are more complex than a simple one-to-one mapping of sound and meaning. The word "dog" or "Hund," "chien," and "perro" can elicit a variety of representations of "dogness" including visual, auditory, olfactory, and tactile representations. But we also use the word "dog" metaphorically in many contexts that do not necessarily elicit representations of dogs: "I'm sick as a dog," "Dog days," "She worked doggedly on her paper," and so forth. Metaphors, similes, and puns all require an understanding of a word in a novel context. Thus, the cognitive abilities associated with flexibility in contextual usage may be indicators of greater complexity than simple one-to-one mappings.

Context was evoked by Smith (1965, 1978) as both pragmatic (explaining variation in usage and comprehension of calls hitherto thought to be fixed in structure, usage, and function) and adaptive. Reproductively successful individuals are likely to be those that can respond quickly to change. A contextually flexible communication system provides the possibility of rapid adjustment of communication to environmental change. An alarm system that not only dealt with traditional predators but adapted quickly to new forms of predators (such as humans) or to other potentially aversive events may allow an animal to respond more adaptively than if it had only a fixed set of correspondences between signal and response.

Contextual flexibility was initially invoked as a way to explain how animals of limited brain size could respond adaptively to multiple situations without needing specific signals for each context. However, contextual flexibility may paradoxically utilize simpler neural processes than contextually inflexible signals. Owren and Rendall (1997) have developed a model of affective conditioning that allows for individual learning of relationships between signals and affect, and thus, as environments change, individuals can acquire new signal–message–meaning correspondences. Saffran et al. (1996) with human infants and Hauser et al. (2001) with cotton-top tamarins have demonstrated that humans and monkeys can learn about statistical regularities in input, a simple but rapid way in which organisms can acquire contingencies between sounds. Thus, an animal in a novel context (such as captivity) can quickly learn to associate alarm calls with caretakers or veterinarians rather than leopards and snakes or can learn to eat novel food, different from those found in the wild, and still give appropriate food calls. These basic processes of conditioning provide a general way to learn rapidly about associations of signals and events in the environment without reliance on a host of hypothetical modality and context-specific modules that have been shaped by evolution. Successful adaptation is more likely to occur when organisms can respond rapidly to changes in environment compared with reliance on hard-wired stimulus–response connections. At the same time communication systems are not totally flexible. It is unlikely, for example, that a food call would be produced in a threatening context, suggesting that there may be a one-to-one mapping between some signals and some functions.

I have not yet defined contextual flexibility, but in what I've presented so far there are two overlapping definitions that I will continue to use. First, contextual flexibility occurs when there appear to be multiple mappings between the situations in which a signal is produced and the responses of recipients. I also consider contextual flexibility to occur when the usage and understanding of a traditional signal is applied in novel circumstances.

Contextual flexibility is found in communication in many species. I will review three areas: (1) flexibility in development, (2) flexibility in adult usage, and (3) influence of social status as a source of flexibility. Many of the examples I present come from research with my colleagues on vocal and chemical signaling in marmosets and tamarins. These are small-bodied, arboreal monkeys from the neotropics that are cooperative breeders. Animals live in small family groups with typically only one reproducing pair. Other group members are reproductively inhibited but share infant care duties—carrying infants, transferring food at time of weaning, and locating and advertising new sources of food. There is a high level of cooperation and coordination among group members, and they have complex vocal repertoires. For example, the cotton-top tamarin has more than thirty-five distinct calls (or calls used in combination). The limited visibility of an arboreal environment, coupled with a need to monitor and coordinate activities among group members, makes marmosets and tamarins ideal species for studying vocal communication and contextual flexibility.

Contextual Flexibility in Development

The context of communication signals can change greatly during development, both with respect to how adults communicate with infants and how infants communicate. Developmental processes affect the structure, usage, and comprehension of signals. In a comprehensive review of developmental processes in primate communication Seyfarth and Cheney (1997) documented that there is relatively little plasticity in primate vocal structure, although vocal structure can be modified within narrow constraints. Snowdon and Elowson (1999) found that pygmy marmosets changed structure of trill vocalizations subtly to match the calls of new mates when they were paired. There is no evidence of innovation in signal structure in primates in contrast to human infants (and many songbirds). More developmental flexibility is found in the use of signals in different contexts and in comprehending signals. I will illustrate these points using examples from cotton-top tamarins and pygmy marmosets.

A characteristic of marmosets and tamarins is food-sharing behavior between adults and infants starting at the time of weaning. We have focused on the vocalizations associated with food sharing (Roush and Snowdon, 2001; Joyce and Snowdon, 2007). Adults often vocalize when they are transferring food to infants, and in these two separate studies we have found that infants are rarely successful in obtaining food unless the adult is vocalizing. We have also found that the calls adults use with infants are rapidly repeated, more intense forms of the food calls that adults use when feeding. An adult animal discovering food emits a few chirp-like vocalizations, but when an adult is willing to transfer food to an infant, it produces a rapid bout of many food chirps in a sequence. We have shown that adults commonly produce these chirp sequences in the presence of infants, but not when they are feeding in the absence of infants (Joyce and Snowdon, 2007). Adults also adapt food sharing and bouts of food calling to the age and competence of infants. Calling and food transfers peak in the third month of life with twin infants, and infants begin to forage more often on their own with increasing age. Twins are energetically more costly than singletons, and we found that adults began to transfer food sooner with twins than with singletons and that twins were able to feed independently at an earlier age than singletons. Adult tamarins use vocalizations differently with young infants, and the trajectory of adult calling and food transfers is different with twins than with singletons, another example of contextual flexibility in call usage that depends on the age of the infants. We find that infants receiving food sharing and hearing vocalizations at an earlier age more quickly acquire independent feeding skills and the appropriate adult call structure than infants receiving food sharing at a later age (Joyce and Snowdon, 2007).

Adult cotton-top tamarins have eight distinct variations of chirp-like calls (Cleveland and Snowdon, 1982). Do infants have the ability to produce these calls at birth, or is there some developmental process of differentiation of chirps? We (Castro and Snowdon, 2000) developed situations in which we could reliably elicit chirps of different types with adults: Presentation of food elicited two types of chirps associated with feeding, presentation of a human in monkey catching clothes elicited the chirp type associated with mobbing, presentation of a sudden noise (alarm clock or personal defense alarm) elicited the chirp associated with alarm, and opening doors between adjacent colony rooms so that animals could hear calls from unfamiliar animals elicited the chirp associated with territorial defense. We systematically presented each of these situations over the first twenty weeks after infants were born to determine how they would respond to each of these contexts. Infants produced chirps in a short sequence of two to three calls that did not have the structure that differentiated the chirp types used by adults. These "protochirps" appeared to be the main vocal signals used by infants, and they were produced in most of the contexts we tested. Interestingly, infants gave significantly fewer chirps in the alarm and mobbing contexts, suggesting that they were responsive to the calls of adults in these contexts even though they were not producing contextually appropriate calls.

At some point during testing each infant gave at least one adult form of chirp that was appropriate to the context, but no infant gave each of the chirp types and no single context elicited an appropriate chirp from each infant. Furthermore, once an infant had produced an appropriate chirp in an appropriate context, the infant rarely produced that call again in subsequent tests. By five months of age, when tamarins were completely weaned, they were not consistently producing contextually differentiated chirp types. The one exception was a chirp used in feeding contexts. The D chirp, which adults give when they acquire food and which forms the basis of the calling bouts used in food transfers, was the form of chirp that appeared earliest in development. The D chirp was produced by the most infants and had the greatest probability of being given again in future tests (Castro and Snowdon, 2000). These results suggest that tamarins must both learn subtle features of call structure and learn how to use them in appropriate contexts, but infant tamarins appeared sensitive to threat situations and inhibited calling in mobbing and alarm contexts.

Bird song has long fascinated those interested in the development of communication because the variability seen in different populations of the same species suggests that song must be learned from others. Indeed, many studies have demonstrated the varieties of song learning (Catchpole and Slater, 1995). Little evidence exists of similar processes in nonhuman primates, raising a question of what has led to such flexible communication in birds and humans in contrast to the less flexible vocal development of nonhuman primates. Variation in vocal structure between populations has been described in two species, chimpanzees (Mitani and Brandt, 1994; Mitani et al., 1999) and Japanese macaques (Green 1975b; Masataka, 1992). In the case of Japanese macaques conditioning of calls by humans during provisioning appears to be involved in the vocal differences observed.

There are three possible explanations for vocal variation across populations: (1) Animals in a population may adapt call structure to local features of the habitat, (2) call structure may differ due to genetic drift that might alter subtle aspects of the vocal production apparatus, and (3) call structure may change in response to social influences, either through development or adult interactions. We have been recording vocalizations from pygmy marmosets at five different locations spanning a 300-km east–west transect in Eastern Ecuador (de la Torre and Snowdon, in review). We have recorded and analyzed twenty trills and J calls from two adults (one male and one female) from an average of three groups at each site and have recorded at least ten long calls from two animals in each group. This sample allows us to analyze sex differences and to see whether groups or populations differ in call structure. Within the home range of each group, we have also measured ambient noise levels and have broadcast tones and rerecorded them at varying distances as described below so that we can correlate any differences in call structure with local habitat acoustics.

Our results show significant differences in the structure of two of the call types in each of the five populations. Comparisons of predicted call structure with habitat acoustics showed only a few predictions were confirmed. Most differences in acoustic variables between populations could not be easily related to differences in habitat acoustics. Either genetic variation between populations or social learning processes may account for the differences not explained by habitat acoustics. In captive marmosets when two animals are paired with each other, they adjust parameters of trill vocalizations to match each other (Snowdon and Elowson, 1999). In dolphins (Tyack and Sayigh, 1997) and greater spear-nosed bats (Boughman, 1997, 1998) individuals also changed vocal structure when encountering new social companions. It is possible that social interactions within a population of wild marmosets also lead to the population-typical vocal structures we have observed. Supportive of the idea that pygmy marmosets may have socially influenced behavioral variation is a recent finding that within the same five populations, pygmy marmosets have a preferred tree species for exudate feeding and that the preferred tree is not necessarily the exudate species most available within the environment. Marmosets in one population ignore a locally abundant tree species that is the most used by marmosets in a different population (Yepez et al., 2005).

We have observed a phenomenon that we call Pygmy Marmoset Babbling (PMB) that appears to share many features with infant babbling (Elowson et al., 1998). Infant pygmy marmosets from the first week of life produce long sequences of vocalizations, often lasting several minutes. These sequences mainly include calls that are recognizable from the adult repertoire. Oller (2000) interprets these as similar to the Expansion Stage in human vocal development, where infants produce repetitive and functionally variable sounds. Pygmy marmoset infants produce a subset of 60% of the call types of the adult repertoire. These calls are often produced in sequences of two to five calls of one type before the infant switches to a different type (showing repetition and rhythmicity). The sequence of sounds produced bears no relationship to the infant's ongoing behavior. For example, a threat call will follow soon after an affiliative call, which in turn follows after a fear call, which follows after a food call. Thus, calls of PMB are not given in a context typical for adults. There appears to be no functional relevance to the PMB, except that PMB increases social interactions between adults and infants. Babbling infant marmosets were more likely to be in contact with other group members when babbling than when not babbling (Elowson et al., 1998). We have observed babbling in wild pygmy marmosets in the Amazon, where PMB has a similar developmental trajectory and duration as observed in captivity (Stella de la Torre, personal communication). Anecdotal reports of babbling in other species of marmosets have been noted.

PMB decreases as marmosets get older, and the proportion of adult forms of vocalization increases with a clear progression toward more precise forms of adult calls. We measured changes in the structure of trill vocalizations, the most common call in the repertoire of adult marmosets as well as in the babbling of young marmosets. With increasing age pygmy marmosets produced more regular and accurate forms of trills, achieving accuracy in different trill parameters at different ages, consistent across individuals. Furthermore, we found a significant relationship between the amount and diversity in babbling during the first five months and the rate with which individuals developed adult structures, suggesting that babbling might function as a form of vocal practice (Snowdon and Elowson, 2001). Marmosets change the context of babbling as they get older. In both captive and field studies, adults use PMB in agonistic contexts to indicate submissiveness. Thus, babbling is a highly variable vocal production that elicits attention from adults, that serves as practice for infants to develop adult vocal structures, and that appears again in subordinate adults using infantile calls to indicate submissiveness.

Contextual Flexibility in Adult Usage

A study of isolate reared squirrel monkeys (Herzog and Hopf, 1984) reported that squirrel monkeys gave species-typical alarm vocalizations on their first exposure to predators, suggesting a hard-wired relationship between stimulus and vocal response. However, considerable evidence from other species suggests that nonhuman primates have contextually flexible responses. In this section we consider several examples of this flexibility.

Are animals sensitive to the presence of recipients when they vocalize? Sherman (1977) reported that California ground squirrels were more likely to give alarm calls when close kin were present. Cheney and Seyfarth (1986) reported that vervet monkeys recognized members of different social alliances. Infant rhesus monkeys give structurally different calls when separated from mothers but in visual contact than when separated out of visual and auditory contact with the mother (Bayart et al., 1990), and Lillehei and Snowdon (1978) found that infant stumptail macaques gave different calls when alone and seeking contact with the mother versus alone and not seeking contact.

Audience effects are also present in chickens. Marler et al. (1986b) found that cocks gave food calls when their mates or novel females were present but inhibited food calls when a rival male was present. When a predator was presented, a cock would give an alarm call if another chicken (male or female) was nearby, but it would not give an alarm call when alone (Karakashian et al., 1988) or in the presence of a bird of a different species (Evans and Marler, 1991). Thus, chickens are not reflexive communicators but adjust calling to the audience present.

In a very different type of audience effect we have found that wild pygmy marmosets in the Amazon in close contact with human ecotourism that had a group member captured by humans vocalized at a much lower rate than marmosets in habitats less influenced by human disruption (de la Torre et al., 2000). A lower rate of calling makes a marmoset less readily detected by humans.

Communication creates a conflict between having an influence on another individual, essential to coordinate social behavior, and being conspicuous to potential predators. Thus, it makes sense to inhibit calls in the absence of conspecifics or in the presence of threats. Are other types of calling affected similarly? Pygmy marmosets have several vocalizations that they give at a high rate through the day. We have described different forms of trill-like calls and have found that marmosets respond antiphonally to playbacks of these calls, recognize each other as individuals, and have a turn-taking convention (Pola and Snowdon, 1975; Snowdon and Pola, 1978; Snowdon and Cleveland, 1980, 1984).

Three forms of pygmy marmoset calls all appear to be used to maintain vocal contact with other group members, but they differ in acoustic properties for sound localization and differ in transmission in natural habitats. de la Torre and Snowdon (2002) studied sound transmission in natural habitats in the Ecuadorian Amazon and related call structure to environmental variables. We studied the home ranges of three groups in each of two habitats. Three types of natural vocalization were used: trills, J calls, and long calls (Type B alerting calls; Pola and Snowdon 1975). We broadcast each of these vocalizations in the natural habitat and rerecorded the calls at distances from 1 m to 80 m. We also broadcast pure tones of different frequencies and tones at different pulse rates. We found rapid degradation of calls over very short distances. The trill was distorted at 20 m and barely detectable at 40 m. The J call showed some deterioration at 20 m and was highly distorted at 40 m. Only the long call was minimally distorted at 40 m, and it was the only call that we could rerecord at 80 m. With increasing distance high frequencies decreased at faster rates than predicted by physical principles of spherical spreading, and with increasing distance there was also increased reverberation, making individual notes less distinct. The trill with a repetition rate of 30–35 Hz was most affected, the J call with a repetition rate of 17–22 Hz was next affected, and long calls at 3 Hz had the least reverberation. All calls are high pitched—above 8 kHz for minimum frequency. Recordings of ambient noise within the habitats of pygmy marmosets indicated significant biogenic noise that decreased above 8 kHz. Thus, the signals used by marmosets utilize a high frequency window above the range of other biological noises.

Do pygmy marmosets vocalize in a way to maximize these aspects of habitat acoustics while minimizing detection from other species? We recorded vocalizations of each type from within each group, detected the next nearest marmoset to the caller, and measured the distance between them. There was a close match between call degradation and the distance over which marmosets used the calls. More than half of the trills were given when animals were within 5 m of others, typically during feeding, foraging, or traveling. Only 5% of trills were observed when animals were more than 10 m apart. More than 80% of J calls were given at distances greater than 5 m, and these calls were rarely observed when animals were separated by more than 20 m. Long calls were given primarily when animals were more than 20 m from others (de la Torre and Snowdon, 2002). Thus, pygmy marmosets appear to use selectively different types of calls depending on their distance from other group members.

Alarm calls and mobbing calls are thought to have evolved to be highly specific to predators. The work on vervet monkey alarm calls described above suggests a close relationship between calling and predators. Captive primates develop in a different ecological environment than wild primates. Does the environmental context of captivity affect the use of alarm or mobbing vocalizations? We have presented captive born cotton-top tamarins with a boa constrictor and found only a mild arousal response to the snake that was not different from the response to a laboratory rat (Hayes and Snowdon, 1990). More recently we have tried to find stimuli that explicitly lead to the mobbing responses described in wild tamarins reacting to predators. In our captive colony we can induce robust mobbing behavior to a human caretaker dressed in veterinary garb as if to capture a monkey. During systematic presentations of a human dressed in veterinary garb, we discovered three new vocalizations from tamarins that were associated with mobbing behavior (Campbell and Snowdon, 2007). We have repeated the Hayes and Snowdon (1990) study with a new generation of captive tamarins and still find no evidence of alarm or mobbing responses to natural predators like snakes but instead find reactions to laboratory-specific threats, suggesting that tamarins learn to give alarm and mobbing calls in conditions appropriate to their current habitats. We have tried to condition tamarins by playing back mobbing calls in association with a live boa constrictor but do not find any evidence of mobbing or evasive reaction developed toward presentation of the snake (Campbell and Snowdon, in preparation). Thus, captive tamarins produce mobbing calls in very different contexts from those used by wild tamarins, suggesting that mobbing call context may be learned and applied to the ecological context of captivity.

We were interested in whether tamarins would learn socially to avoid a noxious food. In contrast to studies on birds and rodents, there have been few demonstrations of social learning to avoid food in primates. We presented family groups of tamarins with a highly preferred food, tuna fish that was mixed with white pepper. Only a third of forty-five tamarins ever sampled the peppered tuna, and few of those that sampled tuna sampled it again on subsequent trials. When we presented tuna without pepper, many animals did begin eating tuna again, but some individuals who never sampled the peppered tuna still avoided eating normal tuna years later (Snowdon and Boe, 2003). What led the tamarins to avoid the tuna? We observed that the animal that first tasted the food often gave alarm calls and exhibited visual signs of disgust (head shaking, chin rubbing, retching). We also observed a greatly reduced rate of food calling to tuna, even after we presented tuna without pepper added. We hypothesized that tamarins were successful in socially learning about noxious foods, due to the vocal and visual communication signals (Snowdon and Boe, 2003). The most important result for the current discussion is the use of alarm calls to noxious foods that the tamarins had not experienced previously. Here is the application of an alarm call in a completely novel context that appears to have been rapidly "understood" by other group members as indicating "danger." This is an example of learning to apply a call functionally appropriate to the new situation.

Addington (1998) studied feeding competition and vocal communication in pygmy marmosets. The excitement squeal (previously identified by Pola and Snowdon, 1975) occurred in 99% of feeding sessions and only 1% of baseline sessions, suggesting that this call functions as a food-associated call for pygmy marmosets. Pygmy marmosets did not use food-associated calls in a strict relationship to food presence. Instead they appeared to call strategically, giving calls more readily when they would not jeopardize their own access to a quality food resource. Marmosets called more to high-quality portable food than to high-quality nonportable food but called more often to low-quality nonportable foods. They also gave more food calls overall when multiple food sources were available than when a single food source was available.

Other research supports the hypothesis that primates give food calls when they can maintain their own access to high-quality foods. We presented pygmy marmosets with a variety of food types of differing preferences to see if we would find the same relationship between food preferences and call rate that we found in cotton-top tamarins (Elowson et al., 1991). For the most part we found a linear relationship between preference and calling rate, except to one of our foods, mealworms that elicited no calling (Snowdon, unpublished data). Gros-Louis (2004), studying wild capuchin monkeys, also reported reduced food calls to animate prey versus fruits. A noisy predator might less easily catch insect prey than a quiet one, but insects are also not a sharable resource, whereas fruits and other foods are typically clumped, allowing multiple animals to feed. Hauser et al. (1993) tested common chimpanzees with different amounts of food and found a higher rate of calling to multiple pieces of watermelon than to single pieces. However, when the chimpanzees were tested with a single watermelon versus a watermelon of similar size cut into twenty pieces, the call rate was reduced to the single watermelon. Chapman and Lefebvre (1990) found that spider monkeys gave food calls only at food patches that were relatively large with high-quality fruit. These results taken together suggest that food calling might be linked to the degree to which the food can be shared with others while ensuring access by the animal that calls. Thus, the message of "food calls" may not have to do with foods per se but communicates a readiness to tolerate proximity or close interactions by others.

Social Status and Contextual Flexibility

Social status has major impacts independent of age on the structure of, usage of, and response to communication signals. Studies of baboons in Botswana have found that dominant males have both longer and higher wahoo barks than subordinate males; however, these higher pitched and longer barks are not a characteristic of the male per se but a function of his status. As the dominant male gains status, his calls become longer and higher pitched, and the male's calls decrease in pitch and duration as his status declines (Fischer et al., 2004).

Cotton-top tamarins, as noted earlier, are cooperative breeders with adult male and female offspring being reproductively suppressed while living as nonbreeding helpers in family groups. Are there any changes in communication with change in social status when an animal moves from being a reproductively suppressed helper to a breeding adult? The tamarins have two forms of vocalization used in feeding contexts—a C chirp when they approach food and make a decision about what to eat and a D chirp when they have picked up food and are eating (Elowson et al., 1991). When we studied the development of the structure and usage of these calls from infancy through adulthood, we found both juveniles and postpubertal subadults produced calls that were similar, but imperfect, versions of calls produced by adults. In addition, unlike adults, younger tamarins overgeneralized, producing calls to manipulable nonfood objects as well. And young tamarins gave a large number of other vocal types not heard in adults while feeding (Roush and Snowdon, 1994).

At first we were surprised to find no evidence of change over a broad developmental range. However, after thinking about the social system of tamarins, we hypothesized that all reproductively subordinate animals, regardless of chronological age, might benefit from vocalizing differently from adults to indicate subordinate status. We tested this hypothesis by recording vocalizations in feeding contexts before and after a tamarin was removed from its family and paired with a novel, opposite-sex animal (Roush and Snowdon, 1999). Within two to three weeks after pairing, regardless of age, tamarins were giving food calls only in feeding contexts, eliminating the other call types, and within one to two months after pairing, the animals produced calls with adult structure. It appears that social status, rather than age alone, may be a context affecting how tamarins communicate. We found a similar result with long calls. Tamarins produce three different types of long calls with one, the Combination Long Call, being observed more often in tamarins living as nonreproductive helpers in families and rarely observed among breeding adults (Cleveland and Snowdon, 1982).

In Addington's (1998) study of feeding competition and calling behavior in pygmy marmosets, described above, she also found that social context affected calling. The excitement squeal appeared as a food-associated call, occurring on almost every trial with food and rarely in control trials. However, Addington also found that a long call was often added to a food-associated call, creating a combined call with long calls always preceding excitement squeals. These combined long call–food call sequences were given almost exclusively by unmated marmosets and rarely appeared among mated marmosets. The reproductive status of a marmoset appears to affect the type of vocalization given. Since long calls carry long distances in the wild (de la Torre and Snowdon, 2002), an unmated animal that discovers food appears to be broadcasting its discovery over a longer range than a mated animal, perhaps to advertise for mates. In the wild, groups are spaced further apart than the distance over which long calls can carry, but unmated adults often forage away from the rest of the group and may more readily encounter a potential mate by adding long calls to excitement squeals.

A final example illustrates contextual flexibility in response to a different modality of communication, olfactory signals. We have developed methods to image brain activity in awake, conscious common marmosets and have been studying neural responses to odors of ovulating and nonovulating females (Ferris et al., 2001, 2004). Odors from ovulating females produce activation of the anterior hypothalamus and medial preoptic areas, known to be critical in sexual arousal and sexual behavior. Odors from ovariectomized females also produced some activation of these hypothalamic areas but much less than odors of ovulating females. Activation in response to both odors was greater than the response to vehicle and no stimulus control. In addition, many other brain areas were differentially activated in response to ovulatory versus nonovulatory odors: the hippocampus associated with memory, the anterior cingulate cortex associated with evaluation and decision making, the putamen and substantia nigra associated with general arousal and motivation, and the temporal cortex and cerebellum associated with integration of sensory function. Although one might interpret the scent marks associated with ovulation as an excellent example of one-to-one mapping to a sexual response function, the involvement of many other brain areas not directly associated with sexual arousal suggests the possibility of contextual flexibility.

We have two studies suggesting contextual flexibility in response to sexual odors in marmosets. In the first study we trained marmosets to expect an opportunity to mate with a novel female in the presence of a lemon odor. The lemon odor was presented for three minutes and then an ovulating female was released from a nest box and copulation occurred. On control trials the vehicle was presented in the presence of an ovulating female but she was not released from the nest box and no copulation occurred. Each male rapidly anticipated the opportunity to mate during lemon trials by increasing nest box exploration compared with control trials. We had imaged these males' neural reactions to the lemon odor prior to sexual conditioning and then imaged them again in response to the lemon odor following conditioning. There was no increased activation in the hypothalamic regions but, instead, increased activation of the nucleus accumbens (involved in motivation) and in the parietal and cingulate cortex (involved in assessment of reward value). Marmosets can be rapidly conditioned to substitute a novel odor for responding sexually to a female, and this conditioning alters neural activity in certain brain areas.

In the second study we evaluated the behavioral and neurohormonal responses to sexual odors (Ziegler et al., 2005). We presented male common marmosets with odors of novel, ovulating females or with vehicle control odors and observed changes in their behavior and took blood samples thirty minutes after exposure to the odors to monitor cortisol and testosterone levels. Males showed increased rates of erection and of licking the odor of an ovulating female and showed nonsignificant increases in sniffing, touching, and scent marking as well as decreased latency to contact the odor. We also found a significant increase in testosterone levels after exposure to the odor of a novel ovulating female (but no changes in cortisol levels). However, the increased testosterone response was seen only in singly housed and paired males. Fathers living with dependent offspring had no change in testosterone levels. Thus, the social status of a male actually influenced his physiological response to a communicative signal.

Summary

The original innate releasing mechanism, fixed-action pattern model of communication did not allow for contextual flexibility. However, current research suggests that contextual flexibility in communication may be relatively common and may provide organisms with a rapid means of adapting to environmental change. Organisms that can apply signals to novel contexts and those that can respond to these novel applications of traditional signals should have greater survival and reproductive success than those that are inflexible in production or response. To date, there is only limited evidence of the mechanisms underlying contextual flexibility, but contextual flexibility may be easily accounted for by basic processes of conditioning and statistical learning. Making use of general learning processes that underlie most other aspects of behavior leads to a more parsimoniously designed nervous system than one depending exclusively on previously adapted modules, specific to signals or contexts.

There is extensive evidence of contextual flexibility among nonhuman primates within adult communication and in changes in context signal relationships during development. Furthermore, at least in some species social status serves as an important context that affects signal production. From the perspective of nonhuman primates, the notions of fixed-action patterns and signed stimuli may still be present in a more sophisticated form in the predator-specific alarm calls of some species, but contextual flexibility in communication may be more the rule than the exception. Although I have emphasized contextual flexibility, most of the results presented refer to flexibility in the usage and the understanding of signals. Compared with songbirds and human infants, the monkeys discussed here do not appear able to create novel signals and they show only limited ability to modify aspects of vocal structure. Nonhuman primates still display many fixed and inflexible features in signal production.

Acknowledgments

I thank my research collaborators Rebecca L. Addington, Carla Y. Boe, Matthew W. Campbell, Nicole A. Castro, A. Margaret Elowson, Craig F. Ferris, Stella M. Joyce, Cristina Lazaro-Perea, Rebecca Roush, Stella de la Torre, Pablo Yepez, and Toni E. Ziegler for their contributions to the research described here. Luisa Arneda, Matthew W. Campbell, Katherine A. Cronin, Shannon Digweed, Liza R. Moscovice, and members of the Workshop provided valuable critical feedback on the manuscript. Research was supported by USPHS Grants MH029775, MH035215, and MH058700 and Hilldale Student–Faculty Research Fellowships.

References

Addington RL. 1998. Social foraging in captive pygmy marmosets (*Cebuella pygmaea*): Effects of food characteristics and social context on feeding competition and vocal behavior. Unpublished Ph.D. dissertation. University of Wisconsin, Madison.

Austin JL. 1962. *How to Do Things with Words*. Oxford: Oxford University Press.

Bauers KA, Snowdon CT. 1990. Discrimination of chirp variants in the cotton-top tamarin. *American Journal of Primatology* 21: 53–60.

Bayart F, Hayashi K, Faull KT, Barchas JD, Levine S. 1990. Influence of maternal proximity on behavioral and physiological responses to separation in infant rhesus monkeys. *Behavioral Neuroscience* 104: 98–107.

Boughman JW. 1997. Greater spear-nosed bats give group distinctive calls. *Behavioral Ecology and Sociobiology* 40: 61–70.

Boughman JW. 1998. Vocal learning by greater spear-nosed bats. *Proceedings of the Royal Society of London B* 265: 227–33.

Campbell MW, Snowdon CT. 2007. Vocal response of captive-reared *Saguinus oedipus* during mobbing. *International Journal of Primatology* 28: 257–70.

Castro NA, Snowdon CT. 2000. Development of vocal responses in infant cotton-top tamarins. *Behaviour* 137: 629–46.

Catchpole CK, Slater PJB. 1995. *Bird Song: Biological Themes and Variations*. Cambridge, England: Cambridge University Press.

Chapman CA, Lefebvre L. 1990. Manipulating foraging group size: Spider monkey food calls at fruit trees. *Animal Behaviour* 39: 891–6.

Cheney DL, Seyfarth RM. 1986. The recognition of social alliances among vervet monkeys. *Animal Behaviour* 34: 1722–31.

Cleveland J, Snowdon CT. 1982. The complex vocal repertoire of the adult cotton-top tamarin (*Saguinus oedipus oedipus*). *Zeitschift für Tierpsychologie* 58: 231–70.

de la Torre S, Snowdon CT. 2002. Environmental correlates of vocal communication in wild pygmy marmosets, *Cebuella pygmaea*. *Animal Behaviour* 63: 847–56.

de la Torre S, Snowdon CT, Bejarano M. 2000. Effects of human activities on wild pygmy marmosets in Ecuadorian Amazonia. *Biological Conservation* 94: 153–63.

Digweed SM, Fedigan LM, Rendall D. 2005. Variable specificity in the anti-predator vocalizations and behaviour of the white-faced capuchin, *Cebus capucinus*. *Behaviour* 142: 997–1021.

Dittus WPJ. 1984. Toque macaque food calls: Semantic communication concerning food distribution in the environment. *Animal Behaviour* 32: 470–7.

Elowson AM, Snowdon CT, Lazaro-Perea C. 1998. Infant "babbling" in a nonhuman primate: Complex vocal sequences with repeated call types. *Behaviour* 135: 643–64.

Elowson AM, Tannenbaum PT, Snowdon CT. 1991. Food-associated calls correlate with food preferences in cotton-top tamarins. *Animal Behaviour* 42: 931–7.

Evans CS, Marler P. 1991. On the use of visual images as social stimuli in birds: Audience effects on alarm calling. *Animal Behaviour* 41: 17–26.

Ferris CF, Snowdon CT, King JA, Duong TQ, Ziegler TE, Ugurbil K, Ludwig R, Schultz-Darken NJ, Wu Z, Olson DP, Sullivan JM Jr, Tannenbaum PL, Vaughn JT. 2001. Functional imaging of brain activity in conscious monkeys responding to sexually arousing cues. *NeuroReport* 12: 2231–6.

Ferris CF, Snowdon CT, King JA, Sullivan JM Jr, Ziegler TE, Ludwig R, Schultz-Darken NJ, Wu Z, Olson DP, Tannenbaum PL, Einspanier A, Vaughn JT, Duong TQ. 2004. Imaging neural pathways associated with stimuli for sexual arousal in nonhuman primates. *Journal of Magnetic Resonance Imaging* 19: 168–75.

Fischer J, Kitchen DM, Seyfarth RM, Cheney DL. 2004. Baboon loud calls advertise male quality: Acoustic features and their relation to rank, age and exhaustion. *Behavioral Ecology and Sociobiology* 58: 140–8.

Green S. 1975a. Variation of vocal pattern with social situation in the Japanese monkey (*Macaca fuscata*): A field study. In *Primate Behavior, Vol. 4*, ed. LA Rosenblum, pp. 1–102, New York: Academic Press.

Green S. 1975b. Dialects in Japanese monkeys: Vocal learning and cultural transmission of locale specific behavior? *Zeitschift für Tierpsychologie* 38: 304–14.

Gros-Louis J. 2004. The function of food-associated calls in white-faced capuchin monkeys, *Cebus capucinus*, from the perspective of the signaler. *Animal Behaviour* 67: 431–40.

Gyger M, Marler P. 1988. Food calling in the domestic fowl, *Gallus gallus*: The role of external referents and deception. *Animal Behaviour* 36: 358–65.

Hauser MD, Newport EL, Aslin RN. 2001. Statistical learning of the speech stream in a non-human primate: Statistical learning in cotton-top tamarins. *Cognition* 78: B53–B64.

Hauser MD, Teixidor P, Field L, Flaherty R. 1993. Food-elicited calls in chimpanzees: effects of food quantity and divisibility. *Animal Behaviour* 45: 817–19.

Hayes SL, Snowdon CT. 1990. Predator recognition in cotton-top tamarins (*Saguinus oedipus*). *American Journal of Primatology* 20: 283–91.

Herzog M, Hopf S. 1984. Behavioral responses to species-specific warning calls in infant squirrel monkeys reared in social isolation. *American Journal of Primatology* 7: 99–106.

Joyce SM, Snowdon CT. 2007. The developmental of food transfer behavior in cotton-top tamarins (*Saguinus oedipus*). *American Journal of Primatology* 69: 955–65.

Karakashian SJ, Gyger M, Marler P. 1988. Audience effects on alarm calling in chickens. *Journal of Comparative Psychology* 102: 129–35.

Lillehei RA, Snowdon CT. 1978. Individual and situational differences in the vocalizations of young stumptail macaques, *Macaca arctoides*. *Behaviour* 65: 270–81.

Macedonia JM. 1990. What is communicated in the antipredator calls of lemur? Evidence from playback experiments with ringtailed and ruffed lemurs. *Ethology* 86: 177–90.

Marler P, Dufty A, Pickert R. 1986a. Vocal communication in the domestic chicken: I. Does a sender communicate information about the quality of a food referent to a receiver? *Animal Behaviour* 34: 188–93.

Marler P, Dufty A, Pickert R. 1986b. Vocal communication in the domestic chicken: II. Is a sender sensitive to the presence and nature of a recipient? *Animal Behaviour* 34: 194–8.

Masataka N. 1992. Attempts by animal caretakers to condition Japanese macaque vocalizations result inadvertently in individual specific calls. In *Topics in Primatology, Vol. 1*, ed. T Nishida, WC McGrew, P Marler, M Pickford, FBM deWaal, pp. 271–8. Tokyo: University of Tokyo Press.

May B, Moody DB, Stebbins WC. 1989. Categorical perception of conspecific communication sounds by Japanese macaques, *Macaca fuscata*. *Journal of the Acoustical Society of America* 85: 837–47.

Mitani JC, Brandt KL. 1994. Social factors influence acoustic variability in the long distance calls of male chimpanzees. *Ethology* 96: 233–52.

Mitani JC, Hunley KL, Murdoch ME. 1999. Geographic variation in the calls of wild chimpanzees: A reassessment. *American Journal of Primatology* 47: 133–51.

Oller DK. 2000. *The Emergence of Speech Capacity*. Mahwah, NJ: Lawrence Erlbaum Associates.

Owings DH, Hennessey DF. 1984. The importance of variation in sciurid visual and vocal communication. In *Biology of Ground Dwelling Squirrels*, ed. JO Murie, GM Michener, pp. 167–201. Lincoln: University of Nebraska Press.

Owings DH, Leger DW. 1980. Chatter vocalizations of California ground squirrels: Predator and social role specificity. *Zeitschift für Tierpsychologie* 54: 163–84.

Owings DH, Virginia RA. 1978. Alarm calls of California ground squirrels (*Spermophilus beecheyi*). *Zeitschift für Tierpsychologie* 46: 58–70.

Owren MJ, Rendall D. 1997. An affect-conditioning model of nonhuman primate vocal signaling. In *Perspectives in Ethology, Vol. 12*, ed. MD Beecher, DH Owings, NS Thompson, pp. 299–346. New York: Plenum Press.

Pola YV, Snowdon CT. 1975. The vocalizations of the pygmy marmoset (*Cebuella pygmaea*). *Animal Behaviour* 23: 826–42.

Roush RS, Snowdon CT. 1994. Ontogeny of food associated calls in cotton-top tamarins. *Animal Behaviour* 47: 263–73.

Roush RS, Snowdon CT. 1999. The effects of social status on food associated calls in captive cotton-top tamarins. *Animal Behaviour* 58: 1299–305.

Roush RS, Snowdon CT. 2001. Food transfers and the development of feeding behavior and food associated calls in cotton top tamarins. *Ethology* 107: 415–29.

Saffran JR, Aslin RN, Newport EL. 1996. Statistical leaning in 8-month-old infants. *Science* 274: 1926–8.

Seyfarth RM, Cheney DL. 1997. Some general features of vocal development in nonhuman primates. In *Social Influences in Vocal Development*, ed. CT Snowdon, M Hausberger, pp. 249–73. Cambridge, England: Cambridge University Press.

Seyfarth RM, Cheney DL, Marler P. 1980. Vervet monkey alarm calls: Semantic communication in a free-ranging primate. *Animal Behaviour* 28: 1070–94.

Sherman PW. 1977. Nepotism and the evolution of alarm calls. *Science* 197: 1246–53.

Smith WJ. 1965. Message, meaning and context in ethology. *American Naturalist* 99: 405–9.

Smith WJ. 1969. Messages of vertebrate communication. *Science* 165: 145–50.

Smith WJ. 1978. *The Behavior of Communicating*. Cambridge, MA: Harvard University Press.

Snowdon CT, Boe CY. 2003. Social communication about unpalatable foods in tamarins (*Saguinus oedipus*). *Journal of Comparative Psychology* 117: 142–8.

Snowdon CT, Cleveland J. 1980. Individual recognition of contact calls in pygmy marmosets. *Animal Behaviour* 28: 717–27.

Snowdon CT, Cleveland J. 1984. "Conversations" among pygmy marmosets. *American Journal of Primatology* 7: 15–20.

Snowdon CT, Elowson AM. 1999. Pygmy marmosets alter call structure when paired. *Ethology* 105: 893–908.

Snowdon CT, Elowson AM. 2001. 'Babbling' in pygmy marmosets: Development after infancy. *Behaviour* 138: 1235–48.

Snowdon CT, Pola YV. 1978. Interspecific and intraspecific responses to synthesized pygmy marmoset vocalizations. *Animal Behaviour* 26: 192–206.

Struhsaker TT. 1967. Auditory communication among vervet monkeys (*Cercopithecus aethiops*). In *Social Communication among Primates*, ed. SA Altmann, pp. 281–324. Chicago: University of Chicago Press.

Tinbergen N. 1959. Comparative studies of the behaviour of gulls (*Laridae*): A progress report. *Behaviour* 15: 1–70.

Tyack PL, Sayigh LS. 1997. Vocal learning in cetaceans. In *Social Influences on Vocal Development*, ed. CT Snowdon, M Hausberger, pp. 208–33. Cambridge, England: Cambridge University Press.

Yepez P, de la Torre S, Snowdon CT. 2005. Interpopulation differences in exudate feeding of pygmy marmosets in Ecuadorian Amazon. *American Journal of Primatology* 66: 145–58.

Ziegler TE, Schultz-Darken NJ, Scott JJ, Snowdon CT, Ferris CF. 2005. Neuroendocrine response to female ovulatory odors depends upon social condition in male common marmosets, *Callithrix jacchus*. *Hormones and Behavior* 47: 56–64.

Zoloth SR, Petersen MR, Beecher MD, Green S, Marler P, Moody DB, Stebbins W. 1979. Species-specific perceptual processing of vocal sounds by monkeys. *Science* 204: 870–2.

Zuberbühler K. 2003. Referential signaling in nonhuman primates: Cognitive precursors and limitations for the evolution of language. *Advances in the Study of Behavior* 33: 265–307.

5 Constraints in Primate Vocal Production

Kurt Hammerschmidt and Julia Fischer

Introduction

Understanding the roots of human language is a fascinating topic that has generated as much interest as controversy (see Jackendoff, 1999; Fitch, 2000; Hauser et al., 2002). A key feature of human speech is its flexibility both in terms of the combination of units that make up statements and the mapping of sound and meaning. A prerequisite for this high degree of flexibility is learning, in terms of both production and comprehension. From a comparative perspective, this raises the question of whether and to what extent other primate species exhibit vocal learning. Much of the original research on primate vocal development focused on whether the ability to produce certain vocalizations was a learned versus an innate process (Latta, 1966; Talmage-Riggs and Mayer, 1972; Winter et al., 1973). More recently, primatologists have realized that it is important to distinguish between the developmental trajectories of the ontogeny of vocal production, call usage, and comprehension of calls (Seyfarth and Cheney, 1997; Janik and Slater, 1997). In a nutshell, while vocal production appears largely innate, learning does play a role in the usage and comprehension of calls, but this is not restricted to the primate order (Kaminski et al., 2004; Fischer et al., 2004).

In the following, we will focus on a series of studies that have explored the vocal ontogeny of nonhuman primates and flexibility in the structure and usage of nonhuman primate calls. We will then compare these findings to the development of preverbal utterances in human children. We subsequently discuss the neurobiology underlying vocal production to examine the reasons for relatively strong constraints that appear to operate on the flexibility in nonhuman primate vocal production.

Ontogeny of Vocal Production in Nonhuman Primates

The earliest studies in this field involved attempts to teach chimpanzees to speak (Hayes and Hayes, 1951; Kellogg, 1968). For instance, Vicki, the chimpanzee (*Pan troglodytes*) raised in a human family, in fact learned to utter a few "words," but the difficulties she had in mastering this task were apparently more striking than her successes (Hayes and Hayes, 1951). The finding that apes or monkeys have difficulties acquiring human speech, however, does not refute the possibility that learning plays an important role in the development of their own species-typical vocal communication. Most of the evidence accumulated about vocal development comes from studies on monkeys (see Seyfarth and Cheney, 1997; Fischer, 2003, for a review), while little is known about the vocal development of apes (Tomasello and Zuberbühler, 2002). In contrast, numerous studies have explored apes' ability to employ signs or linguistic symbols (Wallman, 1992) as well as gestures (Tomasello et al., 1997; see Call, this volume). This imbalance in what we know about monkeys versus apes in these two domains of communication is only beginning to be redressed (Crockford et al., 2004; Marshall et al., 1999).

In their classic paper about the ontogenetic development of squirrel monkey (*Saimiri sciureus*) calls, Winter and colleagues (1973) investigated whether species-specific auditory input is necessary to develop a species-specific vocal repertoire. They compared squirrel monkey infants that were raised normally with infants raised in acoustic isolation. Their qualitative description of the vocal repertoire yielded no difference between the two groups. A quantitative description of two call types ("isolation peep" and "cackling") also failed to show differences related to the rearing conditions. In another study, Winter and colleagues found no structural differences between calls of normally raised squirrel monkeys and squirrel monkeys that were deafened shortly after birth (Talmage-Riggs and Mayer, 1972). The authors concluded that neither hearing a conspecific vocalization nor the ability to hear at all is a necessary prerequisite to develop a species-specific vocal repertoire. Further support for the idea that primate vocal production is largely genetically determined came from hybridization experiments in gibbons (Brockelman and Schilling, 1984; Geissmann, 1984). Descendants of cross-fostered *Hylobates lar* and *Hylobates pileatus* produced songs that corresponded neither to the paternal species nor to the maternal species but were in between. Furthermore, cross-fostering experiments in macaques have shown that rhesus monkey (*Macaca mulatta*) and Japanese macaque (*M. fuscata*) infants showed no modification in the species-specific use of their "coos" and "grunts" (Owren et al., 1992, 1993).

In a review paper, Seyfarth and Cheney (1997) reported an increasing number of studies illustrating learning effects in vocal behavior. They suggested that this might be partly due to the improved acoustic analysis methods developed in the years before. This view was supported by further studies suggesting that monkey calls are not completely determined genetically. For instance, Sugiura (1998) found that coo calls produced as a response to a playback coo are more similar to the playback coo than to spontaneously produced coos. Similarly, Snowdon and Elowson (1999) observed that if two pygmy marmosets are paired, they modify their trill vocalizations in such a way that the calls of both partners become more similar to each other than they were before pairing.

In the light of these findings, Hammerschmidt and colleagues decided to repeat the previous study by Winter and colleagues with an improved methodology and extend the work by analyzing additional call types. They studied the vocal development of six squirrel monkeys over the first two years of their life. Four of the animals were normally raised, one animal was deprived of adult species-specific calls, and one animal was congenitally deaf. The study showed that in all twelve call types under investigation, age-related changes in one or the other acoustic parameter occurred. The degree to which vocal structures changed was call type specific, with some calls showing more substantial changes than others. Most changes occurred within the first four months, but some occurred up to an age of ten months. Most of the call types showed high intra-individual variability. This variability did not decrease during development but remained throughout the two-year study period.

In addition to this high intra-individual variability, the calls also exhibited substantial interindividual variability (Hammerschmidt et al., 2001). Table 5.1 shows the percentage of the acoustic parameters measured that revealed statistically significant individual differences across the twelve call types. As can be seen, each animal differed from the others significantly on a variety of parameters for at least one call type. It is notable that the acoustically deprived and congenitally deaf squirrel monkey did not stand out from the normally raised animals in terms the degree to which they differed from each other or the group as a whole but instead showed a mean percentage of parameters varying from the other animals for given call types that was within the range seen for the normally raised animals. This is an additional indication that the ontogenetic changes found were mainly maturational. In sum, the more advanced methodology did not substantially alter the earlier view (Winter et al., 1973) that squirrel monkeys develop a species-specific vocal repertoire without the opportunity to hear an adult model.

Interestingly, call types like "yap" and "chuck" exhibited greater age-related changes. Maybe this is not so surprising because recent studies have shown that call types with complex frequency modulation require an additional neural coordination, located in the ventrolateral pontine brain stem (Jürgens, 2000; Hage and Jürgens, 2006a, b). These frequency-modulated calls could be blocked after a glutamate antagonist injection, whereas calls with only weak or moderate frequency modulation, like "caws," "shrieks," or "purrs," remained unaffected (Jürgens, 2000).

One surprising finding was that peak frequency increased significantly with increasing body size. Generally, larger body size is associated with longer vocal folds and hence a lower fundamental frequency, which in turn often results in lower peak frequency. However, the pitch of a call is determined not only by the length and mass

Table 5.1
Individual differences in call types between normally raised, acoustically isolated, and congenitally deaf animals

	Animals					
Call types	N 1	N 2	N 3	N 4	Acoustic isolated	Congenitally deaf
Caw		9		36		
Bawl						
Shriek	12					
Yap						
Chuck					29	7
Twitter	23		15			
Cackle					9	
Long peep			9			
Staccato peep						18
Short peep						
Ascending peep		9				18
Descending peep		13		13		
Mean	**2.9**	**2.6**	**2**	**4.1**	**3.2**	**3.6**

The scores give the percentage of call parameters showing statistically significant differences between the respective animals on each indicated call type. "Mean" gives the mean percentage across all 12 call types differing for each individual from the other animals. N 1–4 represent normally raised infants. Depending on the acoustic structure of calls, different sets of call parameters were calculated: Peak frequency, distribution of frequency amplitudes, dominant frequency bands, frequency range, and percentage of noisy structures. For yaps, chucks, twitters, and peeps the important measure is the modulation of peak frequency, the frequency with the highest amplitude. In these pure tone-like calls the peak frequency corresponds to the fundamental frequency. For calls with more distributed frequency amplitudes, like caws, bawls, and shrieks, the statistical distribution of the frequency amplitudes and the modulation of dominant frequency bands are the important measures.

of the vocal folds but also—and quite considerably so—by their tension. The tension is mainly controlled by the cricothyroid muscle (Müller-Preuss and Jürgens, 1978), but a complex coordination of several other muscles is necessary to produce high-pitched frequencies (Kirzinger and Jürgens, 1994). This coordination might require some training until its full effectiveness is reached, suggesting that ontogenetic changes must not necessarily be based on learning processes dependent on adult models.

In the squirrel monkey study it was not possible to pin down exactly which factors influenced changes in call structure during development. Therefore, Hammerschmidt and colleagues studied the developmental modifications of rhesus macaque "coo" calls. In an earlier study Newman and Symmes (1974) reported that young rhesus monkeys raised in isolation produced abnormal-sounding coo calls as adults. The coo call is a useful call type for studying ontogenetic changes because this call type

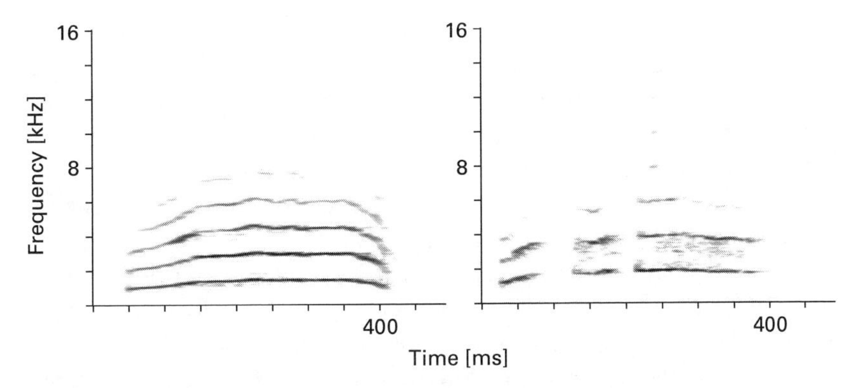

Figure 5.1
Spectrograms of coo calls from two subjects (approximately 1 week old). The first spectrogram shows a normal coo, and the second spectrogram shows a coo with amplitude gaps. Frequency range is 16 kHz, frequency resolution 39 Hz, and time resolution 5 milliseconds.

can be elicited reliably from birth on for several months under constant conditions. Hammerschmidt et al. (2000) recorded twenty rhesus macaque infants from the first week of life until the age of five months. Infants were either raised with their mothers in normal breeding colonies or separated from their mothers at birth and housed in a nursery with other age-matched peers. The calls were recorded during a short five- to fifteen-minute separation from conspecifics.

Adult-like "coo calls" could be observed from the first week of life (see figure 5.1). However, the "coo calls" underwent several changes: With increasing age, the calls dropped in pitch, which is to say that the fundamental frequency and the related harmonics decreased in frequency. In addition, the calls showed reduced variability in the course of their fundamental frequency and call amplitude, while the call duration increased slightly. Neither rearing condition nor sex had any apparent influence on age-related changes in "coo" structure.

The factor that provided the best explanation for age-related changes in call parameters was weight. Except for one call parameter, all other significant correlations could be excluded if they were controlled for weight. As Fitch (1997) showed in a study on rhesus macaques, body size, vocal tract length, and weight were highly correlated in adult animals. Thus, weight, as a proxy measure for body growth and changes in vocal tract characteristics, accounted for nearly all the changes in vocal structure that could be found up to the age of five months. Only the changes in one parameter ("mean gap length") still correlated significantly with age. The parameter "mean gap length" describes the portion of amplitude gaps in a call (see figure 5.1). To produce a constant amplitude throughout a call, it is necessary to establish a specific tension of the vocal folds together with a specific lung pressure. Without a correct combination no audible sound can be produced (Häusler, 2000). As mentioned above, mean gap length was the only parameter that could not be partialed out by weight. Instead, it decreased significantly with age. The authors surmised that the change in mean gap length indicated an increase in practice in call production (Hammerschmidt et al., 2000). Whether auditory input plays a role in the development of coo calls with a constant fundamental frequency course within a call cannot be answered by this study. In any case, rhesus macaques do not appear to require an adult model to produce species-specific "coo" calls.

This view is supported by learning experiments. In a conditioning study, Larson and colleagues (1973) tried to train rhesus macaques to respond to visual signals with a specific vocalization. The monkeys were able to increase the individual duration of a "coo" call step-by-step, being reinforced for every little increase. However, the maximum call duration of the lengthened "coo" calls remained within the normal range of "coo" calls in the population under study.

In sum, these studies showed that age-related changes in the acoustic structure of nonhuman primate calls do exist. However, most of these changes seem to be related to simple maturational factors, such as growth or practice that is needed to achieve the correct interplay between the motoneurons that innervate respiratory, laryngeal, and supralaryngeal muscles. In conclusion, it seems unlikely that the application of sophisticated analytical tools will fundamentally change current views on vocal development (Fischer, 2003). It is possible that some divergent statements can be reconciled if earlier studies are put into their proper historical perspective. For instance, those researchers that described calls as being "adult-like from birth on and showing no signs of modification" (Seyfarth and Cheney, 1997) would probably have had no trouble estimating the age of a caller on the basis of the call characteristics alone, despite the fact that, essentially, the call *types* were the same across different age classes. At a time when comparatively little evidence had been accumulated, the extent to which vocal production appeared to be innate may have come as a surprise, particularly in light of the close relatedness of humans and nonhuman primates. When it became clear that monkeys neither go through sensitive phases (in terms of their vocal production) nor gradually acquire the structure of their vocalizations, subtle changes became more interesting.

Pygmy Marmoset "Babbling"

Whereas the previous studies have addressed the issue of whether nonhuman primate calls exhibit plasticity in terms of their acoustic structure, less is known about plasticity in terms of the usage of calls. Several studies suggest that infants vocalize more frequently than adults (reviewed in Snowdon and Elowson, 2001; see Snowdon, this volume). Whereas some of this vocal activity may be related to the weaning process (Hammerschmidt et al., 1994; Hammerschmidt and Todt, 1995), other studies suggest that this increased vocal activity may provide vocal practice. One outstanding case is the vocal activity of young pygmy marmosets, *Cebuella pygmaea.* Elowson, Snowdon, and colleagues (Elowson et al., 1998a; Elowson et al., 1998b) observed that infant pygmy marmosets go through a phase of intense vocal output during which they produce long bouts of calls, most of which resemble calls of the species' repertoire. These call bouts consist of rhythmic and repetitive strings of syllables that begin to occur within the first two weeks of life. Vocal activity peaks just prior to weaning at the age of seven to eight weeks. The authors likened this behavior to human infant babbling (Elowson et al., 1998a) but did not differentiate between different types of human infant babbling. Typical human infant babbling, called canonical babbling, begins between six and ten months of age and consists of repetitive strings of well-formed syllables (Koopmanns-van Beinum and van der Stelt, 1986; Oller, 1980, 2000). If at all, pygmy marmoset babbling can be likened to precanonical babbling. In human infants, precanonical babbling begins at three month of age and lacks well-formed syllables. Unlike human words, which typically refer to an entity or class of entities, human babbling seems to have no such referents and is thought to constitute emotional expression and to promote control over one's vocal output via auditory feedback. Elowson and colleagues pointed out that both human and pygmy marmoset babbling lack apparent reference, promote interactions with caregivers, and might provide vocal practice (Elowson et al., 1998a; Snowdon and Elowson, 2001; but see Snowdon, this volume).

It may be worth noting that there are also some important differences between pygmy marmoset and human infant babbling: For one, canonical babbling in human infants can be clearly differentiated from earlier vocal behavior. More importantly, children assemble the parts that make up their early words from their prelinguistic phonetic repertoire, that is, from the syllables that form canonical babbling (Locke, 1993; Vihman, 1996). In other words, whereas babbling in children is a crucial part of the child's development of speech, pygmy marmosets decrease the amount of vocal activity after their babbling phase, while the structure of the elements largely remains the same. Finally, not only pygmy marmoset babbling but also the large majority of adult marmoset vocalizations fail to provide information about external objects or events and are in this sense "content free." Despite some parallels between pygmy marmoset and human infant babbling, some important differences remain.

Acoustic Convergence

As mentioned before, at least some nonhuman primates retain a high degree of variability in their vocalization throughout their life. It was hypothesized that this variability may be the substrate for subtle modification of the call structure later in life. One of the earliest reports of convergence processes came from Green (1975). Green observed that three populations of Japanese macaques revealed differences in the acoustic structure of a call type produced during feeding. He suggested that these differences were established due to a so-called "behavioral founder effect," where one subject responded to food provision with a spontaneous call that was imitated by other group members. In an attempt to document this process in more detail, Sugiura (1998) conducted the playback study mentioned above. He found that female Japanese macaques in one population responded to the contact call playbacks with more similar calls than they uttered spontaneously. He concluded that these monkeys must have had some control over their vocal output and had been able to adjust it accordingly.

Chimpanzees have also been reported to exhibit vocal learning. Male chimpanzees (*Pan troglodytes verus*) living in three adjacent communities in the Tai forest of the Ivory Coast differ in the structure of their "pant hoots" more strongly than either of these differs from a more distant community 70 km away. Neither habitat nor genetic structure could account for the observed pattern, and thus the authors concluded that the chimpanzees might actively modify the pant hoot structure, either to sound more like the other males in their own group or to sound different from males in neighboring groups (Crockford et al., 2004).

Mitani and colleagues (1992) had also examined the differences in the acoustic structure of chimpanzee pant-hoots between two populations. Although genetic or anatomic differences between populations could not be ruled out, the authors assumed that the differences in the pant-hoots were mediated by learning. In a second study, Mitani and colleagues recorded pant-hoots of two populations of wild chimpanzees in East Africa (Mitani et al., 1999). They were able to assign the majority of adult calls to the proper population. In contrast to the earlier study, the authors assumed that this acoustic variation could be related to differences in habitat structure and body size of the chimpanzees. To rule out genetic relatedness as a variable in chimpanzee call convergence, Marshall and colleagues did a study in which they investigated whether two groups of chimpanzees differed in the acoustic structure of their pant-hoots (Marshall et al., 1999). The animals had different origins. The authors found an overall similarity in the pant-hoots, suggesting a strong innate component of this call type. Nevertheless, they found some significant differences and reported that a male introduced a novel element to these pant-hoots that subsequently was used by five other males from the same colony. This novel element was produced by blowing air through pursed lips. However, this new element was not voiced, suggesting that no special neural control of vocal apparatus was involved in the production of this element.

A third example of acoustic differences at the group level comes from Barbary macaques (*M. sylvanus*). Fischer and colleagues (1998) reported that alarm calls given by members of two populations of Barbary macaques revealed significant differences between sites. Whereas the general structure was the same at both sites, detailed acoustic analysis revealed significant differences between calls from the two populations. Playback experiments in which calls from their own or the other population were broadcast suggested that the differences were perceptually salient (Fischer et al., 1998). Because neither genetic differences nor habitat structure could account for the observed differences, Fischer and colleagues concluded that the general acoustic structure was fixed whereas there was a restricted potential for acoustic modification.

To summarize, several studies have suggested that nonhuman primates have the potential for minor modification of their call structure. As far as the mechanism is concerned, several authors have suggested that the process of call convergence can be attributed to vocal accommodation (Mitani and Brandt, 1994; Fischer et al., 1998; Snowdon and Elowson, 1999; summarized in Fischer, 2003). In speech, humans make minor adjustments in their vocal output so that it sounds more like the speech of the individual they are talking to. This subconscious process is termed speech accommodation (Giles, 1984). Speech accommodation includes a wide range of subtle adaptations, such as altering the speed, the length of pauses and utterances, as well as the intonation and pitch of utterances. They also include changes in vocabulary and syntax. Locke (1993) observed similar adjustments in the speech acquisition process of infants and termed it vocal accommodation.

In light of the finding that nonhuman primates have only limited voluntary control over the vocal apparatus (Sutton et al., 1973; Deacon, 1991; Jürgens, 1992; Alipour et al., 1997) it remains puzzling how accommodation is accomplished. Possibly vocal accommodation can be achieved by articulation, which seems to be easier to control voluntarily than the laryngeal apparatus. If vocal accommodation accounts for the observed modification in call structure, then auditory feedback must somehow play a role, despite the fact that it does not seem to be a prerequisite for developing a species-specific repertoire (Hammerschmidt et al., 2001). More precisely, matching requires that the model be rapidly stored, and an individual's own motor output planned in accordance with the stored template (Heyes, 2001). With regard to long-term effects of call convergence, an alternative mechanism to vocal accommodation has been suggested by Peter Marler (Marler, 1991). He proposed that some call variants are selectively reinforced by social stimulation ("action-based learning"). In this case, contextual learning would account for the observed changes in call characteristics. Again, however, the precise mechanisms mediating action-based learning are not known.

Ontogeny of Vocal Production in Human Infants

Although speech is the central channel of human communication, there is another vocal behavior used for communication, which occurs much earlier than speech and which persists throughout life: nonverbal vocal expressions, like "crying," "laughing," "squealing," or "wailing." A number of studies have compared the vocal ontogeny of normally hearing and hearing-impaired infants (Stoel-Gammon and Otomo, 1986; Koopmans-van Beinum and van der Stelt, 1986; Oller and Eilers, 1988; Oller, 2000). These studies mostly have focused on the development of phonation and articulation with respect to features of mature speech and have attended less to the physical structure of infant vocalizations.

Scheiner and colleagues conducted a longitudinal study to investigate the vocalizations of normally hearing and profoundly hearing impaired infants with the aim to characterize and compare the preverbal vocal utterances with respect to their acoustic structure and their development during the first year of life. Since such preverbal utterances are hardly describable with linguistic methods, they decided to use a multiparameter acoustic analysis to examine the physical structure of the utterances in terms of time, frequency, and energy characteristics. In this kind of analysis this study is also comparable to studies of nonhuman primates. Scheiner and colleagues were able to distinguish ten expiratory vocal types, which could be differentiated on the basis of a small number of acoustic parameters. Seven vocalization types appeared within the first two months, and two vocalization types emerged between two and four months (see figure 5.2).

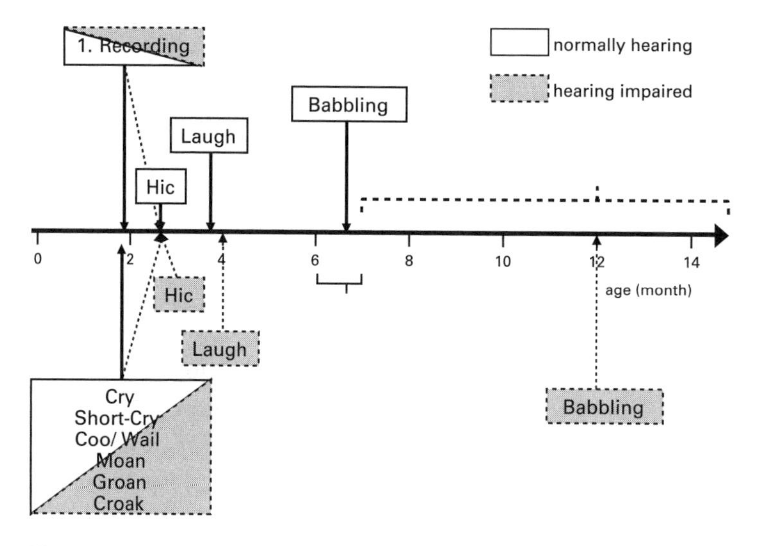

Figure 5.2
Emergence of infant vocalizations separated for normal hearing (NH) and hearing impaired (HI) infants. The figure indicates the age at which at least half of the infants uttered certain vocalizations. The age of first recording differs slightly for NH and HI infants.

In both normally hearing and hearing impaired infants preverbal utterances emerged at the same time, with the exception of canonical babbling (Scheiner et al., 2004). Canonical babbling in hearing impaired infants either was delayed in onset, compared to normally hearing infants, or did not appear at all during the recording period. This finding was in line with previous studies comparing the onset of canonical babbling in normal hearing and hearing impaired infants (Oller and Eilers, 1988; Eilers and Oller, 1994; van Beinum et al., 2001).

Scheiner and colleagues had a sufficient number of calls for four vocalization types ("cry," "short cry," "moan," and "wail") to test for developmental changes during the first years. Within these call types, they found only minor changes in the acoustic structure. These age-related changes were mainly characterized by a shift of call energy from higher to lower frequencies, an increase in harmonic-to-noise ratio, and an increase in the homogeneity of the vocalizations (Scheiner et al., 2002, 2004). Normally hearing and hearing impaired infants did not differ regarding these changes. These results showed that human infants, like nonhuman primates, need no adult model to produce species-specific nonverbal vocalizations. The nonverbal utterances of infants seem to be innate in similar ways to the vocal types of nonhuman primates. Ontogenetic changes are minor and seem to be mainly maturational.

Primate Vocalizations and Their Relation to Affective States

Another correspondence between nonverbal vocalizations in humans and nonhuman primates is that they both function to communicate the affective state of the signaler. As mentioned above, the nonverbal vocal expression of emotions appeared long before the development of language, and the importance of this communication system persists, although the verbal aspects of speech became more important in human communication.

Because there have been so few studies focused on the acoustic structure of preverbal vocalizations, it is not well-known how infant vocalizations are related to their underlying emotional states (but see Oller and Griebel, this volume). Only cries have been investigated extensively, but with controversial results. Some studies have found that infant cries contain discrete information about emotional state (e.g., Wasz-Höckert et al., 1968), while other studies have suggested that infant crying is a graded signal, mirroring a continuum of emotional states reaching from arousal to urgency (e.g., Protopapas and Lieberman, 1997). Still others have not found any changes in cries related to emotional state (e.g., Murry et al., 1975). This disagreement about the expression of infant emotions could be partly due to difficulties in determining the emotional state of an infant.

In the same study mentioned in the chapter above, Scheiner and colleagues (2002) found that all nonverbal vocalizations could occur in positive as well as in negative emotional contexts. Negative emotions were characterized by a significantly higher rate of "crying," "hic," and "ingressive vocalizations." Positive emotions showed a significantly higher rate of "babble," "laugh," and "raspberry." A change from positive to negative emotional state was accompanied by significant structural changes. All vocalization types (cry, coo/wail, and moan) examined in detail showed higher call duration, frequency range, and peak frequency for negative as opposed to positive expressions. Within vocalizations expressing particular positive emotions (joy, contentment) or particular negative emotions (unease, anger) Scheiner and colleagues could not find any significant differences.

A comparison of normally hearing and hearing impaired infants showed the same structural differences in the vocalizations in relation to positive and negative emotions (Scheiner et al., 2004, 2006). The only difference in relation to emotions between normally hearing and hearing impaired infants was in the composition of vocal sequences. Vocal sequences of normally hearing infants varied in the composition of their vocal types significantly with emotional states, whereas in hearing impaired infants they failed to do so. These results suggest that the acoustic structure of single vocalizations is to a large extent independent of auditory feedback, while the composition of vocal sequences may be influenced by auditory input (Scheiner et al., 2006).

The only nonhuman primate species for which a correlation between the structure of vocalization and degree of aversiveness/pleasantness of the concomitant emotional state has been established is the squirrel monkey (Jürgens, 1979). In this animal, the species-specific vocal repertoire could be elicited by activating specific brain regions responsible for the control of these calls. Activation was carried out by electrical stimulation of implanted intracerebral electrodes. By giving the animals the opportunity to switch on and off the vocalization-eliciting stimulation themselves, a quantitative measure could be obtained indicating the degree to which the stimulation was avoided or approached. Depending upon the stimulation site, different self-stimulation scores could be obtained.

An acoustic analysis of the calls uttered during this study showed a relation between different degrees of aversion and acoustic structure. Different call types of particular call categories revealed specific structural changes in relation to the emotional state (aversiveness or pleasantness) in which they were produced. Calls with a higher degree of aversion showed an increase in pitch measured as an increase in peak frequency, distribution of frequency amplitudes, and dominant frequency bands. In addition, they revealed a larger amount of nonharmonic energy (noise). Out of the numerous pitch-related parameters, peak frequency seemed to be the most important acoustic variable, because this variable was the only one that differed in all call categories in relation to the degree of aversion (Fichtel et al., 2001).

There are several other studies supporting the view that peak frequency could be an important acoustic variable to communicate aversiveness. Schrader and Todt (1998) found a positive correlation between the plasma level of adrenaline and the rate of "squeal-grunts" and a negative correlation between plasma cortisol and normal "grunts" in domestic pigs (*Sus scrofa*) subjected to stress through social isolation. In comparison to "squeal-grunts," "grunts" exhibit a lower peak frequency. Weary and Fraser (1995) studied the vocalizations of piglets that had been removed from their mothers immediately before suckling. They compared these vocalizations with those recorded from piglets that had been removed from their mothers immediately after suckling and found that the latter had lower peak frequencies than the former. After a disturbance at their sleeping site, Barbary macaques uttered series of shrill barks. Immediately after the disturbance, these calls had a higher peak frequency than calls uttered at the end of the call series (Fischer et al., 1995). All three papers support the view that the peak frequency is an important variable in the vocal expression of aversive states.

One objection to this interpretation could be that peak frequency could also be influenced by the arousal of the caller. In general, it is difficult to estimate the arousal of an animal. Some researchers, like Rendall (2003), have assigned different levels of arousal to different context conditions. Doing so, he found differences in the fundamental frequency and formant characteristics in the grunts of baboons related to the proposed state of arousal. In other cases, repetition rate or loudness of vocalizations can be good approximations of the arousal level (Fischer et al., 1995). In a recent study of the acoustic correlates of affective prosody in humans, Hammerschmidt and Jürgens (2007) showed that peak frequency significantly increased with loudness and aversion, but other vocal parameters, like fundamental frequency, increased significantly only with loudness. As a consequence, similarly loud hedonistic "prosodies" had a lower peak/fundamental frequency coefficient than aversive "prosodies." In other words, while loudness constitutes a good proxy measure for arousal, peak frequency can be used to differentiate between aversive and hedonistic states independent of arousal level as long as other arousal-correlated acoustic features are available.

Moreover, a playback study on squirrel monkeys indicated that subtle shifts in peak frequency were perceptually salient to listeners (Fichtel and Hammerschmidt, 2003). The same was true for red-fronted lemurs (*Eulemur fulvus*; Fichtel and Hammerschmidt, 2002). The fundamental nature of aversive or pleasant states, furthermore, suggests that their behavioral expressions might be phylogenetically ancient—possibly so old that related species share common acoustic features in the vocal expression of these states.

Neurobiological Foundations of Nonhuman Primate Vocal Behavior

All terrestrial mammals use the same components to produce their vocalizations. The lungs provide the airstream and thus the energy necessary to produce the sound. The airstream then passes the larynx and activates oscillations of the vocal folds that generate the sound. The vocal tract, finally, acts as a filter with specific resonant frequencies, enhancing certain spectral features of the sound while filtering out others (for a more recent review in vertebrate vocal production, see Hauser et al., 2002; Fitch et al., 2002). Despite this common principle in vocal production, some mammals exhibit notable morphological diversity in their sound production organs. A classic example is the huge larynx and hyoid apparatus of howler monkeys (*Alouatta spp.*). It fills the entire space between jaws and sternum and enables the howler monkeys to produce loud, low-pitched vocalization (Keleman and Sade, 1960; Schön-Ybarra, 1988). A different kind of diversity concerns vocal tract elongation. Humans (Lieberman and Blumstein, 1988), some deer (Fitch and Reby, 2001), and large cats (Hast, 1989) like lions and leopards elongate their vocal tract by lowering their larynx. The acoustic consequence is lower formant frequencies in their vocalization, which has been interpreted as an acoustic means to exaggerate acoustically their body size (Fitch, 2003). Despite such morphological specializations, most terrestrial mammals exhibit a common neurobiological circuitry, with the human species being the only exception.

Brain stimulation studies in squirrel monkeys have shown that there are many areas from which it is possible to induce these vocalizations (Jürgens and Ploog, 1970). The wide distributions of vocalization-eliciting sites and their predominately limbic localization, in addition to their long latency and fast habituation, have suggested that the elicited calls in most cases have represented secondary reactions. In other words, they were stimulation-induced motivational states rather than primary motor responses. To distinguish between the different kinds of responses of these brain areas, Jürgens (1976) conducted a study in which he was able to describe the aversive or hedonic emotional states of the squirrel monkey accompanying the electrically elicited vocalizations. The result showed that in most brain areas, vocalizations were indeed accompanied by a specific hedonic or aversive state. These areas include the septum, nucleus accumbens, preoptic area, hypothalamus, and amygdala. However, Jürgens found two exceptions, the anterior cingulate cortex in the forebrain and the posterior periaqueductal gray (PAG) with the laterally bordering tegmentum in the midbrain (Jürgens, 1976). In both areas the elicitation of vocalization was not correlated with a constant behavioral response like in the other areas. Therefore, the two areas must be able to trigger the vocalizations in a more direct way and function as primary vocalization-eliciting areas (Jürgens, 1998).

The anterior cingulate cortex seems to play an important role in volitional vocal control. Sutton et al. (1974) trained rhesus monkeys to emit a "coo" call to obtain food. The animals were required to produce a "coo" of prolonged duration and specific loudness in response to a signal cue light. After bilateral removal of the anterior cingulate cortex, the animals lost the ability to master this task. The same ablation did not interfere with a conditioning task in which the animals had to press a lever instead of producing a vocalization in order to obtain a reward. MacLean and Newman (1988) studied the effects of bilateral ablation in squirrel monkeys. Their results confirmed the findings for rhesus monkeys. They found that squirrel monkeys were no longer able to produce "isolation peeps," whereas other vocalizations, which seem to be no less clearly related to internal trigger mechanisms, remained unaffected. The role of the anterior cingulate cortex in the volitional control of emotional vocalizations does not seem to be limited to nonhuman primates. In a clinical study Jürgens and von Cramon (1982) described a patient who suffered a bilateral lesion in the anterior cingulate cortex. After recovery from the initial state of akinetic mutism this patient showed a long-lasting impairment in the volitional control of emotional intonation. Partiot and colleagues (1995) used positron emission tomography (PET) to measure brain activity while human volunteers silently imagined emotional situations. This imagination was accompanied by a strong activation of the anterior cingulate cortex. This suggests that the anterior cingulate cortex is involved not only in the vocal expression of emotional states but also in the control of emotional states more generally.

The second important area for vocal control is the PAG of the midbrain. Apart from the fact that its electrical stimulation leads to vocalizations in several species (for a review, see Jürgens, 1994, 1998), there are a number of other observations that implicate the PAG in vocal control. Lesioning experiments in squirrel monkeys have shown that the destruction of the PAG causes mutism (Jürgens and Pratt, 1979). Mutism after PAG lesions has also been reported for cats (Adametz and O'Leary, 1959), dogs (Skultety, 1968), rats (Chaurand et al., 1972), and humans (Botez and Carp, 1968). After the destruction of the PAG in squirrel monkeys, none of the numerous forebrain areas, the stimulation of which had elicited vocalizations, kept this vocalization-inducing capability (Jürgens and Pratt, 1979). All these areas have direct projections to the PAG. Thus, it seems likely that the PAG serves as a relay station for all descending vocalization-controlling limbic pathways (Jürgens, 1998). This conclusion is in line with recent retrograde tracing studies (Dujardin and Jürgens, 2005; Dujardin and Jürgens, 2006) that showed that vocalization-eliciting sites of the PAG receive a widespread input, coming from cortical and subcortical areas of the forebrain, large parts of the midbrain, pons, and medulla oblongata.

The PAG has also been extensively explored in macaques (Larson, 1991; Larson and Kistler, 1984; Larson and Kistler, 1986). Besides the same vocalization-eliciting function described for squirrel monkeys, the authors postulated that the PAG is also directly involved in the motor coordination of vocalization, a view that Jürgens (1998, 2002) does not support. Electrical stimulation of the PAG yields natural sounding, species-specific vocalizations. If the PAG were also the vocal pattern generator, one should expect that changes in stimulation parameters would have an influence on the acoustic structure of elicited calls. This did not occur. An increase in stimulation intensity caused an increase in call intensity but no changes in vocal structure. In addition, slight changes in electrode position had no effects on the acoustic structure of elicited vocalizations (Jürgens and Ploog, 1970; Jürgens, 1998).

Vocalizations are complex motor patterns made up of laryngeal, respiratory, and supralaryngeal (articulatory) components. The motoneurons controlling the vocal folds are located in the nucleus ambiguous. Those controlling the cricothyroid muscle are found in the rostral part of the nucleus ambiguous. This muscle controls the length of the vocal fold and hence plays a role in the regulation of fundamental frequency (Larson et al., 1987). The respiratory components of vocalizations are brought about by joint activity of thoracic and abdominal muscles and their corresponding motoneurons (Jürgens, 2002). The articulatory component is less well studied, but it is clear that some squirrel monkey and macaque call types are accompanied by jaw and lip movements. These jaw and lip movements are controlled by motoneurons located in the *N. trigemini* and *N. facialis*. In addition, there are motoneurons in the *N. hypoglossus* that are responsible for tongue movements. A remaining question is which structure could coordinate the activity of these different nuclei involved in vocal production. As mentioned above, some researchers have favored the PAG to be the possible control structure. In a study with squirrel monkeys, most of the neurons in the PAG fired only before the start of vocalizations and did not show any vocalization-correlated activity (Düsterhöft et al., 2004). In addition, for call types with complex frequency modulation Hage and Jürgens found a discrete area in the reticular formation just before the olivary complex that showed vocalization-correlated activity. These neurons showed an increase in neural activity just before and during vocalization. None of these neurons were active during mastication, swallowing, or quiet respiration. In addition, the neurons of this area showed significant correlations with the syllable structure of these vocalizations (Hage and Jürgens, 2006a, b).

In addition, an adequate model of vocal control must also take into account a somatosensory feedback component. Experiments in squirrel monkeys and cats have shown that transection of the internal branch of the superior laryngeal nerve, a fiber bundle transmitting somatosensory information from the larynx to the brain, causes a deterioration of vocalizations (Thoms and Jürgens, 1981; Shiba et al., 1995). When the transection was carried out above the cervical vagus, thus eliminating somatosensory feedback of larynx and lungs together, no further vocalization could be evoked from the PAG. The cited studies showed that the central vocalization mechanism depends upon input from the laryngeal and pulmonary proprioceptors. If this input is lacking, vocalizations are blocked (Jürgens, 1998).

In contrast to nonhuman primate vocalizations, the motor cortical larynx area plays an important role in voluntary vocal fold control in humans. Bilateral lesions in this region lead to a complete loss of voluntary phonation, while nonverbal emotional vocalizations, such as laughing, crying, or moaning, are preserved (for a review, see Jürgens, 2002). This connection does not exist in nonhuman primates (Kuypers, 1958; Simonyan and Jürgens, 2003). These findings are supported by the fact that dysfunction of the bilateral motor facial cortex has been found to have no influence on the ability of nonhuman primates to vocalize (Jürgens et al., 1982; Groswasser et al., 1988). Brain imaging studies in humans, furthermore, reported activation of the facial motor cortex (including the larynx area) during speaking and singing (Bookheimer et al., 2000; Perry et al., 1999). In contrast to humans, electrical destruction of the laryngeal motor cortex in the monkey did not affect vocal communication (Kirzinger and Jürgens, 1982). In addition, Jürgens and colleagues studied anterograde projections of the motorcortical tongue area to the hypoglossal nucleus and neighboring structures in rhesus monkey, squirrel monkey, saddle-back tamarin, and tree shrew. The findings suggested a phylogenetic trend in the projections of the motorcortical tongue area from nonhuman primate mammals via nonhuman primates to man in the sense that the corticomotoneuronal connection is strengthened toward man (Jürgens and Alipour, 2002; Jürgens, 2003).

In sum, the vocal pathway consists of three different subsystems. The first one is responsible for the initiation of vocalizations. The initiation can come from the anterior cingulate cortex or from different limbic brain areas as a consequence of different external or internal stimuli. The PAG serves as a collector or relay station for all these descending vocalization-controlling pathways, integrating the incoming information and triggering a specific innate vocal pattern. These neural circuits can be found in nonhuman primates and humans. The second subsystem is responsible for the voluntary motor control that is necessary to speak or sing—a fact that corresponds with the finding that this pathway can be found in humans only. The system comprises the motor cortex with its connections to the cerebellum, thalamus, as well as putamen and pyramidal pathway. The direct connection between motor cortex and the motoneurons controlling the larynx muscles is particularly important. A direct connection between motor cortex and the motoneurons is not necessary to utter genetically predefined vocal patterns. In addition, there are connections between the limbic cortex and the motor cortex. The third subsystem comprises the formatio

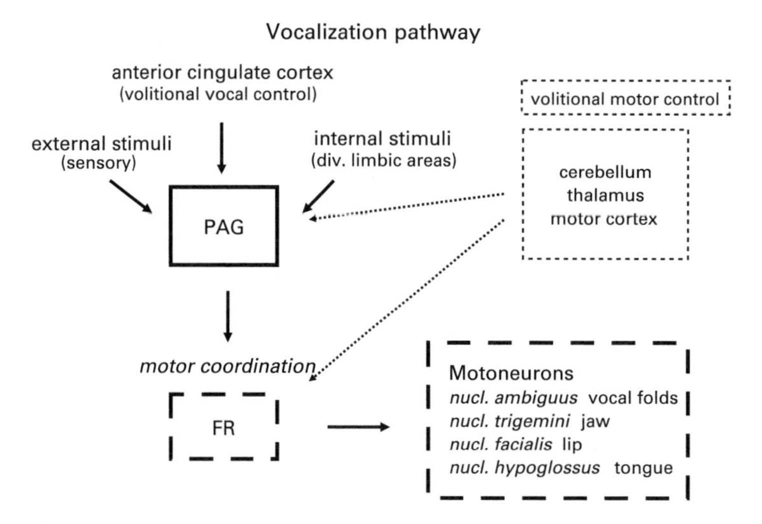

Figure 5.3
Vocalization pathway modified after Jürgens (1992). PAG, periaqueductal gray; FR, formatio reticularis; nucl., nucleus.

reticularis of the lower brain stem and the different motoneurons that innervate the respective muscles for vocal fold, lip, jaw, and tongue movements (see figure 5.3). This system is dependent on input from the PAG or, in the case of human speech or singing, input from the motor cortex. These neural circuits can be found in nonhuman primates and humans. In sum, the same neurobiological circuits that are responsible for the *innateness* of terrestrial mammal vocal production also exist in nonhuman primate and human nonverbal vocal production systems. That is, the neurobiological evidence explains to a large degree the lack of vocal flexibility in nonhuman primates. In contrast, such flexibility can be found in birds (see Hausberger et al., this volume), some marine mammals (see Schusterman, this volume), and human speech.

Flexible Responses to Vocalization in Nonhuman Primates

The previous sections on modifications in the production and/or usage of calls showed that nonhuman primates apparently exhibit little, albeit significant plasticity in some of their calls. In contrast, studies that have examined the development of the comprehension of, and correct responses to, calls have indicated that subjects undergo pronounced changes in development. For instance, vervet, *Chlorocbus aethiops*, infants gradually develop appropriate responses to the different alarm calls given to the main predators, leopards, martial eagles, and pythons (Seyfarth et al., 1980a, 1980b). Vervets also attend to the alarm calls of other species, such as the alarm calls of the superb startling, *Spreo superbus*, which are given in response to ground predators such as leopards (Cheney and Seyfarth, 1985). Vervet infants in an area with higher rates of starling alarm calling responded to playbacks of these calls at an earlier age than did infants living in areas with lower rates of starling alarm calling (Hauser, 1988). Chacma baboon, *Papio h. ursinus*, infants began to discriminate between different variants of their loud calls at an age of about six months (Fischer et al., 2000). These findings raise the question of whether at an age of about half a year, primate infants across a variety of taxa develop the ability to discriminate among different call types. However, a recent study on Barbary macaque infants suggested that experience with the call type and salience of the call also play a role. In this species, infants begin to respond significantly more strongly to their mother's calls than to other females' calls from an age of about two and a half months. Interestingly, in the first few weeks of life they hardly ever orient to any sounds in their environment, suggesting that both maturation and learning determine the onset of appropriate responses (Fischer, 2004).

The view that experience, and not only maturation, mediates the development of infants' responses is supported by the observations that both vervets and baboons respond to the alarm calls of birds, ungulates, and other primate species (Cheney and Seyfarth, 1985; Hauser, 1988; Cheney and Seyfarth, 1990; Zuberbühler, 2000). It seems unlikely that these responses have a purely genetic basis. Finally, in a cross-fostering experiment in which Japanese and rhesus macaque infants were raised by members of the other species, adoptive mothers learned to attend to the calls of their foster offspring even though the infant was a member of another species (Owren et al., 1993).

How do infants learn the appropriate responses to a given vocalization? On the one hand, the observation that young vervets that first had looked at an adult were more prone to respond appropriately suggests that they may mimic the adult's behavior. On the other hand, juvenile Barbary macaques much more frequently ran away or climbed into trees after playbacks of conspecific alarm calls than adults (Fischer et al., 1995; Fischer and Hammerschmidt, 2001). To date, the evidence for "teaching" and other forms of active information transmission remains anecdotal (for discussion, see Hauser, 1996). McCowan and colleagues have suggested that the reason(s) juveniles respond more readily than adults could be either due to infant error, or, more plausibly, because it is adaptive to do so. Young monkeys are more vulnerable to dangers because they are smaller and, possibly, less attentive to their surroundings (McCowan et al., 2001). Thus, not responding to an alarm call could be more costly for young monkeys than for adults.

Summary

Ontogenetic studies revealed age-related changes in the acoustic structure of nonhuman primate calls. However, most of these changes seem to be related to simple maturational factors, such as growth, or the ability to produce a constant subglottic air pressure while vocalizing. Studies with cross-fostered or acoustically deprived animals have shown that an acoustic model is not necessary to produce calls that fall within the species-specific range. The conclusion that the general neural motor patterns are mainly innate is in line with conditioning experiments, which show that only basic acoustic features, like duration or the repetition of elements, can be trained. Overall, the structure of nonhuman primate vocalizations is largely innate. Only subtle changes in vocal structure are possible. While this is highly different from human speech, it resembles findings in the fixed nonverbal acoustic communication of human infants.

These findings are supported by neurobiological studies that have shown that monkeys, in contrast to humans, lack a direct connection of the motorcortex to the laryngeal motoneurons. In addition, comparative studies have found a phylogenetic trend in the projections of the motorcortical tongue area. This is of interest because the tongue is the most important organ in the differentiation of phonemes. Studies have found an increasing strength of the corticomotoneuronal connection from nonprimate mammals to nonhuman primates and humans. The inability to generate new vocal patterns does not exclude a certain degree of vocal plasticity. Nonhuman primates can show a life-long high degree of variability in their vocal types. This variability seems to be the foundation for convergence processes, like vocal accommodation or action based learning, which could lead to a certain degree of vocal differentiation within the same species. However, the neurobiological mechanisms of these convergence processes are not known.

Outlook

It is still unclear why nonhuman primates' vocal production is so strongly constrained, while vocal plasticity has evolved independently in several taxa. Not only humans but also dolphins and parrots are able to imitate vocally. A unique case is songbirds' vocal learning. Songbirds have evolved a highly specialized neural circuitry to store auditory information. As nestlings, but in some species also later in life, they memorize the song sung by their fathers or other males in the vicinity. Months later they begin to match their vocal output to the auditory template until they eventually produce songs that match their learned models with high accuracy. Because of their specific dedication, however, the mechanisms mediating song learning might not be very informative for an understanding of the evolution of speech.

Yet song learning in birds does raise an important question about the adaptive value of vocal imitation. Vocal imitation allows incorporation of vocal patterns produced by others into a repertoire. Imitation may contribute to bonding between individuals (Janik, 2000) and may lead—as in the case of bird song—to the development of elaborate acoustic ornaments. Despite their appeal, none of these arguments provide an explanation of how vocal imitation in humans got started in the first place. Fitch has proposed that speech evolved in the interaction between mother and infant in a relatively close-knit community (Fitch, 2004). Yet, this scenario mainly deals with the undisputed adaptive value of vocal imitation but fails to provide an explanation at the proximate level.

While it seems obvious that voluntary vocal production comes with huge benefits, further research is needed to address the constraints that limit the reorganization of the brain in such a way that vocal imitation and voluntary control can be achieved. One first approach to tackle this question is to clarify the role of the FOXP2 protein, a transcription factor, in the vocal production system of terrestrial mammals. A mutation in the *FOXP2* gene leads to disruptions in speech production in humans (reviewed in Marcus and Fisher, 2003). The gene, mapped onto chromosome 7q31, codes for a forkhead protein, a transcription factor presumed to be important for embryonic and postembryonic development. Analyses of the molecular phylogeny revealed that this gene is highly conserved, with extremely few replacement nucleotide substitutions. In fact, mice and humans differ only in three amino acid substitutions in the FOXP2 protein, two of which occurred after the split of the human and chimpanzee lineage (Enard et al., 2002). Mice whose (murine) *Foxp2* gene was knocked out have a number of deficits in terms of their motor and also vocal behavior (Shu et al., 2005). An analysis of the vocal, cognitive, and motor abilities of mice carrying the human $Foxp2^{hum/hum}$ variant is currently under way (Enard et al., 2007). While the *FOXP2* gene should not be viewed as a "speech gene," it provides an important entry point to tracing the target genes and neural pathways underlying the production of sequential motor actions, a crucial prerequisite for the production of speech and language. On a somewhat different note, it is also important to keep in mind that the selective pressures apparently affecting the sender of the signals do in fact operate on the entire communicative system, in its minimal form both sender and recipient. It might in fact be the case that smart listeners lift some of the pressure off senders because they not only are able to perceive fine-grained differences among calls but also are able to make rich interpretations of calls—in other words, to attribute meaning to them.

References

Adametz J, O'Leary JL. 1959. Experimental mutism resulting from periaqueductal lesions in cats. *Neurology* 9: 636–42.

Alipour M, Chen Y, Jürgens U. 1997. Anterograde projections of the cortical tongue area of the tree shrew (*Tupaia belangeri*). *Journal of Brain Research* 38: 405–23.

van Beinum FJK, Clement CJ, van den Dikkenberg-Pot I. 2001. Babbling and the lack of auditory speech perception: A matter of coordination? *Developmental Science* 4: 61–70.

Bookheimer SY, Zeffiro TA, Blaxton TA, Gaillard W, Theodore WH. 2000. Activation of language cortex with automatic speech tasks. *Neurology* 55: 1151–7.

Botez MI, Carp N. 1968. Nouvelles données sur le problème du mécanisme de déclenchement de la parole. *Revue Roum Neurol* 5: 153–8.

Brockelman WY, Schilling D. 1984. Inheritance of stereotyped gibbon calls. *Nature* 312: 634–6.

Chaurand JP, Vergnes M, Karli P. 1972. Substance grise centrale du mésencéphale et comportement d'agression interspécifique du rat. *Physiol Behav* 9: 475–81.

Cheney DL, Seyfarth RM. 1985. Social and nonsocial knowledge in vervet monkeys. *Philosophical Transactions of the Royal Society of London B* 308: 187–201.

Cheney DL, Seyfarth RM. 1990. *How Monkeys See the World*. Chicago: University of Chicago Press.

Crockford C, Herbinger I, Vigilant L, Boesch C. 2004. Group specific calls in chimpanzees: A case for vocal learning? *Ethology* 110: 221–43.

Deacon TW. 1991. Anatomy of hierarchical information processing. *Behavioral and Brain Sciences* 14: 555–6.

Dujardin E, Jürgens U. 2005. Afferents of vocalization-controlling periaqueductal regions in the squirrel monkey. *Brain Research* 1034: 114–31.

Dujardin E, Jürgens U. 2006. Call type-specific differences in vocalization-related afferents to the periaqueductal gray of squirrel monkeys (*Saimiri sciureus*). *Behavioural Brain Research* 168: 23–36.

Düsterhöft F, Häusler U, Jürgens U. 2004. Neuronal activity in the periaqueductal gray and bordering structures during vocal communication in the squirrel monkey. *Neuroscience* 123: 53–60.

Eilers RE, Oller DK. 1994. Infant vocalizations and the early diagnosis of severe hearing impairment. *Journal of Pediatrics* 124: 199–203.

Elowson AM, Snowdon CT, Lazaro-Perea C. 1998a. Infant 'babbling' in a nonhuman primate: Complex vocal sequences with repeated call types. *Behaviour* 135: 643–64.

Elowson AM, Snowdon CT, Lazaro-Perea C. 1998b. 'Babbling' and social context in infant monkeys: Parallels to human infants. *Trends in Cognitive Sciences* 2: 31–7.

Enard W, Gehre S, Hammerschmidt K, Brückner MK, Gigert T, Hölter SM, Kallnik M, Becker L, Groszner M, Müller U, Gailus-Durner V, Fuchs H, Mouse Clinic Consortium, Klopstock T, Wurst W, Fisher SE, Arendt T, Hrabe de Angelis M, Fischer J, Schwarz J, Pääbo S. 2007. A mouse model for human-specific changes in FOXP2, a gene important for speech and language. Society for Neuroscience 2007, San Diego, CA.

Enard W, Przeworski M, Fisher SE, Lai CSL, Wiebe V, Kitano T, Monaco AP, Paabo S. 2002. Molecular evolution of FOXP2, a gene involved in speech and language. *Nature* 418: 869–72.

Fichtel C, Hammerschmidt K. 2002. Responses of redfronted lemurs (*Eulemur fulvus rufus*) to experimentally modified alarm calls: Evidence for urgency-based changes in call structure. *Ethology* 108: 763–77.

Fichtel C, Hammerschmidt K. 2003. Responses of squirrel monkeys to their experimentally modified mobbing calls. *Journal of the Acoustical Society of America* 113: 2927–32.

Fichtel C, Hammerschmidt K, Jürgens U. 2001. On the vocal expression of emotion: A multi-parametric analysis of different states of aversion in the squirrel monkey. *Behaviour* 138: 97–116.

Fischer J. 2003. Developmental modifications in the vocal behavior of nonhuman primates. In *Primate Audition*, ed. AA Ghazanfar, pp. 109–25. Boca Raton: CRC Press.

Fischer J. 2004. Emergence of individual recognition in young macaques. *Animal Behaviour* 67: 655–61.

Fischer J, Call J, Kaminski J. 2004. A pluralistic account of word learning. *TICS* 8: 481.

Fischer J, Cheney DL, Seyfarth RM. 2000. Development of infant baboons' responses to graded bark variants. *Proceedings of the Royal Society of London B* 267: 2317–21.

Fischer J, Hammerschmidt K. 2001. Functional referents and acoustic similarity revisited: The case of Barbary macaque alarm calls. *Animal Cognition* 4: 29–35.

Fischer J, Hammerschmidt K, Todt D. 1995. Factors affecting acoustic variation in Barbary macaque (*Macaca sylvanus*) disturbance calls. *Ethology* 101: 51–66.

Fischer J, Hammerschmidt K, Todt D. 1998. Local variation in Barbary macaque shrill barks. *Animal Behaviour* 56: 623–29.

Fitch WT. 1997. Vocal tract length and formant frequency dispersion correlate with body size in rhesus macaques. *Journal of the Acoustical Society of America* 102: 1213–22.

Fitch WT. 2000. The evolution of speech: A comparative review. *TICS* 4: 258–66.

Fitch WT. 2003. Primate vocal production and its implication for auditory research. In *Primate Audition*, ed. AA Ghazanfar, pp. 87–108. Boca Raton: CRC Press.

Fitch WT. 2004. Kin selection and "mother tongues": A neglected component in language evolution. In *Evolution of Communication Systems: A Comparative Approach*, ed. K Oller, U Griebel, pp. 275–96. Cambridge, MA: MIT Press.

Fitch WT, Neubauer J, Herzel H. 2002. Calls out of chaos: The adaptive significance of nonlinear phenomena in mammalian vocal production. *Animal Behaviour* 63: 407–418.

Fitch WT, Reby D. 2001. The descended larynx is not uniquely human. *Proceedings of the Royal Society of London B* 268: 1669–75.

Geissmann T. 1984. Inheritance of song parameters in the gibbon song analysed in two hybrid gibbons (*Hylobates pileatus and Hylobates lar*). *Folia Primatologica* 42: 216–35.

Giles H. 1984. The dynamics of speech accommodation. *International Journal of the Sociology of Language* 46: 1–155.

Green S. 1975. Dialects in Japanese monkeys: Vocal learning and cultural transmission of locale-specific vocal behavior? *Zeitschrift für Tierpsychologie* 38: 304–14.

Groswasser Z, Korn C, Groswasser-Reider I, Solzi P. 1988. Mutism associated with buccofacial apraxia and bihemispheric lesions. *Brain and Language* 34: 157–68.

Hage S, Jürgens U. 2006a. Localization of a vocal pattern generator in the pontine brainstem of the squirrel monkey. *European Journal of Neuroscience* 23: 840–4.

Hage S, Jürgens U. 2006b. On the role of the pontine brainstem in vocal pattern generation: A telemetric single-unit recording study in the squirrel monkey. *Journal of Neuroscience* 26: 7105–15.

Hammerschmidt K, Ansorge V, Fischer J, Todt D. 1994. Dusk calling in Barbary macaques (*Macaca sylvanus*): Demand for social shelter. *American Journal of Primatology* 32: 277–89.

Hammerschmidt K, Jürgens U. 2007. Acoustic correlates of affective prosody. *Journal of Voice* 21: 531–40.

Hammerschmidt K, Freudenstein T, Jürgens U. 2001. Vocal development in squirrel monkeys. *Behaviour* 138: 1179–204.

Hammerschmidt K, Newman JD, Champoux M, Suomi SJ. 2000. Changes in rhesus macaque coo vocalizations during early development. *Ethology* 106: 873–86.

Hammerschmidt K, Todt D. 1995. Individual differences in vocalizations of young Barbary macaques (*Macaca sylvanus*): A multi-parametric analysis to identify critical cues in acoustic signalling. *Behaviour* 132: 381–99.

Hast M. 1989. The larynx of roaring and non-roaring cats. *Journal of Anatomy* 163: 117–21.

Hauser MD. 1988. How infant vervet monkeys learn to recognize starling alarm calls: The role of experience. *Behaviour* 105: 187–201.

Hauser MD. 1996. *The Evolution of Communication*. Cambridge, MA: MIT Press.

Hauser MD, Chomsky N, Fitch WT. 2002. The faculty of language: What is it, who has it, and how did it evolve? *Science* 298: 1569–79.

Häusler U. 2000. Vocalization-correlated respiratory movements in the squirrel monkey. *Journal of the Acoustical Society of America* 108: 1443–50.

Hayes KJ, Hayes C. 1951. The intellectual development of a home-raised chimpanzee. *Proceedings of the American Philosophical Society* 95: 105–9.

Heyes CM. 2001. Causes and consequences of imitation. *TICS* 5: 253–61.

Jackendoff R. 1999. Possible stages in the evolution of the language capacity. *TICS* 3: 272–9.

Janik VM. 2000. Whistle matching in wild bottlenose dolphins (*Tursiops truncatus*). *Science* 289: 1355–7.

Janik VM, Slater PJ. 1997. Vocal learning in mammals. *Advances in the Study of Behavior* 26: 59–99.

Jürgens U. 1976. Reinforcing concomitants of electrically elicited vocalizations. *Experimental Brain Research* 26: 203–14.

Jürgens U. 1979. Vocalization as an emotional indicator: A neuroethological study in the squirrel monkey. *Behaviour* 69: 88–117.

Jürgens U. 1992. On the neurobiology of vocal communication. In *Nonverbal Vocal Communication*, ed. H Papousek, U Jürgens, M Papousek, pp. 31–42. Berlin: Springer.

Jürgens U. 1994. The role of the periaqueductal grey in vocal behaviour. *Behavioural Brain Research* 62: 107–17.

Jürgens U. 1998. Neuronal control of mammalian vocalization, with special reference to the squirrel monkey. *Naturwissenschaften* 85: 376–88.

Jürgens U. 2000. Localization of a pontine vocalization-controlling area. *Journal of the Acoustical Society of America* 108: 1393–6.

Jürgens U. 2002. Neural pathways underlying vocal control. *Neuroscience and Biobehavioral Review* 26: 235–58.

Jürgens U. 2003. Phylogenese der sprachlichen Kommunikation. In *Psycholinguistik*, ed. G Rickheit, T Herrmann, W Deutsch, pp. 33–57. Berlin: Walter de Gruyter.

Jürgens U, Alipour M. 2002. A comparative study on the cortico–hypoglossal connections in primates, using biotin dextranamine. *Neuroscience Letters* 328: 245–8.

Jürgens U, Kirzinger A, von Cramon D. 1982. The effects of deep-reaching lesions in the cortical face area on phonation: A combined case report and experimental monkey study. *Cortex* 18: 125–39.

Jürgens U, Ploog D. 1970. Cerebral representation of vocalization in the squirrel monkey. *Experimental Brain Research* 10: 532–54.

Jürgens U, Pratt R. 1979. Role of the periaqueductal grey in vocal expression of emotion. *Brain Research* 167: 367–78.

Jürgens U, von Cramon D. 1982. On the role of anterior cingulate cortex in phonation: A case report. *Brain and Language* 15: 234–48.

Kaminski J, Call J, Fischer J. 2004. Word learning in a domestic dog: Evidence for "fast mapping." *Science* 304: 1682–3.

Keleman G, Sade J. 1960. The vocal organ of the howling monkey (*Alouatta palliata*). *Journal of Morphology* 107: 123–40.

Kellogg WN. 1968. Communication and language in home-raised chimpanzee—Gestures, words, and behavioral signals of home-raised apes are critically examined. *Science* 162: 423.

Kirzinger A, Jürgens U. 1982. Cortical lesion effects and vocalization in the squirrel monkey. *Brain Research* 233: 299–315.

Kirzinger A, Jürgens U. 1994. Role of extralaryngeal muscles on phonation of subhuman primates. *Journal of Comparative Physiology A* 175: 215–22.

Koopmans-van Beinum FJ, van der Stelt JM. 1986. Early stages in the development of speech movements. In *Precursors of Early Speech*, ed. B Lindblom, R Zetterstrom, pp. 37–50. New York: Stockton Press.

Kuypers HG. 1958. Some projections from the pericentral cortex to the pons and lower brainstem in monkey and chimpanzee. *Journal of Comparative Neurology* 110: 221–55.

Larson CR. 1991. On the relation of PAG neurons to laryngeal and respiratory muscles during vocalization in the monkey. *Brain Research* 552: 77–86.

Larson CR, Kempster GB, Kistler MK. 1987. Changes in voice fundamental frequency following discharge of single motor units in circothyroid and thyroarrytenoid mucles. *Journal of Speech and Hearing Research* 30: 551–8.

Larson CR, Kistler MK. 1984. Periaqueductal gray neuronal activity associated with laryngeal EMG and vocalization in the awake monkey. *Neuroscience Letters* 46: 261–6.

Larson CR, Kistler MK. 1986. The relationship of periaqueductal gray neurons to vocalization and laryngeal EMG in the behaving monkey. *Experimental Brain Research* 63: 596–606.

Larson CR, Sutton D, Taylor EM, Lindeman R. 1973. Sound spectral properties of conditioned vocalization in monkeys. *Phonetica* 27: 100–10.

Latta J. 1966. Vocal repertoire of the squirrel monkey (*Saimiri sciureus*), its analysis and significance. *Experimental Brain Research* 1: 359–84.

Lieberman P, Blumstein SE. 1988. *Speech Physiology, Speech Perception, and Acoustic Phonetics*. Cambridge, England: Cambridge University Press.

Locke JL. 1993. *The Child's Path to Spoken Language*. Cambridge, MA: Harvard University Press.

MacLean PD, Newman JD. 1988. Role of the midline frontolimbic cortex in production of the isolation call of squirrel monkeys. *Brain Research* 450: 111–23.

Marcus GF, Fisher SE. 2003. FOXP2 in focus: What can genes tell us about speech and language? *Trends in Cognitive Sciences* 7: 257–62.

Marler P. 1991. Song-learning behaviour: The interface with neuroethology. *TINS* 14: 199–205.

Marshall AJ, Wrangham RT, Arcadi AC. 1999. Does learning affect the structure of vocalizations in chimpanzees? *Animal Behaviour* 58: 825–30.

McCowan B, Franceschini NV, Vicino GA. 2001. Age differences and developmental trends in alarm peep responses by squirrel monkeys (*Saimiri sciureus*). *American Journal of Primatology* 53: 19–31.

Mitani JC, Brandt KL. 1994. Social factors influence the acoustic variability in the long-distance calls of male chimpanzees. *Ethology* 96: 233–52.

Mitani JC, Hasegawa T, Gros-Louis J, Marler P, Byrne RW. 1992. Dialects in wild chimpanzees? *American Journal of Primatology* 27: 233–43.

Mitani JC, Hunley KL, Murdoch ME. 1999. Geographic variation in the calls of wild chimpanzees: A reassessment. *American Journal of Primatology* 47: 133–51.

Müller-Preuss P, Jürgens U. 1978. Projections from different limbic vocalization areas in the squirrel monkey. In *Recent Advances in Primatology, Vol. 1: Behaviour*, ed. DJ Chivers, WJ Herbert, pp. 803–5. New York: Academic Press.

Murry T, Hollien H, Muller E. 1975. Perceptual responses to infant crying: Maternal recognition and sex judgements. *Journal of Child Language* 2: 199–204.

Newman JD, Symmes D. 1974. Vocal pathology in socially deprived monkeys. *Developmental Psychobiology* 7: 351–8.

Oller DK. 1980. The emergence of the sounds of speech in infancy. In *Child Phonology, Vol. 1: Production*, ed. G Yeni-Komshian, J Kavanagh, C Ferguson, pp. 93–112. New York: Academic Press.

Oller DK. 2000. *The Emergence of the Speech Capacity*. Mahwah, NJ: Lawrence Erlbaum Associates.

Oller DK, Eilers RE. 1988. The role of audition in infant babbling. *Child Development* 59: 441–9.

Owren MJ, Dieter JA, Seyfarth RM, Cheney DL. 1992. "Food" calls produced by adult female rhesus (*Macaca mulatta*) and Japanese (*M. fuscata*) macaques, their normally-raised offspring, and offspring cross-fostered between species. *Behaviour* 120: 218–31.

Owren MJ, Dieter JA, Seyfarth RM, Cheney DL. 1993. Vocalizations of rhesus (*Macaca mulatta*) and Japanese (*M. fuscata*) macaques cross-fostered between species show evidence of only limited modification. *Developmental Psychobiology* 26: 389–406.

Partiot A, Grafman J, Sadato N, Wachs J, Hallett M. 1995. Brain activation during the generation of nonemotional and emotional plans. *Neuroreport* 6: 1397–1400.

Perry DW, Zatorre RJ, Petridrides M, Alivisatos B, Meyer E, Evans AC. 1999. Localization of cerebral activity during simple singing. *Neuroreport* 10: 3453–8.

Protopapas A, Lieberman P. 1997. Fundamental frequency of phonation and perceived emotional stress. *Journal of the Acoustical Society of America* 101: 2267–77.

Rendall D. 2003. Acoustic correlates of caller identity and affect intensity in the vowel-like grunt vocalizations of baboons. *Journal of the Acoustical Society of America* 113: 3390–3402.

Scheiner E, Hammerschmidt K, Jürgens U, Zwirner P. 2002. Acoustic analyses of developmental changes and emotional expression in the preverbal vocalizations of infants. *Journal of Voice* 16: 509–29.

Scheiner E, Hammerschmidt K, Jürgens U, Zwirner P. 2004. The influence of hearing impairment on the preverbal vocalizations of infants. *Folia Phoniatrica et Logopaedica* 56: 27–40.

Scheiner E, Hammerschmidt K, Jürgens U, Zwirner P. 2006. Vocal expression of emotions in normally hearing and hearing-impaired infants. *Journal of Voice* 20: 585–604.

Schön-Ybarra M. 1988. Morphological adaptation for loud phonation in the vocal organ of howling monkeys. *Primate Report* 22: 19–24.

Schrader L, Todt D. 1998. Vocal quality is correlated with levels of stress hormones in domestic pigs. *Ethology* 104: 859–76.

Seyfarth RM, Cheney DL. 1997. Some features of vocal development in nonhuman primates. In *Social Influences on Vocal Development*, ed. CT Snowdon, M Hausberger, pp. 249–73. Cambridge, England: Cambridge University Press.

Seyfarth RM, Cheney DL, Marler P. 1980a. Vervet monkey alarm calls: Semantic communication in a free-ranging primate. *Animal Behaviour* 28: 1070–94.

Seyfarth RM, Cheney DL, Marler P. 1980b. Monkey responses to three different alarm calls: Evidence of predator classification and semantic communication. *Science* 210: 801–3.

Shiba K, Yoshida K, Miura T. 1995. Functional roles of the superior laryngeal nerve afferents in electrically induced vocalisation in anesthetized cats. *Neurosci Res* 22: 23–30.

Shu W, Chob JY, Jiang Y, Zhangc M, Weis D, Gregory A, Elder GA, Schmeidler J, De Gasperid R, Gama Sosa MA, Rabidou D, Santucci AC, Perl D, Morrisey E, Buxbaum JD. 2005. Altered ultrasonic vocalization in mice with a disruption in the Foxp2 gene. *PNAS* 102: 9643–8.

Simonyan K, Jürgens U. 2003. Efferent subcortical projections of the laryngeal motorcortex in the rhesus monkey. *Brain Research* 974: 43–59.

Skultety FM. 1968. Clinical and experimental aspects of akinetic mutism. Report of a case. *Archives of Neurology* 19: 1–14.

Snowdon CT, Elowson AM. 1999. Pygmy marmosets modify call structure when paired. *Ethology* 105: 893–908.

Snowdon CT, Elowson AM. 2001. "Babbling" in pygmy marmosets: Development after infancy. *Behaviour* 138: 1235–48.

Stoel-Gammon C, Otomo K. 1986. Babbling development of hearing impaired and normally hearing subjects. *Journal of Speech and Hearing Disorders* 51: 33–41.

Sugiura H. 1998. Matching of acoustic features during the vocal exchange of coo calls by Japanese macaques. *Animal Behaviour* 55: 673–87.

Sutton D, Larson C, Lindeman RC. 1974. Neocortical and limbic lesion effects on primate phonation. *Brain Research* 71: 61–75.

Sutton D, Larson C, Taylor EM, Lindeman RC. 1973. Vocalizations in rhesus monkeys: Conditionability. *Brain Research* 52: 225–31.

Talmage-Riggs G, Mayer W. 1972. Effect of deafening on vocal behaviour of squirrel monkey (*Saimiri sciureus*). *Folia Primatologica* 17: 404–20.

Tomasello M, Call J, Warren J, Frost GT, Carpenter M, Nagell K. 1997. The ontogeny of chimpanzee gestural signals: A comparison across groups and generations. *Evolution of Communication* 1: 223–59.

Tomasello M, Zuberbühler K. 2002. Primate vocal and gestural communication. In *The Cognitive Animal*, ed. C Allen, M Bekoff, GM Burghardt, pp. 293–99. Cambridge, MA: MIT Press.

Thoms G, Jürgens U. 1981. Role of the internal laryngeal nerve in phonation: An experimental study in squirrel monkey. *Experimental Neurology* 74: 187–203.

Vihman MM. 1996. *Phonological Development: The Origins of Language in the Child.* Oxford: Blackwell Publishers.

Wallman J. 1992. *Aping Language.* Cambridge, England: Cambridge University Press.

Wasz-Höckert O, Lind J, Vurenkoski V, Patanen T, Valanne E. 1968. *The Infant Cry: A Spectrographic and Auditory Analysis.* London: Spastics International Medical Publications in association with William Heinemann Medical Books Ltd.

Weary DM, Fraser D. 1995. Calling by domestic piglets—Reliable signals of need. *Animal Behaviour* 50: 1047–55.

Winter PP, Handley D, Schott D. 1973. Ontogeny of squirrel monkey calls under normal conditions and under acoustic isolation. *Behaviour* 47: 230–9.

Zuberbühler K. 2000. Interspecies semantic communication in two forest primates. *Proceedings of the Royal Society of London B* 267: 713–8.

6 Contextual Sensitivity and Bird Song: A Basis for Social Life

Martine Hausberger, Laurence Henry, Benoît Testé, and Stéphanie Barbu

Language Evolution: The Comparative Approach as a Tool

As mentioned by Sterelny (this volume), it is not likely that we can ever have direct evidence of the evolutionary trajectory of language. Plausible explanations can arise from archaeological evidence, but even then controversies appear between claims of a sudden appearance of the language capacity and the idea of a slow, gradual process involving a variety of anatomical, physiological, and neurological changes (e.g., MacWhinney, this volume). Another traditional way of broaching the question of language evolution is to compare our vocal system with the vocal systems of our closest relatives, which are the great apes and, to a lesser extent, monkeys. Despite the important studies aiming at teaching language to apes, it appears in the main that only limited language-like performance can be achieved. Moreover, it is generally admitted that nonhuman primates have fixed innate vocal signals (Seyfarth and Cheney, 1997), although evidence has accumulated lately of more flexibility than assumed until recently (Snowdon, this volume). Nevertheless, it seems that the comparative approach, based on homology, leads more to clues about how human language diverged at an early stage from the vocal repertoire of closely related species than to why and how it evolved the way it did.

Phylogeneticians will argue that in order to understand evolution, both homology (commonly inherited similarity) and homoplasy (phylogenetically scattered, independently acquired similarity) have to be considered (Martins, 1996; Sanderson and Hufford, 1996; Deleporte, 2002). Convergence, a process where similar functional adaptations occur in different lineages in response to similar (e.g., socioecological) conditions, indicates a general evolutionary trend in organisms showing comparable biological capacities. Comparing species with convergent adaptations may help in understanding the general selective pressures favoring a given type of evolutionary trajectory at the behavioral functional level, even if underlying structures and mechanisms are not identical in detail.

This may be particularly true for language evolution. Parallels between language development and bird song learning are commonplace according to the literature in animal communication (e.g., Marler, 1970; Doupe and Kuhl, 1999), where researchers increasingly find indications of convergence in vocal and contextual flexibility in social birds and marine mammals (Schusterman, this volume; McCowan et al., this volume; Snowdon and Hausberger, 1997; Janik and Slater, 1997).

Language is a social act (Locke and Snow, 1997), and as such it seems obvious that social pressures may have been crucial in its evolution (see Sterelny, this volume). Social markers may have been important for intragroup cohesion and regulation of intergroup encounters (Griebel and Oller, this volume). It has been suggested that vocal sharing in social species of birds can act as "passwords" (Feekes, 1982; Brown and Farabaugh, 1997), possibly enabling group members to aggregate and tolerate each other in large assemblies (Hausberger, 1997). It can be expected that vocal plasticity may have been selected for in highly social species with social mobility (Snowdon and Hausberger, 1997). A similar coevolution between social organization and vocal development may have led to these similarities, which involve a high degree of plasticity occurring even in adults (Snowdon and Hausberger, 1997).

However, vocal communication is not restricted to vocal production: As in all communication, usage is equally important. Although no other species has achieved the high level of sophistication of human symbology or grammar, many animal studies have underlined the importance of producing the right signal at the right moment in order to be socially integrated (e.g., West et al., 1997). Paying attention to social partners is an important feature of social organization: It can involve audience effects such as those observed in nonhuman primates and birds (Snowdon, this volume) but also listening to and observing the responses of the other to one's own vocal signals (Ten Cate, 1987).

Obviously this must have been a key in the development of cooperative complex tasks such as planning hunting actions or building huts in the early stages of human culture (MacWhinney, this volume). Communication skills may have arisen in such contexts. Interestingly, parallels can be found with the socially directed attention found in gestural communication of apes but also in other mammals (Call, this volume). Social communication therefore implies appropriate usage, both in terms of contextual adjustment and social attention.

If language pragmatics evolved with social organization, one may expect to find some convergence in vocal usage in different animal species with similar social lives. Could elementary rules of communication be shared across phylogenetic taxa or even regularities be found as a result of a possible automatization of language during evolution (Sterelny, this volume)?

We will here consider human communication skills as a reference and look for possible convergences in animal species, with an emphasis on a social songbird species, which, as mentioned earlier, provides an especially important model for such comparisons.

Vocal Communication in a Social Context: A Communication Contract?

Vocal communication involves perception and production of signals but also proper usage of these signals. These different aspects do not show the same developmental trajectories and tend to appear in their adapted form at different ages in both humans and animals (e.g., Locke and Snow, 1997; Owren and Goldstein, this volume). As vocal communication, and language in particular, is a social act, usage especially has to conform to the social situation and results from learning through interactions both in humans (Locke, 1993) and animals (e.g., vervet monkeys; Seyfarth and Cheney, 1997). Inappropriate usage may lead to aggression as has been shown in cowbirds raised without proper adult influence (West et al., 1997). Research on human vocal communication developing in the field of social psychology as well as in sociolinguistic and language philosophy converges toward the assumption that communication relies upon a set of rules and principles that determine form and effects. As Krauss and Chui (1998, p. 43) argue, "To communicate, participants implicitly adhere to a set of conventions.... Listeners expect speakers to adhere to these rules, and communicators utilize these expectations when they produce and comprehend messages."

According to the social psychologist Ghiglione (1986), individuals, when communicating, establish a contract, defined by a certain number of rules and principles. Failure to respect these rules and principles can induce a breach of the communication contract and can signify the end of the exchange between interlocutors. Ghiglione defines four principles that constitute the core of his "theory of communication contract." The first principle is "pertinence," which predicts that the interlocutors must evaluate whether the context is appropriate for establishing an interaction. The second is "reciprocity," where the other must appear as a valid interlocutor to engage in an exchange. The third, "influence," predicts that any act of communication aims at influencing the other. And the fourth, "contract-based rules," implies that the rules of communication are known and accepted by each interlocutor. Examples of application of these principles in human communication, and especially in conversations, are abundant in the literature. Of course, examples where these principles apply in animal communication are common (e.g., Otter et al., 1999; Mennill and Ratcliff, 2004; Hall et al., 2006), showing further parallels between human and animal vocal communication. Within this general framework, it seems of special interest to consider contextual sensitivity especially where vocal interactions are involved.

Reciprocity in vocal interaction, for example, may depend on the characteristics of the callers: Adult vervet monkeys may respond to alarm calls of juveniles but only if they correspond to the proper situation (Seyfarth and Cheney, 1986); female Campbell monkeys are more likely to respond vocally to the social calls of older individuals than to those of younger ones (Lemasson, personal communication) or to those of group members than to those of strangers (Lemasson et al., 2005).

Although there is considerable evidence of influence through vocal interaction in animals as seen, for example, when one group of birds repels another (e.g., Brown and Farabaugh, 1997) or when spacing is modulated with loud calls in nonhuman primates (Gautier and Gautier-Hion, 1977), much less emphasis has been put on the possible "regulatory" effect of vocalizations on vocalizations. For instance, one of the major effects of the roosting calls of maternal hens seems to be inhibition of the chicks' cheeping and excitation before perching for the night (Guyomarc'h, 1974). Still more striking is the example of barnacle geese pairs in which females may either encourage the male in its display (i.e., triumph ceremony) by producing loud calls similar to his or, on the contrary, stop the whole process by producing soft calls that stop the male's behavior and loud calling in a few seconds (Hausberger and Black, 1990). Interestingly, a correlation was found between the frequency of encouraging responses and the pairs' age: Females tend to stop males' displays and calling more in older, well-established pairs (Bigot et al., 1995), a finding that leads to tempting comparisons with humans.

The Existence of "Rules" in Animal Communication and Their Degree of Context Sensitivity

The existence of rules in animal communication has been controversial. According to Ghiglione's framework, communication is a contract and to be mistaken about the contract is equivalent to not communicating or communicating badly (Ghiglione, 1986). Observations of Kerbrat-Orrechioni (1984, p. 225) illustrate the point: "... when they prepare themselves to dance or to play with one another, the people involved must agree as to the kind of game or of dance and the rules: partners must agree as to the personal rules that govern the particular verbal game they intend to play." While most authors would agree on the cultural/social bases for such agreements on rules in humans, different processes may be involved, including implicit as well as explicit learning, and some degree of automatization may not be totally excluded (see Sterelny, this volume).

In animals, evidence of a sort of shared "agreement" on "rules" is mainly found in the temporal organization of vocal interactions. Duetting birds are particularly remarkable, as two birds can combine, superimpose, or alternate their notes in such a way that it is impossible to distinguish what emerges from each singer (Farabaugh, 1982). Duets may be initiated by male or female or both depending on the species

(Hall, 2004; Langmore, 1998). In many cases, moreover, the other bird can produce the partner's part if the latter is absent, which means that each bird has in memory the partner's entire part. Nonoverlapping in vocal interactions (except chorusing) seems a very common phenomenon in social species and has been described, for example, in killer whales (Miller et al., 2004) but also in territorial birds such as the white throated sparrow (Wassermann, 1977). In case of overlap, lesser skylark and nightingale males stop singing (Gochfeld, 1978; Naguib, 1990) while robins or black capped chickadees get very excited (Dabelsteen et al., 1997; Mennill and Ratcliff, 2004).

Vocal responses to social calls are given in very short intercall intervals (less than one second in most cases) in female squirrel monkeys (Biben et al., 1986) and Campbell monkeys (Lemasson et al., 2005) but with rare occurrences of overlap (e.g., one out of thirty-eight cases in vervet monkeys; Hauser, 1992), suggesting for some authors a phenomenon of turn taking in some primate species. Although all these examples may be the result of chained responses cued and triggered automatically by timing of the partner's response, there are cases suggesting that more than simple chaining may be involved. Thus, trills given by adult pygmy marmosets present clear orders of turn taking so that among a group of adults, it is rare to find an animal giving a second trill before each of the other group members has trilled (Snowdon and Cleveland, 1984). In this case, the pattern observed results from social conventions rather than endogenous calling rhythms because all animals do not call at the same temporal interval. The alternation of calling allows group members to monitor whether all monkeys are still within earshot and to determine whether an animal is missing. The alternation of calling among all members of the group appears to be adaptive. Alternating calling has also been claimed to be adaptive for rhesus and vervet monkeys (Hauser, 1992). Adults of both species show declination in the fundamental frequency during the production of two and three call bouts, a pattern that guides turn-taking decisions. Interestingly, young vervet monkeys do not exhibit a consistent pattern of fundamental frequency change and are interrupted more often than adults. The above-mentioned examples where call responses depend on who is calling (vervet monkeys, Campbell's monkeys) also suggest that more than just automatic responses are involved.

Starling Song: Contextual Sensitivity and Communication Rules?

Starling song is a complex vocal system that is very much subject to social influence and appears as a potential candidate in the search for social regulation of communication in nonhumans (Hausberger, 1997). Starlings are sensitive to context. Indeed, they sing different song types according to the number of birds present, nest proximity, and foraging or roost situation (Adret-Hausberger, 1982). They may or may not engage in singing according to who their closest neighbor is: Outside the breeding season, females mostly sing if their closest neighbor is another female (Henry and Hausberger, 2001).

Starlings, like other songbirds, readily respond to the whistled songs of their neighbors and engage in song-matching interactions. However, interaction is much less common if the song heard belongs to another dialectal population: Male starlings respond less often, with a longer latency, and may even fly off if they hear a foreign dialectal variant instead of the local one (Adret-Hausberger, 1982) or if the songs are altered so that the local identity or the song type cannot be identified (Hausberger et al., 2000). The fact that the other individual is or is not accepted as a valid interlocutor according to its dialectal identity supports the reciprocity principle.

Whether or not starlings try actively to influence others when singing is difficult to assess, as song is generally not associated with any aggressive behavior or fighting, but male starlings do indeed change their song quantity and quality based on whether they are paired or not. Thus, removal of the female during the breeding season induces a drastic and rapid change in male song quality, with an increase in warbling song sequences (i.e., resulting in a long and melodious song), including particular elements such as high-pitched trills that decrease again after pair formation. In contrast, an increase of the typical male universal whistles (i.e., class 1 whistles; Hausberger, 1997) is observed after pair formation (Henry et al., 1994). This is in accordance with the suggestion that warbling song is important in attracting females and class 1 whistles in spacing between males (recalling to mind the loud calls of primates; Adret-Hausberger and Jenkins, 1988). Observations do confirm that high pitched trills are present at courting (Verheyen, 1980) and that copulations occur during warbling bouts (Feare, 1984). The warbling repertoire seems to be involved in mate choice (Adret-Hausberger et al., 1990; Eens et al., 1991; Mountjoy and Lemon, 1991). Starlings may adapt their choice of song types based on the type of response they expect from the receiver: attraction of females or male spacing.

But are there "rules" that seem to be shared by interlocutors? Starlings do respect some general "rules" in their singing interactions. This point will be developed in the following section. Mostly, starlings tend to abide by a pattern of alternation, as do most other animals, as mentioned before, when interacting from nest to nest.

Conclusion: Contextual Sensitivity in Starling Song?

Starlings do take into account the general context and who is nearby before singing or not and selecting what to sing. They seem to have elementary "rules" that all adults tend to respect in a usual context. Singing may not occur or song quality may differ if one of these principles is violated. For instance, the male universal class 1 whistles never or rarely occur in captivity: The inappropriate context and the absence of a valid interlocutor (i.e., male nest neighbor) mean that two of the principles at least (pertinence, reciprocity) are not fulfilled and prevent this type of song from occurring (Henry, 1998).

Similar findings have been described in other species. Thus, male cowbirds must produce their songs in the right context (e.g., the right song according to the sex of their closest neighbor) in order to influence partners in the right way and to be successful in both attracting females and being socially acceptable for the other males (West and King, 1980). They fail to meet these requirements if they did not have the right social experience during development. One further common point seems therefore that both humans and animals have to learn these principles through social experience. The importance of social play in this acquisition deserves consideration, as it may be involved in both cases (Kuczaj and Makecha, this volume). Comparative research on social play could certainly contribute to understanding how timing and synchronization can be acquired. Finally, in both cases, the nonrespect of the principles induces a breach of the communication contract and may signify the end of the exchange. The fact that the ability to follow the rules depends upon experience during development suggests, however, some degree of flexibility between and within individuals.

Context-Bound Rules or Contextual Flexibility?

Analyzing conversations in humans is an ideal way to approach the study of social communication rules. Exchanges have to be organized if everyone wants to acquire the information transmitted. Indeed, speech generally requires that participants take turns, since one cannot efficiently speak and listen simultaneously (Locke, 1993). Conversation rules have been therefore largely studied with regard to one particularly important feature: turn-taking organization. Turn taking implies that time is left for the other to take part in the conversation and presupposes attention to the others present that can be detected through a variety of nonverbal cues such as eye contact, posture, and so forth (France et al., 2001). According to Conti-Ramsden (1990, p. 267), "a turn is a series of one or more utterances by a speaker with or without accompanying gestures or one or more nonverbal acts performed by a speaker strung together without a pause." The termination of a turn is marked by a silence (Miura, 1993).

In fact, turn taking can be considered as a very general feature of social interactions such as games or traffic at intersections and combines the particularity of being both context free and "capable of extraordinary context-sensitivity" (Sacks et al., 1974, p. 699). This balance between fixed principles and contextual flexibility is illustrated in Craig and Washington (1986), who, on the basis of Sacks et al.'s (1974) model, have identified eight characteristics in turn-taking organization: (1) Only one person speaks at a time, (2) the number of participants may vary, (3) the order for speaker turns is variable, (4) turn size is not fixed, (5) the content of speaker turns is spontaneous, (6) simultaneous speech is infrequent and brief, (7) techniques exist for repairing turn exchange errors, and (8) turn allocation techniques are used to regulate the exchange.

Turn-Taking Organization in Conversation: A Universal Feature?

Turn taking is indeed found in a variety of cultures such as Thai (Moerman, 1977), Creols of New Guinea (Sankoff, in Sacks ct al., 1974), or Dogons (Calame-Griaule, 1965). In the Dogon society, overlap with someone's speech is a serious impoliteness that could lead to psychological and physiological trouble. Indeed, words that cannot be spoken because turn-taking opportunities are abridged are assumed to be "repressed in the spleen," which is "the seat of grudge and humiliations" (Calame-Griaule, 1965). Accumulating repressed words can make one sick, according to the Dogon. In general, repair mechanisms exist where the speaker who has committed errors and violations such as interrupting the other speaker will prematurely stop the ongoing sentence or perform ritualized behaviors.

Turn Taking in Conversation as an Example of Contextual Flexibility

Although the general features of human conversation seem rather universal and context bound, a variety of factors may induce some flexibility. For instance, the *number of interlocutors* has an influence on turn order, turn size, and length of pauses (Sacks et al., 1974). Individuals can also choose to violate whatever conventions exist, either because they are angry or obnoxious, because there is something urgent to inject in the conversation, or because their social status allows it.

Social status is an important feature influencing the nature of turn taking. In some cultures, the order in which individuals speak in a group is strictly determined by seniority. "In public, the rule for servants, females and other inferiors, is to speak when spoken to but otherwise to maintain silence in public" (e.g., in Burundi culture; Albert, 1964, p. 40). In occidental cultures, overlaps are mostly initiated by high-status individuals (e.g., Carletta et al., 1998) and more often by men than by women, a pattern that is often considered an implicit sign of dominance (e.g., Zimmermann and West, 1975) or alternatively of different gender subcultures (Maltz and Borker, 1982). Charadeau and Ghiglione (1997) have reported that in a television political show, two persons were able to occupy 50% of the speech time while the ten other people present were "confined" in the remaining 50%. Moreover, these two persons showed clear signs of nonattention to the others' turn contents.

Culture induces differences in the perception of what is in agreement with turn-taking rules and what is not. For instance, the minimal pause after which interlocutors feel they have been interrupted is very short in French and American people (respectively three tenths and five tenths of a second), whereas pauses must last at least one second for Alaskan Athabascans in order for them to consider that their turn has not been respected. On the other hand, silence becomes embarrassing after a few seconds in France, whereas Swedish Lapons expect only five or six exchanges per hour (Kerbrat-Orecchioni, 2001). Intercultural conversations are especially interesting in this case, as they require adjustment and flexibility. However, it seems that speakers from one of the cultures in any such interaction tend to take advantage of the situation by dominating the conversation (Kerbrat-Orecchioni, 2001).

As mentioned above, another important feature in turn-taking tendencies is *individual history*. Turn taking has to be learned and is considered one of the most important abilities that children have to achieve (Locke, 1993). Mother–infant interactions in early infancy appear especially influential (Rutter and Durkin, 1987). Mothers seem to control children rather than facilitating their learning, although controversies persist on this point (Miura, 1993). Neglected children tend to not respect turn-taking rules (Black and Logan, 1995) and socially accepted children have better skills in initiating and maintaining verbal exchanges than children who are rejected by peers (Hazen and Black, 1989). Communication skills such as being able to follow a conversational organization appear therefore as both a product of proper social influence during development and a way of becoming socially well integrated.

In conclusion, turn taking in human conversations is a strikingly general feature but must be acquired through adult modeling and is quite sensitive to context.

Contextual Flexibility of Turn-Taking Rules in Animals? The Starling Case

As mentioned earlier, alternating vocal interactions have been described in a large number of songbird and cetacean species. The term "conversation" has even been used in nonhuman primates (Snowdon and Cleveland, 1984; Hauser, 1992). According to Snowdon (1982, p. 233), these examples indicate that rule-governed communication is not unique to humans, although "in no way do they approach the complexity of human rules." But whatever the real foundations of these general principles, the functional similarities across humans and animals are striking. Understanding how animals succeed in reaching a balance between general features of communication and the contextual flexibility necessary to adapt in special cases may help in understanding language evolution.

Here again, European starlings are especially interesting: They are highly social, nest in colonies that they visit every day all year round in sedentary populations, forage in groups of varying sizes, and can spend the night in communal roosts where several hundred to several thousand birds can be present (Feare, 1984). In all these situations, song is present and the interactions numerous.

Two main categories of songs have been distinguished (Hausberger, 1997): warbling, which is a long, continuous series of motifs sung mostly at low intensity, and whistles, which are loud, discrete, and short vocalizations. Some whistles are universal in males and used in vocal interactions at a distance. They are produced while the bird is sitting, and head movements and glances all around are frequent, showing the bird's attention to its environment. Warbling sequences do not give rise to

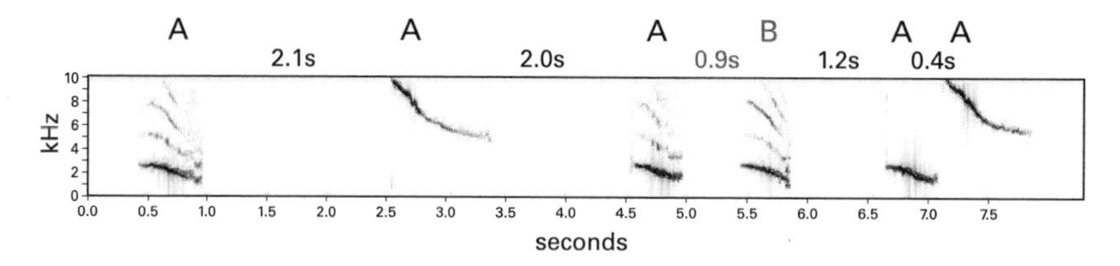

Figure 6.1
Sonogram showing a vocal interaction between two birds: Male A and Male B. Male B answers Male A using the same type of whistles (song-matching interaction). Turn taking is respected: Male B produces his whistle in the interval between two whistles by Male A.

interactions, and starlings do not respond to playback when warbling. Their posture is then characteristic with the bill pointing toward the sky, the whole body tensed, sometimes accompanied by wing movements. The whole context gives an impression of excitation (Verheyen, 1969; Hartby, 1969), while the animal seems to pay little attention to what happens in the surroundings. Thus, starlings, based on the song they produce, may or may not favor alternating interactions. In the case of whistles, time is left between successive sequences (Hausberger, 1991; see figure 6.1) and attention may be paid to the response, whereas in the case of warbling, the bird does not leave time for nonoverlapping responses and does not show clear attention to its environment or to any possible reply.

Interestingly, the song category chosen, and therefore the opportunity for nonoverlapping interactions, shows clear contextual flexibility. One particularly important feature in determining song choice appears to be social context. Whereas isolated birds tend to favor the production of warbling songs and have a tendency otherwise to produce only one whistle per sequence, starlings living in large colonies tend to have long sequences of whistles and only a few sequences of warbling. Our actual data indicated that a correlation existed between the number of neighbors and the tendency to whistle instead of warbling. Starlings favor songs that correspond to turn taking when confronted with neighbors in a colony. In social isolation, they instead favor a "self-centered continuous song."

Temporal rules also apply as male starlings produce whistles in succession with two-to-six-second intervals, which leaves time for replies to occur. However, our actual research indicates that this universal feature is again sensitive to context: The presence of another singer in the neighborhood induces longer intervals, while no difference occurs if the neighbor remains silent. In any case, overlapping seems indeed to be considered as "breaking the rules," since 90% of the cases of overlaps induce an ending of the ongoing interaction.

Like humans, starlings must learn to follow the "rules," and competency depends upon individual histories. Normal development in starlings, as in many songbirds, consists of a succession of stages from subsong (a continuous unstructured song), to plastic song (an intermediate stage where the first adult elements can be detected), and to the final stage of adult song. During the early stages of development, starlings' songs are therefore continuous, and when several young birds sing, they all tend to overlap. This may reflect a lack of attention to social partners, which is to say their songs may not in such cases be social but instead may be a form of solitary play. The first whistles and first alternating interactions appear when the birds are seven to nine months old (Adret-Hausberger, 1989). Hand-raised birds living with adults develop both discontinuous and continuous songs just as wild birds do, whereas young birds raised without contact with adults tend to produce at adulthood only continuous songs and they continue overlapping. More interesting still is the fact that inexperienced hand-raised starlings housed in pairs or in groups of peers do not develop any discontinuous songs and keep overlapping even if they can hear adult songs through loudspeakers (Poirier et al., 2004). Mere maturational processes may therefore be excluded as the determinants of the song development pattern, since the direct presence of adults appears necessary in order to develop the ability to produce songs that are adapted to vocal exchanges (Poirier et al., 2004; Bertin et al., 2007). Adults seem to canalize vocal structuring in young birds.

In some particular contexts, however, overlapping is frequent. This is especially the case in large flocks and roosts where chorusing seems to be a general feature in all songbird species (Catchpole and Slater, 1995). In this context, both continuous and discontinuous songs are produced, but vocal interactions using whistles can still give cues such as the dialectal identity of the singer, and these can be used as a "password" (see Hausberger, 1997). Frequent overlapping seems to be counteradaptive in this context where dialect identification may be important. Nevertheless, in a very noisy environment, birds are still able to detect partner vocalizations by focusing attention on them (a cocktail-party effect; Aubin and Jouventin, 1998). However, we can also suppose either that the number of birds present is so high that overlapping cannot be avoided or that the general level of excitation prevents control over the precise timing of production. Although this question requires further investigation, the latter hypothesis is interesting in that it would suggest that conforming to temporal rules requires a quiet state of attention and that emotional arousal may lead to change in the singing style.

Rules and Contextual Flexibility: A Basic Requirement for an Efficient Communication System?

Animals, like humans, seem to have some requirements before engaging in a vocal interaction. Starlings, and possibly a variety of other social species, seem to show a sort of communication contract that makes successful communication possible. Failure to abide by even a single prerequisite may prevent or stop social interaction.

The extent to which human/animal comparisons are possible remains to be discussed in order to provide an evolutionary perspective. The organization of interactions or conversations in turn taking is certainly an interesting basis for such investigations.

Turn Taking in Interactions: A Shared Need in Human and Animal Communication?

According to Schegloff (2000, p. 1), "the orderly distribution of opportunities in social interaction is one of the most fundamental preconditions for viable social organization." For conversation analysts, turn taking in conversation is a fundamental characteristic of human societies (Sidnell, 2001) and has even been considered as a language universal by Miller (1963). Sidnell (2001, p. 1263) considers that "it may be a species-specific adaptation to the contingencies of human social intercourse." Although mutual comprehensibility may be the basic reason such an organization has developed (Duncan, 1972), more traditional cultures such as the Dogon mentioned earlier claim to have physiological responses to being interrupted (Calame-Griaule, 1965). "Politeness" might well be a uniquely human characteristic of principles that might originally be founded in basic communication and biological needs such as intelligibility and shared attention.

Such needs are also present in animal communication. Indeed, in many species, alternation is favored over overlapping, intervals between successive vocalizations give time for responses to occur, and overlapping engenders one of several outcomes: the end of interaction, a "repair" act, or excitation of the individual that has been interrupted. Furthermore, variations can be observed based on the number, identity, and behavior of individuals present, the level of excitation, or the individual history. Therefore, in a number of cases, vocal interactions in animals do conform to the basic requirements of turn-taking organization, both in its general context-free aspects and in contextual flexibility.

Developmental Pathway of Turn Taking: From Continuous Self-Centered Vocalizations to Interactions

Like humans, animals must learn how to communicate, and "turn taking" appears to be acquired through adult modeling. It is striking that in bird and some primate species, including humans, early stages of vocal development are characterized by long, continuous sequences (subsong, babbling) and that the pattern of alternating the vocalizations with those of the partner has to be learned through interactions with adults (Rutter and Durkin, 1987). In starlings, also, adults seem to canalize the vocal production of the young toward an appropriate patterning (Poirier et al., 2004).

Songbirds and parrots show periods of solitary vocal play at different stages of vocal learning during which they seem to engage in "rehearsal," using long sequences of sounds, of what they have heard in social contexts (Pepperberg et al., 1991). These periods of "solitary play" remind us of those mentioned for babies (Locke, 1993). At these times, the young individual is obviously self-centered, which may make it totally concentrated on auditory feedback, thus improving vocal performance (Pepperberg et al., 1991; Konishi, 1989). On the other hand, "social play" requires appreciating the other's intention along with having patience to leave time for responses from the other (Kuczaj and Makecha, this volume).

From Continuous to Temporally Organized Vocalizations: Social Context and Attention to Partners

Adult starlings may present both strategies based on social context: Long, continuous sequences of uninterrupted songs predominate in social isolation, while the production of clear, distinct, and interspersed structures predominates when neighbors are present in the colony. The latter structures provide information on the singer's individual and group identity (Hausberger, 1997), and therefore listening to the other may provide useful information regarding who it is and what its intentions may be.

Similarly, infants of babbling age vocalize more when alone than when in the company of another person (Delack, 1976). Interestingly, differences have been found between normally developing and Down's syndrome infants in mother–child interactions. When the mother remains silent, both groups vocalize with increasing frequency over age. However, when the mother talks, Down's syndrome children continue to vocalize past the age of thirteen to sixteen weeks, when vocal frequencies decline in normally developing children (Berger and Cunningham, 1983). Down's syndrome infants also overlap more frequently with their mother's speech than do normally developing infants (Jones, 1977).

Allocation of attention is therefore involved here, and the progression between the continuous vocal flow toward a segmented sequencing may well reflect, both in young animals and humans, the development of an ability to progress from self-centered attention toward a partner's directed attention.

From Continuous to Temporally Organized Vocalizations: Emotional State and Vocal Control

In starlings, clear, temporally organized sequences of discontinuous songs are characteristic of quiet contexts, and excitation seems to be associated with a disruption of rhythm. In humans, increased arousal can alter the breathing pattern and probably the vocal patterning (McFarland, 2001). Interestingly, the continuous warbling of starlings is associated with potentially emotional situations such as courting, copulation, or close aggression (Verheyen, 1980), whereas the clear sequences of loud, discontinuous whistles are more involved in long, "calm," distance exchanges.

Furthermore, warbling is mostly processed in the right hemisphere, which is generally considered to be involved in sustaining diffuse attention and dealing with informational complexity (Rogers and Andrew, 2002). On the other hand, whistles are mostly processed by the left hemisphere, which tends to sustain attention to a stimulus to which a response is planned (Goldberg and Costa, 1981; George et al., 2005). Interestingly, Liégeois-Chauvel et al. (1999) have found that the left hemisphere in the human auditory cortex is involved in the processing of the temporal patterning of speech.

Therefore, development of attention toward one's own vocalization and the partner's response may well have been a key in the development of turn-taking types of exchanges (see also Owren and Goldstein, this volume). The disruption of vocal patterning in high arousal states suggests that this patterning requires some control over the vocal system. In humans, this control develops in the context of adult influence (Oller and Griebel, this volume). Neglected children, as is the case in young starlings deprived of adult influence, show deficits in this ability to leave time for the other to respond (Black and Logan, 1995). Their lack of control over vocal timing (phonation) may reflect immaturity and/or high intensity of emotions.

In humans, the powerful in a society may be those who are calm and in control, who listen to others and are listened to, and who show appropriate contextual sensitivity and adaptability. Such control is an indication of their mature communication skills, acquired, as in animals, through appropriate social experience.

Conclusion

In conclusion, animal communication may provide a useful reminder that the evolution of most basic principles of communication rules and contextual sensitivity may not have required higher cognitive processes. Do we need to refer to complex rules in order to explain turn-taking organization in humans? Regularizing the system of interaction through turn taking may allow automatization in language processing and may have been needed with the increase of its complexity (Sterelny, this volume). On the other hand, animals have evolved processes that cannot be entirely explained by mere chaining and triggering. Regulations of vocal interactions, their contextual sensitivity, and dependency upon social experience indicate further convergence with human communication.

Acknowledgments

This research was supported by the Centre National de la Recherche Scientifique ("Origine de l'Homme, du Langage et des Langues" program). We are grateful to

Huguette Schuelke for her help with literature and to Pascal Marchand, Alban Lemasson, Marie-Annick Richard, and Pierre Deleporte for useful discussions and comments. We are especially grateful to U. Griebel and D. K. Oller for inviting us to contribute to this volume and thank them both for a stimulating meeting and for their very useful comments on the manuscript. Comments by Robert Lachlan and Stan Kuczaj have been also of great help.

References

Adret-Hausberger M. 1982. Social influences on the whistled songs of starlings. *Behavioral Ecology and Sociobiology* 11: 241–6.

Adret-Hausberger M. 1989. Song ontogenesis in starlings *Sturnus vulgaris*: Are song and subsong continuous? *Bird Behaviour* 8: 8–13.

Adret-Hausberger M, Jenkins PF. 1988. Complex organization of the warbling song in the European starling *Sturnus vulgaris. Behaviour* 107: 138–56.

Adret-Hausberger M, Güttinger HR, Merkel FW. 1990. Individual life history and song repertoire changes in a colony of starlings (*Sturnus vulgaris*). *Ethology* 84: 265–80.

Albert E. 1964. Rhetoric, logic and poetics in Burundi: Culture patterning of speech and behavior. *American Anthropologist* 66, special issue December: 33–54.

Aubin T, Jouventin P. 1998. Coctail-party effect in king penguin colonies. *Proceedings of the Royal Society of London B* 265: 1665–73.

Berger J, Cunningham CC. 1983. Development of early vocal behaviors and interactions in Down's syndrome and nonhandicapped infant–mother pairs. *Developmental Psychology* 19: 322–6.

Bertin A, Hausberger M, Henry L, Richard-Yris MA. 2007. Adult and peer influences on starling song development. *Developmental Psychobiology* 49: 362–74.

Biben M, Symme D, Masataka N. 1986. Temporal and structural analysis of affiliative vocal exchanges in squirrel monkeys. *Behaviour* 98: 259–73.

Bigot E, Hausberger M, Black JM. 1995. Exuberant youth: The example of triumph ceremonies in Barnacle geese (*Branta leucopsis*). *Ethology Ecology & Evolution* 7: 79–85.

Black B, Logan A. 1995. Links between communication patterns in mother–child, father–child, and child–peer interactions and children's social status. *Child Development* 66: 255–74.

Brown ED, Farabaugh SM. 1997. What birds with complex social relationships can tell us about vocal learning: Vocal sharing in avian groups. In *Social Influences on Vocal Development*, ed. CT Snowdon, M Hausberger, pp. 98–127. Cambridge, England: Cambridge University Press.

Calame-Griaule G. 1965. *Ethnologie et Langage: La Parole Chez les Dogon.* Paris: Gallimard.

Carletta J, Garrod S, Frase-Krauss H. 1998. Placement of authority and communication patterns in workplace groups: The consequences for innovation. *Small Group Research* 29: 531–59.

Catchpole CK, Slater PJB. 1995. *Bird Song: Biological Themes and Variations*. Cambridge, England: Cambridge University Press.

Charaudeau P, Ghiglione R. 1997. *La Parole Confisquée. Un Genre Télévisuel: Le Talk Show*. Paris: Dunod.

Conti-Ramsden G. 1990. Maternal recasts and other contingent replies to language-impaired children. *Journal of Speech and Hearing Disorders* 55: 262–74.

Craig HK, Washington JA. 1986. Children's turn-taking behaviours: Socio-linguistic interactions. *Journal of Pragmatics* 10: 173–97.

Dabelsteen T, Mc Gregor PK, Holland J, Tobias JA, Pedersen SB. 1997. The signal function of overlapping singing in male robins. *Animal Behaviour* 53: 249–56.

Delack J. 1976. Aspects of infant speech development in the first year of life. *Canadian Journal of Linguistics* 21: 17–37.

Deleporte P. 2002. Phylogenetics and the aptationist program. Commentary/Andrews et al.: Adaptationism—How to carry out an exaptationist program. *Behavioral and Brain Sciences* 25: 514 5.

Doupe AJ, Kuhl PK. 1999. Birdsong and human speech: Common themes and mechanisms. *Annual Review of Neuroscience* 22: 567–631.

Duncan SH. 1972. Some signals and rules for taking speaking turns in conversations. *Journal of Personality and Social Psychology* 23: 283–92.

Eens M, Pinxten R, Verheyen RF. 1991. Male song as a cue for mate choice in the European starling. *Behaviour* 116: 211–38.

Farabaugh SM. 1982. The ecological and social significance of duetting. In *Acoustic Communication, Vol. 2: Song Learning and Its Consequences*, ed. DE Kroodsma, EH Miller, pp. 85–124. New York: Academic Press.

Feare C. 1984. *The Starling*. Oxford, England: Oxford University Press.

Feekes F. 1982. Song mimesis within colonies of *Cacicus cela.* A colonial password? *Zeitschrift für Tierpsychologie* 58: 119–52.

France EF, Anderson AH, Gardner M. 2001. The impact of status and audio conferencing technology on business meetings. *International Journal of Human–Computer Studies* 54: 857–76.

Gautier JP, Gautier-Hion A. 1977. Vocal communication in Old World primates. In *How Animals Communicate*, ed. TA Sebeok, pp. 890–964. Bloomington, IA: Indiana University Press.

George I, Cousillas H, Richard JP, Hausberger M. 2005. State-dependent hemispheric specialisation in the songbird brain. *Journal of Comparative Neurology* 488: 48–60.

Ghiglione R. 1986. *L'Homme Communiquant*. Paris: Colin.

Gochfeld M. 1978. Intraspecific social stimulation and temporal displacement of songs in the lesser skylark. *Zeitschrift für Tierpsychologie* 48: 337–44.

Goldberg E, Costa LD. 1981. Hemisphere differences in the acquisition and use of descriptive systems. *Brain and Language* 14: 144–73.

Guyomarc'h JC. 1974. Les vocalisations des Gallinacés. Structure des sons et des répertoires. Ontogenèse motrice et acquisition de leur sémantique. Thèse de Doctorat d'état, Université de Rennes.

Hall ML. 2004. A review of hypotheses for the functions of avian duetting. *Behavioral Ecology and Sociobiology* 55: 415–30.

Hall ML, Illes A, Vehrencamp SL. 2006. Overlapping signals in banded wrens: Long-term effects of prior experience on males and females. *Behavioral Ecology* 17: 260–9.

Hartby E. 1969. The calls of the starling (*Sturnus vulgaris*). *Dansk Ornitologisk Forenings Tidsskrift.* 62: 53–81.

Hausberger M. 1991. Organization of whistled song sequences in the European starling. *Bird Behaviour* 9: 81–7.

Hausberger M. 1997. Social influences on song acquisition and sharing in the European starling (*Sturnus vulgaris*). In *Social Influences on Vocal Development*, ed. CT Snowdon, M Hausberger, pp. 128–56. Cambridge, England: Cambridge University Press.

Hausberger M, Black JM. 1990. Do females turn males on and off in barnacle goose social display? *Ethology* 54: 232–8.

Hausberger M, Leppelsack E, Richard JP, Leppelsack HJ. 2000. Neuronal bases of categorization in starling song. *Behavioural Brain Research* 114: 89–95.

Hauser MD. 1992. A mechanism guiding conversational turn-taking in vervet monkeys and rhesus macaques. In *Topics of Primatology, Vol. 1: Human Origins*, ed. T Nishida, FBM de Waal, W McGrew, P Marler, M Pickford, pp. 235–48. Tokyo: Tokyo University Press.

Hazen NL, Black B. 1989. Preschool peer communication skills: The role of social status and interaction context. *Child Development* 60: 867–76.

Henry L. 1998. Captive and free living European starlings use differently their song repertoire. *Revue d'Ecologie* 53: 347–52.

Henry L, Hausberger M. 2001. Differences in the social context of song production in captive male and female European starlings. *Comptes Rendus de l'Academie des Sciences, Paris* 234: 1167–74.

Henry L, Hausberger M, Jenkins PF. 1994. The use of repertoire changes with pairing status in male European starlings. *Bioacoustics* 5: 261–6.

Janik VM, Slater PJB. 1997. Vocal learning in mammals: Advances in the study of behaviour. *Advances in the Study of Behavior* 26: 59–99.

Jones OHM. 1977. Mother–child communication with prelinguistic Down's syndrome and normal infants. In *Studies in Mother–Infant Interaction*, ed. HR Schaffer, pp. 379–401. New York: Academic Press.

Kerbrat-Orecchioni C. 1984. Les négociations conversationnelles. *Verbum* VII: 223–243.

Kerbrat-Orecchioni C. 2001. Les cultures de la conversation. In *Le Langage: Nature, Histoire et Usage*, ed. JF Dortier, pp. 209–15. Paris: Editions Sciences Humaines.

Konishi M. 1989. Birdsong for neurobiologists. *Neuron* 3: 541–9.

Krauss RM, Chui C. 1998. Language and social behavior. In *The Handbook of Social Psychology*, ed. DT Gilbert, ST Fiske, G Lindzey, pp. 41–88. Boston, MA: McGraw-Hill.

Langmore NE. 1998. Function of duet and solo song of female birds. *Tree* 13: 136–40.

Lemasson A, Hausberger M, Züberbühler K. 2005. Socially meaningful vocal plasticity in adult Campbell's monkeys (*Cercopithecus campbelli*). *Journal of Comparative Psychology* 119: 220–9.

Liégeois-Chauvel C, de Graaf JB, Laguitton V, Chauvel P. 1999. Specialization of left auditory cortex for speech perception in man depends on temporal coding. *Cerebral Cortex* 9: 484–96.

Locke JL. 1993. *The Child's Path to Spoken Language*. Cambridge, MA: Harvard University Press.

Locke JL, Snow C. 1997. Social influences on vocal learning in human and nonhuman primates. In *Social Influences on Vocal Development*, ed. CT Snowdon, M Hausberger, pp. 274–92. Cambridge, England: Cambridge University Press.

Maltz DN, Borker RA. 1982. A cultural approach to male–female miscommunication. In *Language and Social Identity*, ed. JJ Gumperz, pp. 196–216. New York: Cambridge University Press.

Marler P. 1970. Bird song and speech development: Could there be parallels? *American Scientist* 58: 669–73.

Martins EP. 1996. *Phylogenies and the Comparative Method in Animal Behaviour*. Oxford, England: Oxford University Press.

Mc Farland DH. 2001. Respiratory markers of conversational interaction. *Journal of Speech, Language and Hearing Research* 44: 128–43.

Mennill DJ, Ratcliff LM. 2004. Overlapping and matching in the song contexts of black-capped chickadees. *Animal Behaviour* 67: 441–50.

Miller GA. 1963. Universals of language. *Contemporary Psychology* 8: 417–8.

Miller PJO, Shapiro AD, Tyack PL, Solow AR. 2004. Call-type matching in vocal exhanges of free-ranging resident killer whales, *Orcina orca. Animal Behaviour* 67: 1099–107.

Moerman M. 1977. The preference for self-correction in Thai conversational corpus. *Language* 53: 872–82.

Mountjoy DJ, Lemon RE. 1991. Song as an attractant for male and female European starlings and the influence of song complexity on their response. *Behavioral Ecology and Sociobiology* 28: 97–100.

Miura I. 1993. Switching pauses in adult–adult and child–child turn-takings: An initial study. *Journal of Psycholinguistic Research* 22: 383–95.

Naguib M. 1990. Effects of song overlapping and alternating in nocturnally singing nightingales. *Animal Behaviour* 88: 1061–7.

Otter K, Mc Gregor PK, Terry AMR, Bulford FRL, Peake TM, Dabelsteen T. 1999. Do female great tits (*Parus major*) assess males by eavesdropping? A field study using interactive song playback. *Proceedings of the Royal Society of London B* 266: 1305–9.

Pepperberg IM, Brese KJ, Harris BJ. 1991. Solitary sound play during acquisition of English vocalizations by an African grey parrot (*Psittacus erithacus*): Possible parallels with children's monologue speech. *Applied Psycholinguistics* 12: 151–78.

Poirier C, Henry L, Mathelier M, Lumineau S, Cousillas H, Hausberger M. 2004. Direct social contacts override auditory information in the song learning process in starlings. *Journal of Comparative Psychology* 118: 179–93.

Rogers L, Andrew RJ. 2002. *Comparative Vertebrate Lateralization*. Cambridge: Cambridge University Press.

Rutter DR, Durkin K. 1987. Turn-taking in mother–infant interaction: An examination of vocalization and gaze. *Developmental Psychology* 23: 54–61.

Sacks H, Schegloff E, Jefferson G. 1974. A simplest systematic for the organization of turn taking for conversation. *Language* 50: 696–735.

Sanderson MJ, Hufford L. 1996. Homoplasy. *The Recurrence of Similarity in Evolution*. San Diego, CA: Academic Press.

Schegloff EA. 2000. Overlapping talk and the organization of turn-taking for conversation. *Language in Society* 29: 1–63.

Seyfarth RM, Cheney DL. 1986. Vocal development in vervet monkeys. *Animal Behaviour* 34: 1640–58.

Seyfarth RM, Cheney DL. 1997. Some general features of vocal development in nonhuman primates. In *Social Influences on Vocal Development*, ed. CT Snowdon, M Hausberger, pp. 249–73. Cambridge, England: Cambridge University Press.

Sidnell J. 2001. Conversational turn taking in a Caribbean English Creole. *Journal of Pragmatics* 33: 1263–90.

Snowdon CT. 1982. Linguistic and psycholinguistic approaches to primate communication. In *Primate Communication*, ed. CT Snowdon, CH Brown, MR Pedersen, pp. 212–38. Cambridge, England: Cambridge University Press.

Snowdon CT, Cleveland J. 1984. "Conversations" among pygmy marmosets. *American Journal of Primatology* 7: 15–20.

Snowdon TC, Hausberger M. 1997. *Social Influences on Vocal Development*. Cambridge, England: Cambridge University Press.

Ten Cate C. 1987. Listening behaviour and song learning in zebra finches. *Animal Behaviour* 34: 1267–8.

Verheyen RF. 1969. Le choix du nichoir chez l'étourneau *Sturnus v. vulgaris. Le Gerfaut* 59: 239–59.

Verheyen RF. 1980. Breeding strategies of the starling. In *Bird Problems in Agriculture*, ed. EN Wright, IR Inglis, CJ Feare, pp. 69–82. London: British Crop Protection Council.

Wasserman FE. 1977. Intraspecific acoustical interference in the white-throated sparrow. *Animal Behaviour* 25: 949–52.

West MJ, King AP. 1980. Enriching cowbirds song by social deprivation. *Journal of Comparative and Physiological Psychology* 94: 263–70.

West MJ, King AP, Freeberg TM. 1997. Building a social agenda for the study of bird song. *Social Influences on Vocal Development*, ed. CT Snowdon, M Hausberger, pp. 41–56. Cambridge, England: Cambridge University Press.

Zimmermann DH, West C. 1975. Sex roles, interruptions and silences in conversation. In *Language and Sex: Differences and Dominance*, ed. B Thorne, N Henley, pp. 105–26. Rowley, MA: Newbury House.

III THE ROLE OF FLEXIBILITY AND COMMUNICATIVE COMPLEXITY IN THE EVOLUTION OF LANGUAGE

Part III addresses human language origins directly. There has been considerable attention paid to the evolution of language in recent times—the chapters of the present volume, however, offer new proposals based on reviews of growing evidence from such diverse fields as human development, paleoanthropology, connectionist modeling, and naturalistic philosophy.

Part III begins with a combined evolutionary and developmental "evo–devo" (see Hall, 1992) perspective by Oller and Griebel based in part on a long-standing research program of Oller and his colleagues in human infant vocalizations. The fundamental claim of the chapter is that developmental evidence may be critical in determining the steps of evolution that led to vocal language, because both development and evolution can be expected to abide by the same "natural logic" of steps in the emergence of progressively powerful communication capabilities. The natural logic proposed by the authors specifies that evolution of vocal contextual flexibility (both primitive signal flexibility and primitive functional flexibility) is a foundation without which progress toward vocal language is impossible.

Owren and Goldstein present a both comparative and developmental perspective, advocating the idea that each phase of vocal development in humans provides a scaffold for the next, and that learning is critical at each phase. They emphasize both voluntary control of vocalization and the instrumental functions of babbling and other vocalizations. The review outlines features of development that lay critical groundwork for human language from the perspective of primate vocal systems and the neurological control mechanisms that afford the possibility of voluntarity and learning, thus adding perspective to other chapters, especially those by Hammerschmidt and Fischer as well as by Oller and Griebel.

MacWhinney tackles the origins of language by appealing to a "quartet of characteristics" deemed crucial to the hominin line's special linguistic evolution: bipedalism, manual dexterity, neoteny, and social bonding. The quartet is argued to constitute hominin-specific conditions that favored growth in both cognitive and vocal flexibility. MacWhinney posits that certain important steps in hominin communication occurred first in the gestural domain. The chapter proposes that a platform for language evolution was created by the quartet, and he outlines steps that may have occurred, invoking the item-based learning approach that he and others have advocated for child language (Bates et al., 1991; Tomasello, 2003). He expands the argument with evidence of the role of perspective taking in providing foundations for grammatical competence, another long-standing realm of research by MacWhinney and colleagues.

Selectional effects of human niche creation on language evolution, effects that are associated with special characteristics of human social systems, are addressed by Sterelny in the last of the chapters in part III. He argues for cognitive preadaptations in the hominin line for assessing signal reliability (honesty), and the idea that grammar is an adaptation that helps contain the cognitive costs of assessing signals for honesty. Further, he reasons that symbol use has cognitive costs that can be reduced by division of labor within human society, such that meanings of individual symbols are distributed across individuals within communities. The arguments of Sterelny address how the social circumstances of humans created special pressures favoring communicative flexibility and how the costs of that flexibility may have been constrained by certain features of language.

References

Bates E, Thal D, MacWhinney B. 1991. The functionalist approach to language and its implications for assessment and intervention. In *Pragmatics of language: Clinical Practice Issues*, ed. TM Gallagher, pp. 133–161. London: Chapman and Hall.

Hall BK. 1992. *Evolutionary Developmental Biology*. London: Chapman and Hall.

Tomasello M. 2003. *Constructing a Language: A Usage-Based Theory of Language Acquisition*. Cambridge, MA: Harvard University Press.

7 Contextual Flexibility in Infant Vocal Development and the Earliest Steps in the Evolution of Language

D. Kimbrough Oller and Ulrike Griebel

Overview on Vocal Flexibility in Development and Evolution

Prelude

The natural laboratory of human development offers indirect evidence about the underlying capabilities that must have been evolved in the hominin line in order for language to have emerged. In this chapter we argue that in all likelihood the first foundational steps of hominin communicative evolution away from the primate background had much in common with stages that are observable in human infant development (Oller, 2000; Oller and Griebel, 2005). These stages pertain to vocal production in particular, since it is vocal production (see review in Snowdon, this volume), rather than perception and auditory processing, that appears to differentiate humans so strongly from nonhuman primates.

The parallelism we advocate between development and evolution is different from the concrete form of the idea attributed to Haeckel (1866) that ontogeny recapitulates phylogeny. Our view conforms to the abstract idea that ontogeny and phylogeny often abide by the same "natural logic," a logic that should be expected to have extremely broad application because it characterizes sequences of transparently increasing complexity or of sequences required by development. For example, in embryology single-cell structures must precede multicellular ones, and also in evolution the existence of single-cell creatures had to precede multicellular ones. The term "natural logic" can be applied in a variety of realms of science. For example, in chemistry, it is obvious that the building of organic molecules involves naturally logical steps combining first atoms, then simple molecules, and then more complex molecules.

Communicative complexity in many cases can be interpreted in terms of the number of capabilities that have to be coordinated in order for some type of communicative act to be performed. For example, the capability to control one word at a time is simpler than the capability to put two words together, because the latter requires both the former plus something more. Similarly, increasing complexity of capabilities can be viewed as requiring increasingly complex neural subsystems to manage the actions in question. For example, the capability to produce one word at a time draws on neural subsystems that must be expanded to make syntactic combination of words possible.

Empirical study of early vocal development, to be reviewed below, has determined the existence of stages that similarly appear to occur in a naturally logical order. Vocal contextual flexibility provides foundations for the power of human language to communicate an indefinitely large variety of functions and ideas precisely and rapidly. Without vocal contextual flexibility, well-recognized features of language would be impossible, including learning new speech sounds by imitation, recombination of sounds to create words, acquisition of the lexicon of any language, generation of sentences composed over the lexicon, and so on.

Definitions of Signal Flexibility and Functional Flexibility

Our chapter focuses on the appearance of vocal contextual flexibility in the first months of life, including features of both signal flexibility and functional flexibility as defined in Griebel and Oller (this volume). The definitions are extended here.

Signal and function have been recognized as conceptually distinct in language since the founder of modern linguistics, de Saussure, drew the distinction between *signifiant* (that which signifies) and *signifié* (that which is signified) in language (de Saussure, 1968; first published in 1916, three years after his death). The distinction can be expanded to encompass all types of communication. The study of signal characteristics independent of the functions they may transmit has been termed "infraphonology," and the study of the functions that can be served by signals has been termed "infrasemiotics" (Oller, 2000). As we research the development and evolution of language, it proves crucial to maintain the distinction between these two aspects of communicative events.

We use the term "signal" in this chapter to refer both to actions that transmit communications systematically and to actions that have the *potential* to transmit systematic communications. The inclusion of potential signals in the definition is important because the young of a species may systematically produce sounds or other potential signals that lay foundations for later communication.

The term "signal flexibility" focuses on the (infraphonological) ability to produce systematic physical variations in actions that can be signals, but the term is not intended to encompass all such variations. Consider, for example, the hissing of the house cat. The sound itself is a signal of aggression, and systematic variations in intensity of the signal can be produced to denote systematic variations in intensity of its communicated function (aggression). However, the hissing of the house cat is stereotyped in form, that is, limited in acoustic variability. The signal has been naturally selected to serve no other purpose than that of expressing aggression, and if it were not limited in variability it could fail to serve its purpose at points of urgency.

The term "signal flexibility" in our usage invokes in particular the ability to vary signal parameters systematically *beyond* the limited types of variations of signals that are naturally selected to serve a particular function, as in the case of hissing in the house cat. When signals vary beyond the limits of naturally selected stereotypy, a many-to-one mapping from signal to function can occur. In notable cases of this sort (especially in very young animals), variable vocal signals occur where the *social* function appears to be to provide a fitness indicator for conspecifics, especially caretakers. A nonsocial function, vocal practice, appears also to be involved. When human infants engage in vocal play (Stark, 1980) or developing songbirds produce subsong (Nooteboom, 1999), they vary the parameters of potential signals systematically, often in the absence of immediate communicative purpose or effect other than possible fitness signaling, and thus provide examples of "signal flexibility" as we intend the term. In accord with our proposal, learning new vocal signals (either through imitation or through vocal exploration) involves this kind of signal flexibility as a prerequisite foundation.

"Functional flexibility" is the (infrasemiotic) ability to produce a given signal type in such a way as to serve different communicative purposes on different occasions. The notion "function" generally refers to coherent classes of social communication. One coherent social communicative function might be called "aggression" or "threat" (as transmitted by the hissing house cat), another "warning" or "announcement of alarm," another "advertisement of fitness," yet another "greeting," and so on. Because there is coherency to these different purposes, it is possible for a particular signal to be naturally selected within a particular species to serve any one of them. When an individual signal is selected to serve a particular purpose and no other, it becomes a "fixed signal" in the terminology of classical ethology (Lorenz, 1951; Tinbergen, 1951). The hissing of the house cat is just such a signal. A fixed signal serves the same kind of purpose on *each* occasion of use—a fixed signal of threat, such as hissing in the house cat, cannot be used as a warning or greeting on some other occasion. The acoustic characteristics of fixed signals are stereotyped because the signals are naturally selected to serve the same purpose on each occasion, and the purpose cannot be consistently served unless the signal itself is unambiguous.

However, not all signals are fixed in these ways. Words or sentences in human languages are signals that are free to transmit different purposes (warning, naming, ridiculing, etc.) on different occasions. Consider the word "vampire." It can be used to warn a disbeliever, to name a flying mammal, to invoke a mythical concept, or to ridicule an ex-spouse, among many other possibilities. Functional flexibility is this ability to use the same signal to serve different functions on different occasions.

Furthermore, each root word in a human language has to be learned, and language learning requires a mechanism flexible enough to adapt to different sound systems in different languages, to different conventions for the relation of each word with its meaning, to different conventions about how words can be modified and combined in sentences, and so on. Both signal flexibility and functional flexibility are thus required in order for the process of learning to proceed, and until a species acquires at least rudimentary flexibility in these domains, evolution in the direction of the communicative power of language is not possible.

The Proposal in Short

We propose, then, that the earliest requirement of hominin communicative evolution beyond the vocal abilities seen in nonhuman primates, abilities that were presumably also present in our distant primate ancestors, was the establishment of vocal contextual flexibility, both signal flexibility and functional flexibility. This proposal takes stock of the vocal limitations of nonhuman primates both in signal flexibility and in functional flexibility (see Hammerschmidt and Fischer, Owren and Goldstein, and Snowdon, this volume, for reviews of empirical data and prevailing interpretations). Owren and Goldstein (this volume) provide a compatible proposal about the importance of scaffolding of stages of learning in the emergence of vocal flexibility and voluntarity in the human infant in contrast with the relative inflexibility that is observed in nonhuman primates.

We reason that in hominin evolution the capabilities for signal flexibility and functional flexibility must have been built in steps, applying first in simple, primitive communications resembling (at least in part) those of human infancy, and applying only much later in the context of the enormously more complex structures of mature human language. Human infancy, then, provides a model, at least in broad outlines, of how foundational steps might be taken to break free from the limitations of fixed signaling. The empirical evidence on early human development to be reviewed here indicates that systematic foundational steps toward vocal contextual flexibility are taken by humans in the first months of life. By the middle of the first year of human life, humans surpass nonhuman primates in vocal flexibility at any age (Oller, 2000).

Naturally Logical Interpretation of the First Steps of Infant Vocal Development

Natural Logic and the Ordering of Stages in Development and Evolution

Our proposal turns on the idea that the ordering of steps of vocal contextual flexibility in the first months of life occurs *by necessity*: The steps abide by a "natural logic," the broad outlines of which can be discerned through evaluation of the patterns of development. Whenever developmental or evolutionary sequences can be explained as occurring by natural necessity, we invoke the term "natural logic" and try to specify the principles of logical dependency of one capability with respect to another that lie behind the natural necessity for sequencing. In some cases, it may be easy to correctly posit a sequence abiding by natural logic from our armchairs—for example, we need no research (at least not any more) to know that independent single cells precede multicellular clusters in both development and evolution. The relative complexity of single- and multicellular structures is unambiguous. However, there are many cases of development and evolution where empirical research illustrates systematic sequencing prior to scientific recognition of the natural logic that may predispose a sequence of development to occur.

Vocal development in human infants provides clear examples that armchair speculation about natural sequences is often of little use. The naturally logical character of the sequences of vocal development that have been revealed were not at all obvious when systematic research on human infant vocalization began in the 1970s (Oller, 1978; Stark et al., 1975; Zlatin-Laufer and Horii, 1977). Furthermore, only after considerable interpretive progress in empirical developmental research did the importance of the posited developmental natural logic come into focus as a basis for evolutionary speculations. We now turn to a summary of stages of human vocal development and the natural logic that appears to predispose their orderly occurrence.

Stages in Emergence of Vocal Contextual Flexibility in Human Infancy

On the infraphonological side of vocal development, human infants manifest emergence of a capability for signal flexibility through the following stages:

1. Spontaneous production of phonation in comfort, usually from the first month of life.

2. Rapid elaboration of spontaneous vocalizations in nonstereotyped displays of vocal raw material, including primitive supraglottal articulations, usually within the first two months.

3. Creation by the infant of gross but quite identifiable categories of (precanonical) vocalization from the elaborated vocal raw material (of stage 2) by the third or fourth month.

4. Further elaboration of vocal categories to include well-formed or "canonical" syllables, which incorporate signal flexibility both in terms of laryngeal (vocal) control and in terms of supralaryngeal, articulatory control by the middle of the first year of life or shortly thereafter.

On the infrasemiotic side of development, infants show systematic growth of functional flexibility as follows:

1. By establishing a foundation for functional flexibility during the same time period as stage 1 of signal flexibility, with spontaneous production of flexible vocalizations that have no fixed social communicative function (unlike, e.g., crying).

2. By utilizing the nonstereotyped vocal raw material of stage 2 both in playful vocalization (play can be viewed *as* a function) when the infant is alone as well as in interactive face-to-face vocalization.

3. By showing free expressivity, where any of the newly developed categories indicated in stages 3 and 4 are used in multiple and often opposite affective contexts (exulting, complaining, or simply engaging in vocal play). Free expressivity is seen by the fourth or fifth month and continues to be elaborated throughout language development.

These patterns of emerging signal and functional flexibility, having been described in longitudinal studies to be cited below, are now the focus of even more intense research in our laboratories (e.g., Buder et al., in press; Oller et al., 2007). It is expected that the methods being developed to quantify the patterns will be applicable to description of the extent of contextual flexibility not only in human infants but in a wide variety of species whose vocal repertoires (both during development and in mature form) provide a backdrop for the understanding of human development and evolution.

The Stages of Vocal Development and Their Naturally Logical Basis

Infraphonological Stage 1 (The Phonation Stage): Spontaneous Production of Phonatory Acts

Spontaneous vocalizations occur from the first month of life. These "quasivowel" sounds, produced in comfort, begin as vocal acts of modal or normal phonation with no articulation and no within-utterance interruption. Quasivowels are produced with no systematic posturing of the supraglottal tract (which can be said to be "at rest"), and consequently these sounds can be viewed as pure phonatory events. Prior research has documented both acoustic and functional characteristics of quasivowels (Boliek et al., 1996; Koopmans-van Beinum et al., 2001; Koopmans-van Beinum and van der Stelt, 1986; Oller, 1980; Stark et al., 1993; Stark, 1978). Quasivowels are not fixed signals, because they are spontaneously produced, with no apparent stimulus that elicits them—they occur when the infant is alone *and* when the infant is attended. Quasivowels do not appear to be driven by strong emotional content, for they tend to occur when the infant is awake but otherwise quiescent. Quasivowels appear to function sometimes as state and fitness indicators to parents, who often seem pleased and pacified to hear the infant vocalizing in this way (instead of crying, the principal other vocalization of the first month). However, the sounds are inherently ambiguous as to function and are not always interpreted as indicators of comfort—they can also be treated as fussing on some occasions.

"Close calls" appear to provide the most appropriate analogy in nonhuman primates to the acoustic characteristics and functions of quasivowels (Snowdon, 2004), but close calls have been relatively little studied in nonhuman primates, partly because they tend to occur at low intensity (they do not have to be loud to be heard by their targeted audiences) and are thus difficult to record (Becker et al., 2003). It has been speculated that close calls are affiliation expressions that may include several (perhaps many) subtypes in some species (see Snowdon, this volume) and may even include substantial contextual flexibility of production. Developing further quantitative description of human quasivowels and the methodology to characterize the relation of the sounds with contexts is important in order to form a foundation for more fruitful cross-species comparison. No one has yet quantified the degree of contextual flexibility occurring in close calls and quasivowels in order to compare nonhuman primates and human infants—a stable basis for scaling that comparison remains to be developed.

We interpret the ability to produce any sound spontaneously as a critical foundation for that sound's further flexible production, either in playful exploration, category formation, or variable communication usage. While quasivowels are primarily socially undirected in the first month of life, they are incorporated into face-to-face interactions with caregivers in stage 2. Later in the first year of life, quasivowels are often utilized by infants to communicate assent or acknowledgment (McCune et al., 1996). The capability for spontaneous production lays a naturally logical foundation for these later developments.

Infraphonological Stage 2 (The Primitive Articulation Stage): Elaboration of Spontaneously Produced Vocalizations to Include Primitive Articulation

Infants usually show rapid elaboration of spontaneous vocalizations into nonstereotyped displays of vocal raw material by the second or third month of life. New sounds such as squeals and growls begin to appear. Quasivowels also show elaboration in the first weeks, becoming increasingly variable in acoustic character, exhibiting both long and short types, louder and softer ones, and glottal interrupts begin to be heard within quasivowels, that is, the breathing cycle is halted during vocalization. By the fifth or sixth week it is clear that infants adjust breathing by taking in extra air before beginning quasivowel-like vocalization (Boliek et al., 1996).

Quasivowels are initially produced as pure phonatory events with no systematic supraglottal posturing or movement—the supraglottal vocal tract tends to be at rest. But by stage 2, supraglottal articulation does begin in "gooing," which involves seemingly uncoordinated movements where the tongue dorsum is brought into contact with other structures erratically. The articulations produce a kind of "primitive syllabification," which seems primitive precisely because it is unpredictable in time and extent (Zlatin, 1975), unlike the well-formed canonical syllables that occur systematically at a later point in development.

During the same period that gooing appears, we also note elaboration of vocalization at the larynx. High-pitched and low-pitched sounds (without articulation), that is squeals and growls, now occur along with sounds in the midrange of normal phonation for human infants. These sounds manifest emergence of elaborate vocal raw material.

The lack of stereotypy, or to put it another way, the tendency for infant vocalizations in stage 2 to include a tremendous range of sound qualities, is the subject of intense investigation currently in our laboratories in Memphis. Complexity of vocalization is the rule, not the exception, by this point in time, in vocalizations produced both when the infant is alone and in interaction (Buder et al., in press; Oller et al., 2007). Lack of stereotypy, along with the fact that nonstereotyped production occurs both when the infant is alone and when the infant is in social interaction, provides a further indication of signal flexibility.

In general, descriptions of nonhuman primate vocal patterns have emphasized stereotypy and ritualization of particular sounds. Further, the descriptions have emphasized unitary functions for each vocal type. While Snowdon subscribes to this point of view in the main, he has also described instances, especially in New World monkeys, that appear to provide challenges to the general view on nonhuman primates (Snowdon, 2004; Snowdon et al., 1997; Snowdon and Hodun, 1981; Snowdon, this volume). Hammerschmidt and Fischer (this volume) also review interesting cases of vocal flexibility in nonhuman primates. However, research has not yet made direct comparisons, utilizing a well-defined common methodology, between human infant and nonhuman primate vocalizations. At present the most comparable data available across species suggest that human infants by stage 2 produce quantitatively more elaborate vocalizations than nonhuman primates at any age.

Still, the most important point about stage 2 is one of natural logic, and that point does not depend on data from nonhuman primates: Stage 2 builds on stage 1, because spontaneously produced quasivowels are elaborated in stage 2 to include more variable raw material both in phonatory and supraglottal articulation.

Infraphonological Stage 3 (The Expansion Stage): Primitive Vocal Category Formation from Vocal Raw Material

In the months following the first appearance of elaborated displays, infants show creation of new categories of vocalization from the raw material of complex sounds developed during the prior period. In this "expansion stage," repetitive sequences of vocalizations having a particular property (e.g., high pitch) are alternated systematically with sequences having another property (midpitch, e.g., or harsh vocal quality).

These sound types and their repetitive occurrence in vocal play have been described spontaneously by parents through more than thirty years of longitudinal research in our laboratories and others' as "squeals," "vowel-like sounds," and "growls." Squeals and growls are often perceived as opposites, and vowel-like sounds are viewed as neutral in vocal quality and pitch with respect to the other two. All three categories tend to be produced with the supraglottal vocal tract open rather than in the at-rest position that is typical of quasivowels.

We recognize categories in stage 3 in several quantitative ways (Buder et al., 2003; Oller et al., 2003). In the first method, we merely ask listeners to judge utterances presented from real samples as having squeal, growl, or vowel-like quality, and we find "good" agreement among untrained judges as measured by Cohen's kappa. A second method we use is to acoustically analyze the same utterances presented in the interobserver agreement studies and to plot them in multidimensional acoustic space. Even with only two dimensions, the categories identified by observers auditorily segregate significantly (Kwon et al., 2006b), and in further work we are seeking to characterize the degree of segregation that can be obtained when additional acoustic dimensions are utilized. A third method invokes lag sequential analysis of utterances produced by infants in various circumstances (alone, in interaction, while in the same room with the parent but separated in space and not obviously interacting). The utterances are first categorized as fitting into one of the three categories, and thereafter it can be shown the categories do not appear at random but are ordered in repetition and/or alternation by the fourth or fifth month (Kwon et al., 2006a). All these kinds of quantification suggest that new vocal categories have been developed by human infants by stage 3. Current work in our laboratories with neural networks provides additional confirmation of acoustic patterns in infant vocalization that suggest category formation by stage 3.

New vocal category formation within an individual at stage 3 depends, in our interpretation, upon the naturally logical foundations of spontaneous production of sounds in stage 1 and their exploration and elaboration within the available acoustic space in stage 2. Westermann's (this volume) modeling suggests that categories can emerge from vocal exploration and self-perception, even without influence from other ambient vocalization. Our reasoning about the importance of exploration as a foundation for development of new categories of action is emphasized also in an important literature on early motor development focused on hand, arm, and leg movements (Thelen, 1981, 1994, 1995). In both vocal and more general motor development, it appears then that new categories of action can appear by self-organization if the infant engages in self-monitored physical exploration.

The formation of new categories appears to be critical to further development of complex communication. Every aspect of language depends on the ability to learn new discrete sound categories. And crucially, these categories must be free of specific function or meaning, because they must constitute a repertoire that can be recombined to form meaningful units (lexical items and sentences) of unlimited number. This principle of language, obvious for generations, has been referred to in recent formulations as "discrete infinity" (Chomsky, 1986; Hauser et al., 2003). It should be emphasized that what the infant accomplishes in stage 3 is not, however, the development of discrete phonological units of a mature sort. These sounds are not phonemes, nor allophones, nor phonetic features of the mature sort found in human languages. They are instead embryological precursors to such elements, infraphonologically significant forerunners that manifest the infant's ability to form categories from the raw material of elaborated, nonstereotyped sounds, categories that have no predetermined function but are free to be utilized in new ways.

Infraphonological Stage 4 (The Canonical Stage): Emergence of Canonical Syllables
Canonical syllables are relatively mature and well formed in the sense that words in spoken languages can be composed of them. They manifest characteristics that are found in the vast majority of syllables in languages, and without them it would not be possible to have indefinitely large lexicons. Canonical syllables emerge under systematic control by the fifth to the tenth month of human life. They consist typically of a nucleus with the supraglottal vocal tract open rather than at rest (i.e., a vowel-like sound), produced in modal or normal phonation, along with at least one margin (i.e., a consonant-like supraglottal articulation). Importantly, the transition between the margin and the nucleus must be smoothly and quickly articulated (nominally within 120 milliseconds).

Occasional canonical syllables can be heard in human infants long before the canonical stage (e.g., in gooing), but in such cases they appear to occur as accidents of the exploratory elaboration of more primitive vocalizations. Our quantitative approach to identification of the canonical stage is based on indications of repetitive occurrence of well-formed syllables. Parents prove to be excellent judges of the onset of canonical babbling (Oller et al., 2001) and are capable of listing the syllable types that infants produce under systematic control once the canonical stage is under way. They systematically ignore the great bulk of the rich raw material of which each canonical syllable is composed (i.e., they ignore within-category variation, which at this stage is still considerable) and instead focus on repetitive features that indicate syllabic control.

That canonical syllables form a crucial, naturally logical foundation in the development of human languages is not controversial. Languages require canonical syllables to form words, and after the onset of the canonical stage in an infant, several months typically pass before meaningful words are produced consistently so that they can be identified by parents (Oller et al., 1998). There is no credible report that we know of that indicates canonical syllable control by any nonhuman primate at any point in time.

Canonical syllables are not, however, the starting point on the developmental path to human language as has been suggested by MacNeilage (1998). The three prior stages have been found to occur in every one of the scores of infants we have studied in longitudinal research; further, the events of the earlier stages are naturally logical prerequisites to canonical syllables (Oller and Griebel, 2008). Canonical syllables require coordinated phonation *and* supraglottal articulation, developed to primitive extents in stages 1 and 2. Further, canonical syllables are vocal *categories* formed of both phonation and systematically well-timed supraglottal articulations: Vocal categories are first formed out of phonatory distinctions without well-timed articulations at stage 3. Canonical syllables are, thus, formed from components that are developed in three prior stages that constitute a naturally logical sequence of increasing complexity and increasing precision of vocal action.

MacNeilage's view also fails to recognize the naturally logical precedence of phonatory over supraglottal articulatory development. He assumes phonation need not be developed first. However, phonation is the primary sound source in syllables—without phonation, supraglottal articulatory movements produce low-intensity sounds that are ill suited to forming the nuclei of syllables. Articulatory movements during syllable production are largely perceived because phonation provides a carrier for formant transition information. Consequently, systematic development of phonatory capabilities comes in stage 1, providing a basis upon which systematic supraglottal articulatory movements can be perceived when they appear in subsequent stages.

Early Infrasemiotic Development: Emergence of Free Expessivity

Perhaps the most notable development in functional flexibility during the first six months of life is that each category developed at stage 3 is produceable in multiple, sometimes opposite, circumstances of affect and that the same sort of functional flexibility is seen with canonical syllables when they emerge at stage 4 (Oller, 1981, 2000; Scheiner et al., 2002).

The significance of the occurrence of particular sound categories with varying affect cannot be undercut by the argument that variations of affect may be (at least partly) a product of varying general states of arousal or external environmental conditions. Fixed signals, by definition, do not allow variation of the connection between affect and vocal signals—for example, an aggressive signal must be negative, an affiliation signal must *not* be negative, and an exultation signal must be positive. Consequently, empirical illustration of the production of the same sound category in differing conditions of affect on different occasions provides a conclusive illustration of functional flexibility.

To provide such illustration, we conduct cross-classification analysis and find that, for example, squeals are used sometimes to express complaint (as indicated by facial expression) but on other occasions to express exultation or delight (as indicated by broad smiling). On other occasions the same vocalization type is used with a totally neutral face in circumstances that can be interpreted as pure vocal play. All these types of cross-classification variations apply to the other vocal types as well—growls and full vowels are also used with multiple expressions (Oller et al., 2003).

There are also clear instances where a panel of judges each independently characterizes a particular vocalization as pertaining to a particular category, for example, a squeal, based on the audio information only, and the facial expression as, for example, an exultation, based on video information only. Another vocalization judged to be of the same category (e.g., a growl) is seen uniformly as a complaint, while another is judged by the panel to be produced with neutral affect. Further, it appears that infants vary day to day, tending to use particular sound categories more heavily one way or another (with positive or negative emotion) on differing days (Oller et al., 2007). All these patterns are viewed as evidence of a kind of free expressivity (Oller, 2000). Continued quantification of free expressivity in infant vocalizations and the development of a general methodology that could be applied in nonhuman primates is a primary goal of our current work.

The natural logic we propose suggests that free expressivity with newly formed categories is a step of development that depends upon the prior step of category formation. We emphasize here that the vocal flexibility occurring in stage 3 is not merely that of random occurrence of states with unsystematically varying sounds. Rather, the flexibility seems targeted by the infant: New systematic categories are first controlled and thereafter utilized to express specific states under the infant's voluntary control.

Summary on Points of Natural Logic in Early Infant Vocal Development

In summary, then, we see naturally logical dependencies that produce necessary sequences for vocal development in at least the following ways within the first half year of life:

1. Simple, stereotyped quasivowels *precede* more variably produced quasivowels and gooing, a naturally logical progression from simple phonatory acts to more complex phonatory and primitively articulated acts.

2. Variably produced quasivowels and gooing *precede* category formation from the variably produced raw material, a sequence that appears to be based on a naturally logical learning dependency (category formation is apparently achieved through active exploration initiated with quasivowels and gooing).

3. Relatively less complex categories (such as squeals and growls), involving phonatory manipulation only, are formed through vocal exploration *before* the more complex categories of canonical babbling, where both phonatory action and supraglottal articulation must be coordinated.

On the infrasemiotic side, after categories are developed, they are recruited to serve multiple communicative functions. Thus, as new flexible potential signals appear in the repertoire of infants, they provide the naturally logical foundation for multifunctional expression with those signals.

All these sorts of developments are infrastructural foundations for capabilities of vocalization that come later in the first year of life. For example, vocal imitation of categories developed by the infant (including canonical syllables) comes later than category formation because there is no possibility of systematically recruiting the production of a voluntary category to match a sound produced by another person without the prior establishment of a capability to produce the category spontaneously. Indeed, clear evidence of infant vocal imitation occurs only after the first half year of life (Kessen et al., 1979). The ability to perform vocal imitation has long been noted as a logical precursor to all other aspects of spoken language and has recently been reemphasized in Hauser, Chomsky, and Fitch (2003). Notably, the authors do not emphasize the apparently logical precursors to imitation that we have outlined here. One possible reason for this omission is that very few of the individuals involved in research on evolution of language have to the present taken note of the data from longitudinal research on infant vocal development, and consequently they have not considered the sequence of early vocal events that appears to be universal in human infants.

Features of Natural Logic as an Interpretive Framework

Interleaving Development of Naturally Logical Capabilities

The developmental sequence indicated above is seen in interleaving of growth of fundamental capabilities in the domains of signal and functional capability, a pattern of overlap in time that continues throughout development. To illustrate interleaving, notice first that emergence of spontaneous production in a particular signal domain—for example, phonation as in infraphonological stage 1—is followed by elaboration of the signal through systematic vocal exploration of variations in phonation in stage 2, and the creation of categories of primarily phonatory nature in stage 3. Notice also that later, control of a new signal domain, movement of the jaw during phonation, begins to emerge, and again proceeds through the same three stages: First coordination of jaw movement and phonation occurs in spontaneous production, then again later in systematic exploration, and finally in the creation of canonical syllabic categories in stage 4. Thus, the occurrence of spontaneous production does not end when the first evidence of vocal exploration begins, but rather growth of the capabilities is interleaved with spontaneous production leading the way and vocal exploratory elaboration following as each new domain of potential signaling (and possibly each new signal type) comes under systematic control by the infant. The growth of these capabilities (spontaneous production, exploration, and category formation) is interleaved across domains of signal control throughout infant vocal development.

Interleaving is also evident in the relation between signal and functional flexibility. Signal flexibility emerges in successive stages (1–4), and functional flexibility emerges in increasingly complex ways repeatedly at each stage to take advantage of the signal developments as they become available. New signals are recruited to serve existing functions as the signals become available, and the communicative functions that can be served become more elaborate as development proceeds, presumably in part because there are new signals available to serve them.

The interleaving of developments in different domains across years implies that the stages of signal and functional flexibility indicated above do not specify discrete begin and end points for each vocal capability. Instead the stages are characterized by increases in *degree* of capability in the designated domains, and development in those domains is interleaved across time. The stages of vocal development provide a heuristic overview of ordering, but the processes they entail show overlapping developmental schedules.

Interleaving is seen in many domains of development. MacWhinney (1982) has pointed out that in the development of productive syntax, rote learning of word strings and analysis of word strings as individual words both play important roles, with rote learning repeatedly leading the way and analysis following for each new item or set of items (words and word strings) that enter the child's repertoire. Interleaving of processes (rote learning and analysis) is evident, since rote learning does not cease its growth when analysis begins but continues expanding to provide a foundation for new analysis possibilities throughout the early development of syntax.

The Natural Logic Alternative to Preformationism: An Analogy to Illustrate Interleaving

Our interpretive approach is distinctly nonpreformationist—we do not assume that mature vocal categories are preordained by an innate language endowment (see, e.g., Pinker, 1994). Key advantages of the natural logic approach over an innatist approach can be illustrated by considering an analogy with embryology. When the seed that will become a tree germinates, it has no trunk, no branches, no leaves, and no bark. The seed does not possess *preformed* mature structures such as these but only precursors to them. One of the first things that is noticed by the casual observer of the tree's growth is that a shoot appears, with no branches and no trunk. The shoot is a precursor to a tree trunk and to every other structure that the tree will possess above ground. As the tree begins to grow, the shoot thickens, and smaller shoots begin to emerge at the leading edge. These are precursors to branches. Eventually, the base shoot expands so much that we feel comfortable calling it a trunk, and the smaller shoots expand and diversify until we feel comfortable calling them branches. When the tree is mature, it has many new structures both internally and externally, including leaves and bark, that are not at all evident in the seed stage or the shoot stage.

There are two key points to consider in the analogy of the tree's growth to the growth of speech capability. First, in both cases the beginning phases include *no* mature structures but only precursors to them that diversify in stages toward a mature form, with no simple discontinuous change from immature to mature—structures are not preformed (as assumed in traditional innatist proposals) but grow and diversify from precursor forms. Second, the growth in each type of structure is interleaved across time. The shoot does not stop growing when the branches begin to emerge, and the main branches do not stop growing when yet smaller branches begin to emerge or when leaves begin to form. Each stage of development of the tree involves growth at the foundations to support growth at the top. Yet there is a naturally logical precedence of the original shoot to any of its branches and of any of the main branches to any of their smaller branches or to the leaves that will emerge from them.

When we posit a natural logic for human vocal development and evolution, we advocate a view that is analogous to the natural logic of the growth of the tree. Spontaneous production of phonation is a logical foundation for all that vocal language will become, just as the shoot that first appears is a logical foundation for the tree. Neither the shoot nor the capability for spontaneous production ceases to grow after its first appearance—nothing could progress without continued growth of the foundational structures at each stage for both language and the tree. Still, the logical precedence of structural elements is clear. There is an interleaved ordering in both cases where the more foundational elements precede the later ones by necessity.

Naturally Logical Diversity in Evolution and Development

The logically necessary capabilities for speech indicated above (spontaneous production, elaboration of produced forms, category formation, etc.) are abstract, leaving considerable room for individual variation in concrete developmental patterns and in possible routes of evolution. Individual differences among infants in vocal development are notable. For example, while all infants we have studied develop phonatory categories of vowel-like sounds contrasted with squeals or growls in infraphonological stage 3, a variety of additional categories and subcategories (raspberries, ingressive sounds, subcategories of growls produced with either vocal fry or harshness, etc.) occur with considerable individual variation during the same period.

The fact that there are notable *similarities* across infants in the types of sounds that occur at particular stages is presumably a product of similarities in the physical structures that must be manipulated to produce sound and in the relative efficiency commonly occurring sound categories possess to serve the functions of early vocal communication. We envision a Darwinian competition among various possible categories such that the categories that are most efficiently produceable and/or effective in transmitting the messages they may bear at the stage in question survive and tend to occur frequently. Good examples of less successful forms during phonatory development are ingressive sounds or sounds produced with vocal tremor—they occur occasionally but tend to be infrequent and to drop out of repertoires quickly.

Just as development shows variability consistent with the proposed natural logic, so, according to our proposal, evolution must have proceeded consistent with the natural logic along paths with a variety of options in terms of particular sound types that might have been used and particular functions they might have served at various points of hominin evolution. The logic provides outlines within which evolution was presumably constrained, but the possible concrete routes of progression consistent with the natural logic were various, and there was presumably competition among the routes such that only those that yielded efficient communicative systems survived. Oller (2000) provides speculative scenarios of evolution based on the broad outlines described here, including a discussion of the types of sounds that plausibly might have occurred as communications at various points in hominin evolution.

Natural Logic and Neurological/Physiological Foundations for Language

In accord with Darwinian thinking, we assume that the competition that led to the capabilities for vocal control seen in modern humans included natural selection for efficient neurological systems to implement the stages of natural logic in development. Hammerschmidt and Fischer (this volume) and Owren and Goldstein (this volume) both review literature on neurological foundations for vocal flexibility in humans and contrast those foundations with those observed in nonhuman primates. We shall not provide a detailed proposal here regarding how the natural logic we propose might be implemented neurologically. However, the proposal is indeed intended to imply specific mechanisms of neurological structure and development to account for the stages of the natural logic, and so we provide a brief outline of the mechanisms we envision consistent with the stages of development and the natural logic.

The primate background provides evidence to support schematic components of vocal communicative control presumed to be present in all primates, including humans. This pan-primate system includes at least the following components:

1. A vocal output pathway primarily associated with the brain stem's periaqueductal gray (PAG; see Hammerschmidt and Fischer, this volume, for references).

2. An emotional interpretation system (presumably subcortical) that derives information from various sensory modalities (Damasio, 1999; LeDoux, 1998), affording the basis for determination of (fixed) signals appropriate to particular circumstances.

3. Pathways from emotional interpretation centers that feed limbic-based control centers to initiate largely involuntary commands to the PAG (see Hammerschmidt and Fischer, this volume).

4. Fairly well-developed pan-primate auditory categorization systems presumably housed in the temporal lobe's supramarginal gyrus, allowing early speech discrimination in humans as well as remarkably similar discrimination in a variety of nonhuman species including nonhuman primates (see evidence in Eimas et al., 1971; Jusczyk, 1992; Kluender et al., 1987; Kuhl and Miller, 1978; Trehub, 1976).

In the human case, it would appear that new subsystems of control must emerge to account for each of the stages of development that we have outlined:

1. The first *new* system not present in nonhuman primates to come online at infraphonological stage 1 may be a frontal cortex control system making spontaneous phonation possible. This system could include not only excitatory but also disinhibitory properties to allow frequent spontaneous production of vocalization. Disinhibition might help account for relatively unfettered production in human infants in contrast with otherwise more strictly constrained vocalizations that usually occur in nonhuman primates, where vocalizations tend to have stereotyped form and to occur only when the immediate circumstances warrant their production. Such a new frontal control system, according to our proposal, at an early stage of human development could make spontaneous production of quasivowels possible.

2. At infraphonological stage 2, expansion of functions of the frontal control system could make possible production of more variable sounds (as we see, e.g., in early quasivowel elaborations in duration, vocal quality, and pitch). We propose that an additional new neurological foundation of stage 2 may create greater *motivation* to explore vocalizations, a motivation that seems lacking in nonhuman primates. In contrast to the differences in vocal exploration between humans and nonhuman primates, both nonhuman primates and human infants show substantial playful exploratory tendencies in nonvocal domains (see Kuczaj and Makecha, this volume; Thelen, 1995).

3. In order to form new categories from newly elaborated voluntary sounds, we suggest that a new categorization feedback loop must be established in infraphonological stage 3, between the newly formed frontal motor control system and the pan-primate auditory categorization system. This connection (perhaps instantiated at least in part in the arcuate fasciculus) is proposed to account for the self-monitoring and sensitivity to ambient vocalizations that appear to play a key role in the human capability to produce new vocal categories.

4. At infraphonological stage 4 we see strong evidence of a new neurological system allowing voluntary *coordination* of phonatory (glottal) and supraglottal articulatory actions (presumably an articulatory frontal cortex control system, to complement the phonatory one). This system may begin to emerge at stage 2 and account for the primitive articulations of gooing, but the lack of systematic coordination, especially in timing of phonation and articulation in gooing, suggests that the system must become much more active in stage 4, where fine coordination of rapid articulation with phonation is the defining feature of the canonical stage.

5. Finally, a system that allows for freely expressive production of newly controlled categories would presumably require a feedback loop between the pan-primate emotional interpretation system and the new human-specific systems controlling the development and production of new vocal categories. This loop, according to our proposal, would allow for the production of new categories to serve a variety of communicative functions.

Genetic and Epigenetic Changes and Naturally Logical Stages

It is uncertain how much genetic change specific to communication would be necessary to establish the foundational capabilities for language. The reason for the uncertainty is that genetic changes could provide naturally logical foundations not specific to communication, foundations making possible greater capabilities for seeking, learning, and problem solving. An organism with these noncommunicative foundations could nonetheless gain advantages supporting self-organization of additional capabilities specific to spoken language (Kent, 1992; Sachs, 1988).

We propose (in accord with a variety of other theorists such as Bates and MacWhinney, 1982, and Tomasello, 2003) that both specific genetic changes corresponding to language-specific capabilities and other epigenetic changes corresponding to capabilities that may have allowed for self-organizing development of various language-necessary capabilities have been involved in the remarkable process of hominin communicative evolution. In accord with this way of thinking, we will be required ultimately, in seeking to understand the evolution of language, to seek ways to differentiate between the genetic changes that ultimately made language possible and the epigenetic developmental processes through which the concrete form of language is built.

Natural Logic in Other Domains of Child Development

Although the term "natural logic" has not been widely used in the sense we intend for it, there are many examples in child development that are consistent with our viewpoint. Consider, for example, that infants do not develop stranger fear until they recognize individual faces (see review in Lafreniere, 2000). The application of natural logic seems straightforward: Infants cannot be fearful of strangers until they can recognize strangers as such.

Similarly, it is straightforward that infants manifest substantial general postural control, including, for example, head and neck control, long before they are able to sit unsupported (Bayley, 1969). The latter development includes the necessity for general postural control, of which head and neck control are examples, and consequently sitting can be seen as a capability that would be impossible until basic head and neck control are established, since these are subcomponents of general postural control.

In some cases the implied natural logic of sequences of development appears to be straightforwardly applicable to evolutionary scenarios. Consider an example from the foundations for "theory of mind": Until infants can track the gaze of others, they do not acquire the capability for "alternating joint attention." In the latter case, they follow another person's gaze toward an object and then quickly shift their gaze back toward the other, sharing attention to the object with another person (Butterworth, 1996; Mundy et al., 1992; Mundy and Willoughby, 1996). The simple tracking of gaze can be seen as a necessary and logical precursor to the more complex act of gaze sharing. Gaze sharing is more complex because it involves both tracking *and* alternation.

Infant initiation of joint attention is seen when infants initiate a communicative event, pointing and engaging in alternating gaze between a designated object and another person—infants often vocalize to help initiate such an event. Initiation of joint attention represents an additional step in the process of development beyond gaze tracking and gaze sharing, and it seems logically to require the foundations of the prior developments because infants would have no reason to anticipate sharing of attention to something they see until they have experienced the following of gaze and subsequent alternation of gaze with another person.

These steps of development suggest a natural progression based on the establishment of primitive, simpler capabilities along with the building of more complex capabilities through elaboration of, or combinations of, the simpler ones. In the case of the gaze examples, comparisons with other primate species provide confirmation of the naturally logical sequence that we have described. In general, the data suggest that gaze tracking occurs in some species without training by humans but that more advanced functions of joint attention are harder to find in nonhumans and may occur only when special training is provided (Povinelli, 2000; Povinelli and Eddy, 1996; Tomasello, 1996).

Consequently, hominin evolution appears to have proceeded in such a way that alternating joint attention and initiated alternating joint attention appeared after gaze tracking was already in place. Speculation about such a sequence in evolution is of particular interest because it seems obvious that without joint attention in fairly elaborate form, it would be impossible to develop fully referential words. The latter depend on an understanding that a word spoken (e.g., "door") can represent a shareable concept (e.g., a particular class of objects, doors). Consequently, we reason that concrete concepts can be shared between individuals through words only if these concepts have been previously shared in the absence of words.

Orderly Sequences in Language Development that Have Already Come to Be Viewed as Corresponding to Evolutionary Sequences

Simple Concatenative Syntax Precedes More Advanced Grammar

Parallel with the sort of reasoning that has been pursued regarding joint attention, there is increasing acceptance of the idea that logical sequences of events in nonevolutionary domains can help to specify evolutionary sequences. Bickerton, for example, has suggested that there existed in hominin history a stage of "protolanguage" in which only the simplest sort of syntax existed (Bickerton, 1981, 1990). Words were combined in telegraphic style, with no inflections (i.e., grammatical markers on root words) or bound morphemes, and constructions were short. His reasoning draws support from pidgin languages that serve as media of exchange among peoples who share no full-fledged language. He points out that such systems are important and communicatively valuable as simple languages of trade even though they include no inflections or bound morphemes. Consequently, it seems reasonable, in accord with Bickerton's reasoning, to suggest that hominin evolution included a protolanguage stage that resembled pidgin languages.

Developmental patterns in humans in the second and third years of life conform to the patterns that would be expected based on the idea that the protolanguage-to-full-fledged-language sequence is required by natural logic. Children go through a telegraphic communication stage during which early words are produced in a way that has much in common with pidgin languages (Bloom, 1970; Brown, 1973). Only thereafter do bound morphemes and other inflectional phenomena take hold. The sequence makes perfect logical sense, because bound morphemes and inflectional elements provide fine-tuning for the syntactico-semantic relations among individual words produced in sequence. What use would such grammatical devices be in the absence of individual root words? And so it seems reasonable to posit a naturally logical sequence in which simple word concatenation syntax precedes the grammar of bound morphemes and other inflectional phenomena and to infer that any developmental or evolutionary sequence that could ever reach a level of grammar would have to have also reached a level of word concatenation syntax or protolanguage.

Bickerton's reasoning suggests that protolanguage could have represented a stable evolutionary stage prior to the appearance of grammar. The reasoning seems to be supported by (1) normal developmental patterns as suggested above, (2) by the fact that there exist language disorders where grammatical capabilities are severely disrupted but where concatenation syntax is present (Johnston, 1988; Leonard and Schwartz, 1985) and, (3) by the fact that apes that learn signs seem to be capable of simple concatenation syntax but not grammar (Gardner and Gardner, 1969; Terrace et al., 1979). These considerations suggest two genetically determined stages in hominin evolution, presumably a pattern of interleaved growth of capabilities with concatenation syntax leading true grammar.

Words Precede Simple Syntax

In a similar line of reasoning, it can be noted that the occurrence of utterances consisting of single truly referential words precedes concatenation syntax in modern children and must have done so in hominin evolution. There exists the illusion of an exception to this principle in the case where children produce short multiword utterances by rote; however, it has become clear that in such cases children have misanalyzed what they have heard in adult speech, such that a short phrase or other short word sequence from the adult language has been understood by the child as an unanalyzed whole (Bloom, 1970). Later in development these unanalyzed types have to be reanalyzed by the child as multiple words that can function independently. As in the case of telegraphic speech, there is stage stability in children, such that words are produced one at a time for a notable period, after which concatenation in simple syntax begins.

This pattern suggests the plausibility of a "word stage" of evolution, where ancient hominins spoke to each other only one word at a time and presumably with a small vocabulary compared to vocabularies of modern languages or even pidgin languages. While the advantages of such communication fell far short of the advantages of modern language, a "word stage" could surely have had enormous communicative benefit, even with no syntax at all.

The small size of presumed vocabularies at this proposed evolutionary stage also corresponds to strong evidence from modern human childhood. Children learn and use single-word utterances for months before beginning to use productive word combinations. And further, vocabulary size shows both a strong positive predictive value with regard to the appearance of early concatenative syntactic forms and further positive predictive value with regard to the emergence of more advanced grammatical forms (Bates, 1996). It seems, then, necessary for children to amass a critical number of words that can be used individually before syntax can take hold. These facts are consistent with a natural logic whereby combinatorial syntax of any kind presupposes control of the individual words that syntactic operations combine.

The notion of interleaving of developmental stages and processes introduced above is also relevant to evolution. The idea of a naturally logical sequence for emergence of primitive syntax, as we propose it, includes the interleaving of the ability to produce and understand individual words with the logically subsequent ability to analytically produce and understand strings composed of words (MacWhinney, 1982, 1988). The processes are overlapped, and the naturally logical sequence is repeatable at increasingly higher levels of complexity as the lexicon grows and the syntax itself becomes more complex. In accord with our general view that development and evolution have important parallels, we propose that interleaving of word and (analytical) multiword stages must have occurred also in the evolution of human language.

Canonical Babbling Precedes Large-Scale Vocabulary

By a similar line of reasoning to that proposed for primitive syntax, it is arguable that canonical babbling (which implies the production of well-formed syllables especially notable in reduplicated sequences such as "baba" or "dada") precedes individual word development in a naturally logical way (Oller, 1978, 1980, 2000; Oller and Griebel, 2005). The argument is supported empirically by the following observations:

1. In normally developing infants, the appearance of a substantial vocabulary of conventional words is always preceded by the development of canonical babbling.
2. Nonhuman primates do not appear to be able to produce voluntarily controlled canonical syllables.
3. The vast majority of words in natural languages consist of canonical syllables.

This evidence has been accorded prominence in the writings of MacNeilage, who argues for the emergence in ancient hominins of specific neural mechanisms to support canonical babbling as a foundation for speech (MacNeilage, 1998; MacNeilage and Davis, 1990, 1993).

A proviso regarding the relation between canonical babbling and word use is important: Some words (although very few relative to the vast size of human vocabularies) in natural languages consist of noncanonical syllables. The formal definition of "canonical syllables" includes the specification that each such syllable must possess at least one supraglottally articulated consonant. However, some words either have no consonants at all or the only consonants they have are glottal, and thus these words do not require supraglottal articulation. Examples of such noncanonical syllables that constitute words in English are "he" and "ah." Children, when they begin to talk, sometimes include a small number of words with noncanonical form. The existence of noncanonical word forms does not, however, undercut the basic logic of the argument that canonical syllable production is a prerequisite to full-scale vocabulary development. Only a very small number of words can be constructed without articulated consonants, and so there is a practical (a naturally logical) requirement that canonical syllables precede *large-scale* vocabulary development as occurs in children. Phonologies abide by a "particulate principle" that affords the possibility of recombination of syllables to produce lexicons of virtually unlimited magnitude (Studdert-Kennedy, 2000).

In accord with the interleaving characteristic of developmental patterns, we see continued increases in the numbers of canonical syllables that children can command in production as the lexicon grows. New words often require new syllables to produce them, and consequently canonical syllable development must continue to support lexical development for many months after the first appearance of words composed of canonical syllables.

The naturally logical sequence of canonical syllable production followed by word production appears to be dependent, then, on large vocabulary size—very small-scale vocabulary development can occur prior to development of canonical syllables, but large-scale development of vocal lexicons cannot occur in the absence of canonical syllables.

Oller (2000) has pointed out that the earliest meaningful and contextually flexible vocal communications in ancient hominins could have been noncanonical in type. The fact that infants only occasionally use noncanonical forms as words suggests the possibility that the communicative value of contextually flexible noncanonical forms in ancient hominins may have been primarily of a simple illocutionary sort, with each form corresponding to an act such as requesting, rejecting, affiliating, and so on, a pattern of usage that also occurs in noncanonical early word usage by modern children. More advanced semantic communications, where words are used to reference concepts with multiple possible illocutionary forces that can be invoked on different occasions, would have occurred later in evolution, according to our reasoning, following from the observations that semantically laden communication also tends to occur later than simple illocutionary communication in modern child development (Bates et al., 1979). Whatever the situation may turn out to be in terms of the relation between the appearance of semantically laden words and canonical or noncanonical forms, the general point at stake here seems to have been largely unchallenged: Development of canonical syllable control must have played a major role in establishing foundations for full-scale vocabulary evolution in hominins.

The Need for an Explicit Enterprise to Develop Natural Logic as an Explanatory Framework

Thus, in at least three widely publicized cases (simple concatenation syntax precedes grammar, single words precede simple concatenation syntax, and canonical syllable control precedes significant vocabulary growth), there appears to be general acceptance of the idea that certain logical relations implied by ordered human development can be reasonably adopted in the construction of evolutionary scenarios. Acceptance of these ideas is consistent with the notion that natural logic provides the connection accounting for parallelism between development and evolution.

Bickerton (1990), Oller (2000), and Jackendoff (2002) have all supplied stage models for evolution of language that are, at least in part, built upon an idea of naturally logical sequencing. Given that these ideas seem to be at least tacitly accepted in many quarters, and explicitly posited in others, it seems important to take the notion of natural logic further. What is needed is a general theory of the natural logic of vocal communication. Issues of natural logic have often been invoked in speculations about why infant development proceeds as it does or about why evolution must have proceeded as it does, but the speculations have tended to invoke ad hoc constraints (Prince and Smolensky, 1993) or ad hoc possible solutions in a communication problem space (Elman et al., 1996).

The first author has argued elsewhere (Oller, 2005) in favor of a fundamental replacement for the ad hoc approach, an overarching theory of natural logic in the emergence of communicative systems. The study of natural logic should become a general effort to characterize possible sequences of evolution (and development) in vocal communication away from the primate background and toward systems of communication with greater power. An initial model of communicative natural logic is provided in Oller (2000). The model outlines properties of capability required to be developed for language, starting with contextual flexibility (both signal and functional flexibility) in the simple forms detailed in this chapter, and forming foundations for such further sequentially developed capabilities as recombinability of syllables, vocal imitation, learning of conventional words, formation of truly semantic word categories with illocutionary flexibility, formation of propositions (simple syntax), and finally steps of more advanced grammatical development.

If we are to succeed in applying a naturally logical model to language evolution, it would seem critical that we start at the very beginning of the break between the hominin line and our nonhominin primate ancestors. We ask then, what were the first logical steps of vocal communication upon which all others must have been built? As we have argued here, a tentative answer is found in interpretation of the patterns of vocalization of the human infant, beginning well before canonical syllable control, and providing foundations for all that vocal language becomes.

Acknowledgments

This work has been supported by a grant from the National Institutes of Deafness and other Communication Disorders (R01DC006099-01 to D. K. Oller, Principal Investigator, and Eugene Buder, Co-Principal Investigator) and by the Plough Foundation. The authors express their appreciation to all the workshop participants for their comments on the work presented here and especially to Brian MacWhinney and Stan Kuczaj for insightful suggestions.

References

Bates E. 1996. *From preverbal communication to grammar in children*, Waseda University International Conference Center, Tokyo.

Bates E, Benigni L, Bretherton I, Camaioni L, Volterra V. 1979. *The Emergence of Symbols: Cognition and Communication in Infancy*. New York: Academic Press.

Bates E, MacWhinney B. 1982. Functionalist approaches to grammar. In *Language Acquisition: The State of the Art*, ed. E Wanner, LR Gleitman, pp. 173–218. Cambridge, England: Cambridge University Press.

Bayley N. 1969. *Bayley Scales of Infant Development: Birth to Two Years*. New York: Psychological Corporation.

Becker ML, Buder EH, Ward JP. 2003. Spectrographic description of vocalizations in captive *Otolemur garnetti*. *International Journal of Primatology* 24: 415–46.

Bickerton D. 1981. *Roots of Language*. Ann Arbor, MI: Karoma.

Bickerton D. 1990. *Language and Species*. Chicago: University of Chicago Press.

Bloom L. 1970. *Language Development*. Cambridge, MA: MIT Press.

Boliek CA, Hixon TJ, Watson PJ, Morgan WJ. 1996. Vocalization and breathing during the first year of life. *Journal of Voice* 10: 1–22.

Brown R. 1973. *A First Language*. London: Academic Press.

Buder EH, Chorna L, Oller DK, Robinson R. In press. Vibratory regime classification of infant phonation. *Journal of Voice*.

Buder EH, Oller DK, Magoon JC. 2003. *Vocal intensity in the development of infant protophones*. Proceedings of the XVth International Congress of Phonetic Sciences, ed. MJ Solé, D. Recasans, J Romero, pp. 2015–2018. Adelaide, Australia: Casual Productions.

Butterworth G. 1996. *Species typical aspects of manual pointing and the emergence of language in human infancy*, Waseda University International Conference Center, Tokyo.

Chomsky N. 1986. *Knowledge of Language: Its Nature, Origin, and Use*. New York: Praeger.

Damasio A. 1999. *The Feeling of What Happens: Body and Emotion in the Making of Consciousness*. New York: Harcourt Brace.

de Saussure F. 1968. *Cours de Linguistique Générale*. Paris: Payot.

Eimas PD, Siqueland E, Jusczyk P, Vigorito J. 1971. Speech perception in infants. *Science* 171: 303–6.

Elman JL, Bates EA, Johnson MH, Karmiloff-Smith A, Parisi D, Plunkett K. 1996. *Rethinking Innateness: A Connectionist Perspective on Development*. Cambridge, MA: MIT Press.

Gardner RA, Gardner BT. 1969. Teaching sign language to a chimpanzee. *Science* 165: 664–72.

Haeckel E. 1866. *Generelle Morphologie der Organismen: 2 volumes*. Berlin: Georg Reimer.

Hauser M, Chomsky N, Fitch WT. 2003. The faculty of language: What is it, who has it, and how did it evolve? *Science* 298: 1569–79.

Jackendoff R. 2002. *Foundations of Language*. Oxford: Oxford University Press.

Johnston JR. 1988. Specific language disorders in the child. In *Handbook of Speech-Language Pathology and Audiology*, ed. NJ Lass, LV McReynolds, JL Northern, DE Yoder, pp. 685–715. San Diego, CA: Singular Publishing Group.

Jusczyk PW. 1992. Developing phonological categories from the speech signal. In *Phonological Development: Models, Research, Implications*, ed. C Ferguson, L Menn, C Stoel-Gammon, pp. 17–64. Parkton, MD: York Press.

Kent RD. 1992. The biology of phonological development. In *Phonological Development: Models, Research, Implications*, ed. C Ferguson, L Menn, C Stoel-Gammon, pp. 65–89. Parkton, MD: York Press.

Kessen W, Levine J, Wendrich K. 1979. The imitation of pitch in infants. *Infant Behavior and Development* 2: 93–9.

Kluender KR, Diehl RL, Killeen PR. 1987. Japanese quail can learn phonetic categories. *Science* 237: 1195–7.

Koopmans-van Beinum FJ, Clement CJ, van den Dikkenberg-Pot I. 2001. Babbling and the lack of auditory speech perception: A matter of coordination? *Developmental Science* 4: 61–70.

Koopmans-van Beinum FJ, van der Stelt JM. 1986. Early stages in the development of speech movements. In *Precursors of Early Speech*, ed. B Lindblom, R Zetterstrom, pp. 37–50. New York: Stockton Press.

Kuhl PK, Miller JD. 1978. Speech perception by the chinchilla: Identification functions for synthetic VOT stimuli. *Journal of the Acoustical Society of America* 63: 905–17.

Kwon K, Buder EH, Oller DK. 2006a. *Contextual flexibility in precanonical infant vocalizations: Its role as a foundation for speech*, International Child Phonology Conference, Edmonton, AB, Canada.

Kwon K, Buder EH, Oller DK, Chorna LB. 2006b. *The role of fundamental frequency characteristics in classifying prebabbling vocalizations*, Annual meeting of the American Speech-Language-Hearing Association, Miami, FL.

Lafreniere PJ. 2000. *Emotional Development: A Biosocial Perspective*. Belmont, CA: Wadsworth Press.

LeDoux J. 1998. *The Emotional Brain*. London: Weidenfeld and Nicolson.

Leonard LB, Schwartz RG. 1985. Early linguistic development of children with specific language impairment. In *Children's Language*, ed. KE Nelson, pp. 291–318. Hillsdale, NJ: Lawrence Erlbaum Associates.

Lorenz K. 1951. Ausdrucksbewegungen höherer Tiere. *Naturwissenschaften* 38: 113–6.

MacNeilage PF. 1998. The frame/content theory of evolution of speech production. *Behavioral and Brain Sciences* 21: 499–546.

MacNeilage PF, Davis BL. 1990. Acquisition of speech production: Frames then content. In *Attention and Performance XIII: Motor Representation and Control*, ed. M Jeannerod, pp. 453–76. Hillsdale, NJ: Lawrence Erlbaum Associates.

MacNeilage PF, Davis BL. 1993. Motor explanations of babbling and early speech patterns. In *Changes in Speech and Face Processing in Infancy: A Glimpse at Developmental Mechanisms of Cognition*, ed. B de Boysson-Bardies, S Schonen, P Jusczyk, PF MacNeilage, J Morton. Dordrecht: Kluwer Academic Publishers.

MacWhinney B. 1982. Basic syntactic processes. In *Language Development, Vol. 1: Syntax and Semantics*, ed. S Kuczaj, pp. 73–136. Hillsdale, NJ: Lawerence Erlbaum Associates.

MacWhinney B. 1988. Competition and teachability. In *The Teachability of Language*, ed. R Schiefelbusch, M Rice, pp. 63–104. New York: Cambridge University Press.

McCune L, Vihman M, Roug-Hellichius L, Delery K, Gogate L. 1996. Grunt communication in human infants (*Homo sapiens*). *Journal of Comparative Psychology* 110: 27–37.

Mundy P, Kasari C, Sigman M. 1992. Joint attention, affective sharing, and intersubjectivity. *Infant Behavior and Development* 15: 377–81.

Mundy P, Willoughby J. 1996. Non-verbal communication, joint attention and social-emotional development. In *Emotional Development in Atypical Children*, ed. M Lewis, M Sullivan, pp. 65–87. New York: Wiley.

Nooteboom SG. 1999. Anatomy and timing of vocal learning in birds. In *The Design of Animal Communication*, ed. MD Hauser, M Konishi, pp. 63–110. Cambridge, MA: MIT Press.

Oller DK. 1978. Infant vocalization and the development of speech. *Allied Health and Behavioral Sciences* 1: 523–49.

Oller DK. 1980. The emergence of the sounds of speech in infancy. In *Child Phonology, Vol. 1: Production*, ed. G Yeni-Komshian, J Kavanagh, C Ferguson, pp. 93–112. New York: Academic Press.

Oller DK. 1981. Infant vocalizations: Exploration and reflexivity. In *Language Behavior in Infancy and Early Childhood*, ed. RE Stark, pp. 85–104. New York: Elsevier North Holland.

Oller DK. 2000. *The Emergence of the Speech Capacity*. Mahwah, NJ: Lawrence Erlbaum Associates.

Oller DK. 2005. The natural logic of communicative possibilities: Modularity and presupposition. In *Modularity: Understanding the Development and Evolution of Natural Complex Systems*, ed. W Callebaut, D Raskin-Gutman, pp. 409–34. Cambridge, MA.

Oller DK, Buder EH, Nathani S. 2003. *Origins of speech: How infant vocalizations provide a foundation.* Miniseminar for the American Speech-Language Hearing Association Convention, Chicago.

Oller DK, Eilers RE, Basinger D. 2001. Intuitive identification of infant vocal sounds by parents. *Developmental Science* 4: 49–60.

Oller DK, Griebel U. 2005. Contextual freedom in human infant vocalization and the evolution of language. In *Evolutionary Perspectives on Human Development*, ed. R Burgess, K MacDonald, pp. 135–66. Thousand Oaks, CA: Sage Publications.

Oller DK, Griebel U. 2008. The origins of syllabification in human infancy and in human evolution. In *Syllable Development: The Frame/Content Theory and Beyond*, ed. B Davis, K Zajdo, pp. 29–62. Mahwah, NJ: Lawrence Erlbaum Associates.

Oller DK, Levine S, Eilers RE, Pearson BZ. 1998. Vocal precursors to linguistic communication: How babbling is connected to meaningful speech. In *The Speech/Language Connection*, ed. R Paul, pp. 1–23. Baltimore, MD: Paul H. Brookes.

Oller DK, Nathani Iyer S, Buder EH, Kwon K, Chorna L, Conway K. 2007. *Diversity and contrastivity in prosodic and syllabic development.* In Proceedings of the International Congress of Phonetic Sciences, ed. Trouvain J, Barry W, pp 303–308. Saarbrucken, Germany: University of Saarbrucken.

Pinker S. 1994. *The Language Instinct*. New York: Harper Perennial.

Povinelli DJ. 2000. *Folk Physics for Apes: The Chimpanzee's Theory of How the World Works.* Oxford: Oxford University Press.

Povinelli DJ, Eddy TJ. 1996. What chimpanzees know about seeing. *Monographs of the Society for Research in Child Development* 61: 191.

Prince A, Smolensky P. 1993. *Optimality Theory: Constraint Interaction in Generative Grammar.* New Brunswick, NJ: Rutgers University Press.

Sachs Y. 1988. Epigenetic selection: An alternative mechanism of pattern formation. *Journal of Theoretical Biology* 134: 547–59.

Scheiner E, Hammerschmidt K, Jürgens U, Zwirner P. 2002. Acoustic analyses of developmental changes and emotional expression in the preverbal vocalizations of infants. *Journal of Voice* 16: 509–29.

Snowdon C. 2004. Social processes in the evolution of complex cognition and communication. In *Evolution of Communication Systems: A Comparative Approach*, ed. DK Oller, U Griebel, pp. 131–50. Cambridge, MA: MIT Press.

Snowdon CT, Elowson AM, Rousch RS. 1997. Social influences on vocal development in New World primates. In *Social Influences on Vocal Development*, ed. CT Snowdon, M Hausberger, pp. 234–48. New York: Cambridge University Press.

Snowdon CT, Hodun A. 1981. Acoustic adaptation in pygmy marmoset contact calls: Locational cues vary with distance between conspecifics. *Behavioral Ecology and Sociobiology* 9: 295–300.

Stark R, Bernstein LE, Demorest ME. 1993. Vocal communication in the first 18 months of life. *Journal of Speech and Hearing Research* 36: 548–58.

Stark RE. 1978. Features of infant sounds: The emergence of cooing. *Journal of Child Language* 5: 379–90.

Stark RE. 1980. Stages of speech development in the first year of life. In *Child Phonology, Vol. 1*, ed. GY Komshian, J Kavanagh, C Ferguson, pp. 73–90. New York: Academic Press.

Stark RE, Rose SN, McLagen M. 1975. Features of infant sounds: The first eight weeks of life. *Journal of Child Language* 2: 205–22.

Studdert-Kennedy M. 2000. Evolutionary implications of the particulate principle. In *The Evolutionary Emergence of Language*, ed. C Knight, M Studdert-Kennedy, JR Hurford, pp. 161–76. Cambridge, England: Cambridge University Press.

Terrace HS, Petitto LA, Sanders RJ, Bever TG. 1979. Can an ape create a sentence? *Science* 206: 891–902.

Thelen E. 1981. Rhythmical behavior in infancy: An ethological perspective. *Developmental Psychology* 17: 237–57.

Thelen E. 1994. Three-month-old infants can learn task-specific patterns of interlimb coordination. *Psychological Science* 5: 280–5.

Thelen E. 1995. Motor development: A new synthesis. *American Psychologist* 50: 79–95.

Tinbergen N. 1951. *The study of instinct*. Oxford: Oxford University Press.

Tomasello M. 1996. *The Gestural Communication of Chimpanzees and Human Children*, Waseda University International Conference Center, Tokyo.

Tomasello M. 2003. *Constructing a Language: A Usage-Based Theory of Language Acquisition*. Cambridge, MA: Harvard University Press.

Trehub S. 1976. The discrimination of foreign speech contrasts by infants and adults. *Child Development* 47: 466–72.

Zlatin-Laufer MA, Horii Y. 1977. Fundamental frequency characteristics of infant non-distress vocalization during the first 24 weeks. *Journal of Child Language* 4: 171–84.

Zlatin M. 1975. Preliminary descriptive model of infant vocalization during the first 24 weeks: Primitive syllabification and phonetic exploratory behavior. Progress Report, National Institutes of Health Research Grants.

8 Scaffolds for Babbling: Innateness and Learning in the Emergence of Contexually Flexible Vocal Production in Human Infants

Michael J. Owren and Michael H. Goldstein

Introduction

As one of the most distinctive features of human behavior, spoken language is widely believed to be critical to our species' notable flexibility in cognition, social interaction, and culture. Understanding the origins of this complex and apparently unique faculty poses significant challenges and has engendered requisite controversy and discussion. Examples of critical issues include whether speech began relatively early or late in human evolution, whether any species other than humans exhibit or can acquire similar capabilities, and whether the evolutionary process involved was gradual and continuous or instead was so abrupt and discrete that it should be considered an evolutionary discontinuity relative to our closest nonhuman primate relatives (hereafter *primates*).

The phenomenon of *babbling* in human infants is often considered particularly illuminating for these kinds of questions. By babbling, we mean prelinguistic production of meaningless, nonspeech sounds that are nonetheless recognizable precursors of later phonemes. In this chapter, these *prelinguistic* vocalizations will be distinguished from *nonlinguistic* sounds such as laughter and crying, which are not considered direct precursors of phonemes or words. We will also distinguish between *early* and *canonical* forms of babbling, with the former referring to sounds with vowel- or consonant-like components that are not yet fully formed, and the latter applied to the speech-like and often reduplicated syllabic combinations of consonants and vowels that infants begin to produce prior to the emergence of words. While the terms babbling and canonical babbling are often used interchangeably, Oller (2000) argues that early babbling is characterized by *quasi-resonant vowels* and *primitive articulation*, the latter composed of slow, uncertain consonant–vowel transitions referred to as *gooing* or *marginal syllables*. In contrast, canonical babbling is marked by fully resonant vowels that are combined with rapid consonant–vowel alternation to form *canonical syllables*.

One common point of view is that babbling is a genetically based, evolutionary adaptation that arose in humans to facilitate production and perception of language-specific phonemes (e.g., Fry, 1966; Pulvermüller, 1999; Pulvermüller and Preissl, 1991; Petitto et al., 2004). This approach tends to emphasize the uniqueness of the behavior, noting that while babbling-like vocalizations occur in many songbirds (e.g., Doupe and Kuhl, 1999; Goldstein et al., 2003), it is rare or absent in nonhuman mammals (cf. Elowson et al., 1998; Knörnschild et al., 2006).

This chapter presents a different perspective, one that views the ontogeny of speech as an activity- and experience-dependent process from the very earliest stages. Specifically, we argue for the importance of learning both in the emergence of prelinguistic vocalization and in the infant's subsequent progression to early babbling. In other words, while many researchers take the onset of canonical babbling as a starting point, we suggest that even nonlinguistic vocalizations play an important role in the vocal development that culminates in the emergence of speech. According to this *babbling-scaffold* view, each stage of vocal ontogeny acts as a bootstrap or scaffold for the next, beginning virtually from birth. Thus, nonlinguistic vocalization facilitates the emergence of early babbling, which in turn is the foundation of canonical babbling. Two particular factors we will point out in the progression of events preceding canonical babbling are the infant's acquisition of volitional vocal control and its sensitivity to the instrumental value of vocalizing.

The hypothesis relies on a variety of kinds of evidence, beginning with the observation that the infant's nonlinguistic vocalizations are quite like primate calls. For example, they emerge in the absence of auditory experience, are largely unarticulated sounds, and are most likely mediated by subcortical, emotion-related brain areas. In both humans and primates, these kinds of sounds will be termed *innate*, in the limited sense that producing them requires little if any auditory experience or motor practice. However, vocal development in humans quickly diverges from this primate-like state, with infants soon showing levels of volitional control over vocal production that are far beyond the capabilities of primates. Here, the evidence indicates that infants first begin to control some aspects of their nonlinguistic vocalizations, with vowel-like sounds in particular providing the bridge from nonlinguistic to prelinguistic communication.

We argue that there are at least two different kinds of learning involved. The first derives from emotion-triggered production of nonlinguistic vocalizations, which exposes the infant's cortex to converging streams of proprioceptive and auditory feedback. The proposal is that limbically mediated sounds act as a developmental scaffold that facilitates acquisition of volitional vocal control by providing sensorimotor experience to the infant's cortex. This process is interwoven with learning that nonlinguistic and early babbling vocalizations have important instrumental value—for instance, in eliciting caregiving responses from others and promoting other kinds of social interactions as well. Overall, the argument is that even the earliest vocal development involves deeply intertwined learning about vocal production and the social roles of vocalization and that both are critical to the infant's progress from nonlinguistic to prelinguistic communication.

To flesh out this argument, we begin by reviewing data from primate vocal behavior, emphasizing an evident dissociation between production and reception of vocalizations in these animals. The key point here is that primates show little flexibility in producing vocalizations and are evidently not able to exert direct volitional control over call acoustics or the circumstances of calling. In contrast, the same animals show great flexibility in responding to sounds. Humans, on the other hand, are very flexible in both, a difference we argue can be traced to more extensive connections between the human cortex and brain stem neurons that innervate peripheral vocal anatomy. However, studies of normally hearing and hearing-impaired infants also show that fine-tuned volitional control over vocal production is an acquired ability in humans, which we propose develops gradually through stages of *laryngeal* and then *supralaryngeal* control. While neither laryngeal nor supralaryngeal production is likely under volitional control in primates, the hypothesis nonetheless includes that the heritage of nonlinguistic vocalization and flexible auditory learning capabilities that humans share with these animals are critical contributors to the developmental events that eventually produce our unique vocal communication abilities.

Primate Vocal Behavior

Vervet Monkey Alarm Calls

Recent years have seen much interest in the question of whether primates and other nonhumans use vocalizations in ways that parallel word use in human language. In spite of these efforts, Seyfarth et al.'s (1980) pioneering demonstration of semantic-like properties in vervet monkey alarm calls arguably remains the most convincing example of language-like function in primate vocalizations. The work was particularly convincing because vervets produce a number of acoustically distinct alarms that the researchers could show are linked to specific, mutually exclusive predators and escape strategies. Vervet responses upon hearing *snake*, *eagle*, and *leopard* calls played from hidden speakers provided requisitely compelling evidence that these sounds have specific representational value.

However, the development of vervet alarm calls is also strikingly different from the emergence of language in humans. Most importantly, Seyfarth et al. found that vervets produce recognizable alarm calls from a very young age and in largely appropriate situations. In other words, vervet infants evidently do not need to learn how to produce alarm vocalizations with appropriate acoustic features, nor do they need to learn in which general circumstances to use the various calls. However, learning was found to be critical in responding to the calls. While able to call "correctly" from an early age, young vervets do not initially show differentiated escape reactions to alarm vocalizations from others. Instead, the youngsters tend to freeze, run to their mothers, or react in ways that can increase rather than decrease their exposure to danger. Infant vervets do acquire predator-appropriate responses relatively quickly, but in this case, experience with the calls and responses of other group members plays a critical role (Cheney and Seyfarth, 1990).

Vocal Production

Little Learning of Vocal Production to Control Acoustics

The deep divide between producing and responding to calls shown by vervets is mirrored by evident innateness in the vocal production of other primates. For example, data from several monkey species show that vocalizers produce recognizable species-appropriate calls from an early age and that developmental effects on vocal acoustics are likely mainly maturational in nature. Evidence on this point is ably reviewed by Hammerschmidt and Fischer (this volume) and will not be repeated here except to note that the studies include some with monkeys that were deaf (Talmadge-Riggs et al., 1972), socially isolated (Winter et al., 1973; Hammerschmidt et al., 2001), or reared in an altered auditory environment (Owren et al., 1992a, 1992b, 1993).

In addition, the same conclusion can be drawn from recent reports of babbling-like vocalizations in two nonhuman mammals, namely pygmy marmosets (*Cebuella pygmaea*; Elowson et al., 1998) and sac-winged bats (*Saccopteryx bilineata*; Knörnschild et al., 2006). Early vocal production in these species resembles human canonical babbling in that infants produce a mix of adult-like sounds that are divorced from their normal calling contexts. In contrast to human babbling, however, the sounds have no syllabic character, and there appears to be little or no acoustic learning involved. Instead, the "babbled" vocalizations are acoustically recognizable as adult-like calls from the beginning, do not arise from more approximate, intermediate forms, and do not seem to require practice.

Lack of Articulation

Primate vocalizations are also different from human speech in that key acoustic features typically reflect *laryngeal* rather than *supralaryngeal* aspects of vocal production. By laryngeal, we mean acoustic characteristics due to vocal-fold vibration patterns, whereas supralaryngeal refers to filtering effects traceable to cavities and tissues above the larynx. Linguistic contrasts in human speech typically rely on flexible positioning of the tongue, mandible, and lips, referred to as the *articulators*. In primates, however, major sound classes within a given repertoire primarily reflect differentiated vocal-fold action (Brown et al., 2003; Owren, 2003). Compared to humans, primates have thinner tongues, larynges positioned higher in the neck, and a relative lack of flexible soft tissues in the supralaryngeal vocal tract. While supralaryngeal filtering effects are likely important at least in some sounds (Owren and Rendall, 1997), primates rarely seem to use active articulation to create functionally significant sound contrasts (Lieberman, 1975; cf. Hauser et al., 1993; Riede and Zuberbühler, 2003).

Limbic Neural Mechanisms

Evidence about the neural circuitry underlying vocal production in primates is also extensively reviewed by Hammerschmidt and Fischer (this volume). A major point is that subcortical structures play a central role—for instance, with electrical stimulation of midbrain (periaqueductal gray; PAG) eliciting fully formed, natural-sounding vocalizations in several species, including squirrel monkeys, rhesus monkeys, gibbons, and chimpanzees (Jürgens, 2002). Natural-sounding vocalizations can also be produced by stimulation in the hindbrain, as well as in limbic-system structures such as hypothalamus and cingulate cortex. However, stimulating these latter regions is less reliable, is slower, and elicits less natural-sounding calls. Stimulating cerebral cortex does not elicit vocalizations, nor does cortical lesioning interfere with vocal production. In addition to limbic pathways, motor neurons innervating the lips, mandible, tongue, and vocal folds have also been found to have connections to the facial area of motor cortex. However, these *corticobulbar* pathways have not been shown to have a role in vocal production.

Vocal Conditioning Effects

Some evidence of plasticity in vocal production has been provided by operant conditioning studies (Pierce, 1985) but may also mainly underscore the primacy of limbic, rather than cortical mechanisms. For example, several laboratory studies have shown that rhesus monkeys working for food reward can show contingent changes in the rate, duration, and intensity of species-typical *coo* calls (Sutton et al., 1974). Masataka (1992) has also reported anecdotal evidence that provisioned, free-ranging Japanese macaques can learn to produce coos for food reward (see also Sugiura, 1998).

However, macaque coos are feeding-related vocalizations that both rhesus and closely related Japanese monkeys produce spontaneously in the context of foraging and the arrival of caretakers with food. Motivational state plays a strong role both in triggering these calls and in shaping call acoustics (e.g., Green, 1975; Hauser and Marler, 1993; Owren et al., 1992b). It therefore seems likely that experiments demonstrating conditioned modification of these vocalizations have succeeded in large part because the calls are ones that are triggered by food-related cues under natural circumstances. In addition, acoustic dimensions such as rate, duration, and intensity are characteristics that are highly likely to be affected by vocalizer arousal and emotion in both primates (Rendall, 2003) and humans alike (Bachorowski and Owren, in press). Volitional control would be much more convincingly demonstrated if primate subjects could be trained to produce affectively unrelated sounds such as threat, alarm, or copulation calls for food reward, but that has not been done and is likely very difficult or impossible.

Brain-lesioning studies with rhesus monkeys have also shown that conditioned effects on coo acoustics critically involve the cingulate cortex, a limbic-system component (Sutton et al., 1974; Trachy et al., 1981). This structure has reciprocal connections both to other limbic areas and to neocortex, monitors overall internal state, and triggers emotion-related vocalization (Vogt and Barbas, 1986). While lesioning the cingulate abolishes conditioned vocalization, spontaneous calling and PAG-elicited sounds are not affected. In contrast, lesions in PAG abolish all vocalization (Jürgens, 2002). In other words, although cortically mediated learning is probably involved in conditioned vocalization, effects on call acoustics are nonetheless still mediated by emotion-related, subcortical regions. Taken together, then, both behavioral and neural evidence indicates that primate cortex cannot exert direct control over vocal output or trigger vocalizations independent of the caller's affective state.

Responding to Vocalizations

Flexibility in Perception, Cognition, and Behavior

While the sophisticated cognitive capabilities of monkeys and apes seem to take a back seat to relatively involuntary affective processes in vocal production, a very different picture emerges when these animals are responding to vocalizations. As in the case of vervets learning to execute particular escape strategies when hearing various alarm calls, primate responses to vocalizations are highly labile and subject to both short- and long-term modification. A listener can respond quite differently to a sound depending on the call-type involved, the context in which it hears the call, and its relationship to, as well as knowledge of, the vocalizer.

Flexibility in responding to sound has in fact been amply demonstrated in decades of studies of captive primates. For example, behavioral tests of auditory processing have been heavily reliant on the ability of monkeys and apes to learn to perform arbitrary motor actions in response to particular acoustic features of interest (Moody, 1995; Niemiec and Moody, 1995; Sinnott, 1995). Captive primates tested in learning tasks can also readily associate sounds with other stimuli or events, a capability that has been shown based on both affective and instrumental responses. The same conclusion can be drawn from field studies, where primatologists playing back species-typical calls and other auditory stimuli to wild monkeys and apes necessarily rely on that flexibility in conducting their experiments (Cheney and Seyfarth, 1990; Hauser, 1996; Rendall, 2004). Overall, results have been impressive in that regard. For example, researchers in laboratory and field have shown that listening animals can attend and respond to very subtle aspects of vocalizations, including those related to group membership, biological kinship, and dominance rank, as well as more individualized factors such as signaling reliability, recent calling behavior, or the emotional tone of recent interactions between vocalizer and listener.

Widespread Neural Involvement

The dramatic discrepancy in flexibility that primates show when producing versus responding to calls is consistent with differences in underlying neural connectivity. As noted, vocal production is primarily mediated by subcortical, limbic structures. In contrast, processing of sounds such as vocalizations includes both subcortical and cortical regions (Kaas et al., 1999). Lesioning and neurorecording studies have thus confirmed that the cortex is intimately involved when animals respond to any of a variety of natural or artificial sounds (Brosch and Scheich, 2003; Newman, 2003; Cheung et al., 2005). While vocal production is thus importantly "limbic" in nature, responding to vocalizations can be said to bring the entire primate brain into play.

Human Vocal Behavior

Based on the material reviewed so far, it seems difficult to make substantive connections between development of speech in human infants and vocal ontogeny in primates. Most importantly, learning and environment clearly play a critical role as the infant moves from prelinguistic sounds to language (e.g., Oller, 2000), while little auditory or motor learning is required for primates to produce acoustically appropriate calls. We suggest that vocal behavior in humans and primates may nonetheless be importantly related through the innate, nonlinguistic vocalizations and sophisticated auditory learning capabilities present in both. The critical difference becomes that significant expansion of direct corticobulbar pathways in humans brings these characteristics into play as scaffolds that facilitate development of volitional vocal control and the emergence of prelinguistic sound production.

Early Vocal Ontogeny

Early vocal behavior in infants has long been a topic of interest to those studying the ontogeny of speech and language—for instance including nonlinguistic crying and prelinguistic sounds that can arguably be linked to later canonical babbling (Oller, 2000). Recent studies by Scheiner et al. (2002, 2004) examined both nonlinguistic and prelinguistic vocalizations produced during the first year of life, in particular comparing development in normally hearing versus hearing-impaired infants. Scheiner and colleagues documented the occurrence of at least twelve sound types, including ten that were compared between the two groups. Nine of these are, in our view, nonlinguistic in nature, including *cry*, *short cry*, *coo/wail*, *moan*, *whoop/squeal*, *groan*, *croak*, *hic*, and *laugh*, whereas the last category comprised prelinguistic *babble* sounds.

Little difference was found in either the acoustics or time of emergence for any of the nonlinguistic vocal types, indicating that they are produced independently of auditory experience. These sounds included both noisy and harmonically structured vocalizations, but as in primates, the acoustic distinctions involved were primarily laryngeal rather than supralaryngeal in origin. Scheiner et al. were also able to sort the sounds by overall type based on whether the vocalizing infant was most likely in a positive versus negative emotional state. Harmonically structured coo/wail sounds could, for instance, be further broken down into coos, occurring in positive contexts, and wails, occurring in negative contexts. The former were quieter and more vowel-like, while the latter were higher pitched and less vowel-like, although remaining tonal in quality.

The development of babble sounds was markedly different, however. This category included a variety of more clearly prelinguistic vocalizations, and these sounds were either delayed or failed to appear in the repertoires of the hearing-impaired infants over the course of the study period. This difference between nonlinguistic and prelinguistic sounds was striking but also consistent with earlier work that has documented lengthy delays in the onset of canonical babbling in hearing-impaired infants (Oller and Eilers, 1988; Eilers and Oller, 1994; Koopmans-van Beinum et al., 2001).

Emergence of Canonical Babbling

Although canonical babbling has often been considered a starting point for speech development, Oller (1980, 1986, 2000) has presented a compelling argument for the occurrence of earlier prelinguistic stages. In his view, canonical babbling is importantly preceded by *quasi-resonant* vowels in a *simple phonation* stage (zero to two months), a *primitive articulation* stage associated with "gooing" and *marginal syllables* (one to four months), and an *expansion* stage (three to eight months) in which a variety of new sounds are produced. During the latter period, infants begin to position their vocal-tract articulators in new ways, produce sequences of articulated sounds, and exhibit slow but recognizable consonant–vowel transitions (*marginal babbling*).

While the role of learning in these early stages is not well documented, the importance of auditory input for subsequent vocal ontogeny is clear from studying hearing-impaired infants. Reviewing the available evidence, Eilers and Oller (1994) suggested that canonical babbling begins only after the infant has accumulated some threshold level of auditory experience. This argument was based on finding a rather remarkable correlation of +.69 between the ages at which impaired infants received hearing aids and the age of canonical babbling onset. While a corresponding regression value was not reported, fitting a regression line to the graphical data shown reveals an approximate slope of +.75 and a canonical babbling delay of at least six months. In contrast to nonlinguistic vocalization, then, emergence of canonical babbling requires a substantial amount of previous auditory experience (see also Lynch et al.,

1989, for evidence that the necessary threshold may also be reached through other sensory modalities). These findings thus belie any argument that canonical babbling should be considered primarily maturational or "genetic." Oller (2000) instead suggests that sounds occurring during all three prelinguistic stages are building blocks for the later speech-like, consonant–vowel transitions of canonical babbling.

Neural Mechanisms of Volitional Production

While spontaneous, nonlinguistic vocalizations in both humans and primates are associated with specific limbic and brain stem structures, human speech involves integrated activity across a variety of regions (Lieberman, 2002). It is therefore common to view linguistic versus nonlinguistic vocalization in humans as involving distinct neural structures (Myers, 1976; Ploog, 1988; Deacon, 1997), with cortical areas playing critical roles in both production and perception of speech (Jürgens, 2002; Lieberman, 2002). It is also likely not coincidental that cortex has a central role in speech and language, given the flexible and markedly volitional nature of this form of communication.

Not surprisingly, then, anatomical studies have revealed important differences between humans and primates in corticobulbar connectivity. As mentioned earlier, while primates do have some direct cortical connections to the brain stem neurons innervating peripheral vocal-production anatomy, it is not clear that these pathways are involved in vocalization, as opposed to other functions (Jürgens, 2002). Connections are less extensive than in humans, giving primates less opportunity for using the larynx, jaw, tongue, and face in an integrated, volitional fashion for the purpose of making sounds (Deacon, 1997). In addition, humans exhibit at least one corticobulbar pathway that is lacking in primates—namely, connections from motor cortex to the *nucleus ambiguus* of the medulla. This pathway in particular is therefore believed to be central to volitional vocal production (Deacon, 1997; Jürgens, 2002).

The Babbling-Scaffold Hypothesis

The observations and empirical evidence outlined to this point present an intriguing pattern of similarities and differences between vocal behavior in humans and that in primates. A human infant's earliest vocalizations, for instance, are clearly primate like, being innately grounded, nonlinguistic, and largely unarticulated. Conversely, primates resemble humans in showing pronounced auditory-learning capabilities and flexible responding to vocalizations they hear from others. Nonetheless, whereas humans exhibit routine volitional control over vocal production, primates do not. At the neural level, this difference likely has much to do with the more extensive corticobulbar connections present in humans.

We interpret these similarities and differences as suggesting that neither canonical nor even early babbling arises de novo during vocal ontogeny. Instead, we propose that the combination of innate, nonlinguistic vocalizations, typically primate auditory-learning capabilities, and uniquely human corticobulbar connections is key. While the first two are shared with other primates, the addition of direct corticobulbar pathways completes an otherwise open "circuit" between the production and reception sides of primate vocalization. Closing this loop creates the possibility of cortically controlled vocal communication, which is arguably realized through a self-organizing cascade of developmental events in which each stage becomes a foundation for the next level of communicative achievement. However, none of these events necessarily require that humans have novel neural substrates or new learning capabilities relative to primates.

The next sections outline an argument for two developmental stages that we propose bring the infant from nonlinguistic to prelinguistic vocal behavior. The first stage involves acquisition of volitional control of laryngeally based vocal production, with nonlinguistic vocalizations suggested to be playing an important role. The second stage concerns development of the rudimentary supralaryngeal control that is the hallmark of early prelinguistic behavior and that is a prerequisite for the subsequent emergence of more speech-like sounds in canonical babbling.

Nonlinguistic Vocalizations, Cortical Maps, and Laryngeal Control

Infants prototypically produce their earliest vocalizations shortly after birth in the form of crying. Other innately grounded, nonlinguistic sounds also emerge early, with Scheiner et al. (2004) reporting a total of six nonlinguistic sound types appearing within the first two months of life. As in primates, crying and other similarly innate vocalizations are likely limbically controlled signals triggered by particular needs or emotional responses. Caregivers are very sensitive to these nonvolitional sounds as illustrated by the responsiveness of both parents and nonparents alike to crying (Green and Gustafson, 1997; Wood and Gustafson, 2001).

However, infants also rather quickly begin to exhibit some control over their nonlinguistic production—for example, pausing during crying in an apparent effort to gauge caregiver reactions (Bell and Ainsworth, 1972). Over time, infants begin to use their sounds instrumentally as an attention-getting device and to coordinate vocalizations with gestures and other skeletal motor actions (Wolff, 1969; Papoušek and Papoušek, 1984; Gustafson and Green, 1991; Lester and Boukydis, 1992). In other words, vocal behavior in human infants shows subtle but unmistakable elements of volitional control from an early age, thereby diverging from its initial, more purely primate-like form. This discrepancy is not likely to be traceable to major, between-species differences in caregiver reactions, as primate mothers are also responsive to infant distress vocalizations (Patel and Owren, 2007). Nor is the difference likely to reflect that primates cannot learn contingent relationships between vocalizing and caregiver response. As discussed earlier, primates show sophisticated learning about the social significance of calls heard from others, and laboratory studies have confirmed that conditioning can lead to affectively mediated changes in vocal production.

Instead, the discrepancy appears to be rooted in the elaboration of corticobulbar connections in humans, which allows infants to gain more direct, cortical control over vocalizations than is possible for primates. It is certainly the case that both primates and human infants receive proprioceptive and auditory feedback to cortical regions when producing affectively triggered, nonlinguistic vocalizations. However, only the human cortex in turn has direct contact with premotor neurons innervating the peripheral vocal anatomy that is producing these sounds.

This is a crucial difference in our view, with the convergence of motor and sound feedback being precisely what is needed to bootstrap the cortical sensorimotor mapping that is critical to language. That mapping has been explored by Westermann (this volume; Westermann and Miranda, 2004) and others, whose models have demonstrated that separate but interconnected networks can produce highly coordinated, reciprocal mapping of perceptual and motor information. Kuhl and Meltzoff (1992) have made a similar point in the context of later speech production, arguing that a cortically mediated link between perceptual and motor experience is critical for mastering the sounds of a given language. They suggest that sensory experience with particular phonemes establishes stored auditory patterns that guide the infant's motor behavior as it works through successive approximations to the target sounds. Our proposal is that this kind of learning begins even earlier, with cortical maps first beginning to form based on the infant's own nonlinguistic production.

Laryngeal Control and Pitch Modulation

The suggestion also includes that this first stage of vocal development most importantly concerns laryngeally based production. Volitional control first appears particularly evident in the infant's use of pitch, which reveals growing mastery of both vocal-fold tension and air pressure. Although vocal pitch is important in primate vocalizations, the psychophysical evidence indicates that humans are especially sensitive to this acoustic dimension (Owren, 2003). The functional importance and salience of pitch variation to human infants has been amply demonstrated—for instance, through their attentiveness to and preference for highly pitch-modulated "infant-directed speech" (Fernald, 1992). Caregivers are also very attentive to pitch in the infant's vocalizations, consistently using pitch patterning as a basis for interpreting their emotional significance for infants as young as two months of age

(Papoušek, 1989). Even for nonlinguistic sounds, adult listeners interpret vocalizations that end with rising pitch as expressing requesting or wanting (D'Odorico, 1984; Flax et al., 1991; Furrow, 1984; Masataka, 1993).

Analyzing the acoustic variation and social significance involved in pitch-modulated vocalization is furthermore heavily reliant on cortical processing—both in humans and primates alike. Once again, however, it is only in humans that the cortex in turn has extensive, direct contact with the brain stem neurons used to control vocal output. In the infant, these connections can create reciprocal contact between its nascent cortically based sensorimotor maps, social learning, and peripheral vocal-production anatomy. Early emergence of volitional control of the larynx in particular figures into this argument because pitch is laryngeally based in vocal production. It primarily reflects the vocal-fold vibration rate, which is in turn a function of laryngeal muscle tensions and air pressure from the lungs.

Pitch also likely represents the avenue of greatest opportunity for infants that are beginning to gain some control over laryngeal production. Due to the perceptual prominence of pitch modulation, any control the infant can gain over pitch acoustics will be very salient both to itself and to caregivers. As discussed earlier, nonlinguistic vocalizations such as crying are also very salient, likely to both parties. However, many are grounded in less controllable, or even chaotic, forms of vocal-fold vibration (Owren, 2003; Robb, 2003). Both for very rudimentary control that is first exerted over limbically mediated vocalizations and for more fully volitional sounds that the cortex is coming to directly control, the infant can be expected to work on readily achievable modulations that are nonetheless the most salient to itself and to others. We suspect that vocal pitch meets both these criteria.

The final point to make here is that the success the infant begins to enjoy in gaining control over its sounds is likely to be highly reinforcing. In other words, any initial achievement in volitional control can only foster greater motivation to explore the available "vocal space." As noted in the next section, such exploration is likely encouraged both by the self-stimulation afforded by sound production and through reciprocal effects on and from the environment. Here again, initial efforts are expected to focus on laryngeally produced modulations, with pitch changes in particular being a prime candidate for early, purposeful exploration. Indeed, by two to three months of age, infants are routinely producing sounds with highly variable and dramatically exaggerated pitch changes, such as squeals (Oller and Griebel, this volume).

From Nonlinguistic to Prelinguistic: Emerging Articulation

While the distinction drawn between nonlinguistic and prelinguistic vocalizations is conceptually useful, we are also proposing that the line is soon blurred by the infant's earliest, rudimentary attempts at volitional vocal control. Achieving even a modest degree of control over nonlinguistic sounds such as crying and squealing means that these vocalizations could also be considered prelinguistic in a sense. However, the distinction remains useful in that nonlinguistic vocalizations emerge at a very early age, including in hearing-impaired infants for whom the absence of auditory input means that fully volitional control and canonical babbling may never develop. Furthermore, it is clear from vocalizations such as adult laughter and crying that cortical and limbic vocal mechanisms never become fully, or perhaps even substantially, integrated. Human adults can inhibit or simulate these kinds of nonlinguistic sounds to some degree, but volitional suppression nonetheless fails in the face of strong emotion. Further, most people are likely unable to produce accurate and convincing nonspontaneous versions of either laughter or crying.

Instead, it is volitional production of vowel-like sounds that prelinguistic infants first master, and that can be considered the primary vehicle of the nonlinguistic-to-prelinguistic transition we are proposing. In their nonlinguistic instantiation, Scheiner et al. (2002) and others refer to these sounds as coos, while for Oller (2000) the quasi-vowels in this early stage exemplify simple phonation. Oller further notes that these sounds are subject to an early, primitive form of articulation, and in this version might better be called goos. With increasing laryngeal control, we suggest that the initially emotion-triggered and limbically controlled vowel-like coos become the substrate of the next important steps in infant vocal development, namely the emergence of rudimentary articulation. This is another significant step, as supralaryngeal modulation has been found to play a prominent role in the more sophisticated social interactions that somewhat older infants enjoy with their caregivers.

The physiology of the vocal tract above the larynx is in fact changing significantly after the first 2 to 3 months of life. Prior to that point, the larynx is positioned high in the throat and overlaps the nasopharynx, which predisposes the infant to breathe nasally (Kent and Vorperian, 2006). In addition, the small size of the oral cavity in very young infants restricts the range of possible tongue movement. By about three months of age, laryngeal descent has opened up the supralaryngeal vocal tract, reduced the previously nasalized character of vowel-like sounds, and significantly increased the diversity of acoustic effects achievable by repositioning the tongue within the oral and pharyngeal cavities.

We suggest that social learning is also playing a critical role at this early babbling stage. Infant vocal behavior is, for example, known to be responsive to contingent, trial-and-error learning long before the onset of traditional cognitive milestones of communication (Locke, 2001). Even simple caregiver responses such as touching and shaking a rattle have been shown to be effective reinforcers of infant vocalizations that lead to increased rates of production (Rheingold et al., 1959; Weisberg, 1963; Routh, 1969; Poulson, 1983).

Associative learning is thus proposed to be at work even in very young infants, likely by first affecting incidental articulation effects occurring in nonlinguistic vocal production. In one prominent example, Goldman (2001) found that caregivers report hearing infants as young as one month old produce their first "mama"-like sound in the context of crying. Emergence of "mama" sounds (or something similar) was found to peak at two to three months of age in this study, with the critical observation being that infants are not producing a fully formed word at this point. Rather, caregivers are interpreting incidental labial contact as an articulatory gesture, typically attributing communicative intent related to wanting or requesting.

While caregivers continue to be highly responsive to all manner of infant vocalizations at this age, coos and "a" sounds figure prominently among the events that elicit increasing proportions of verbal responses over the first three to four months (Keller and Schölmerich, 1987). In addition, caregivers have been found to be responsive specifically to vocal acoustics associated with speech-like supralaryngeal effects. Infants who produce more fully resonant sounds, for example, are rated as being more attractive or appealing by adults (Papoušek, 1989; Bloom and Lo, 1990; Beaumont and Bloom, 1993; Bloom et al., 1993; see also Goldstein et al., 1998).

Young infants are in turn very sensitive to sounds and responses from caregivers. For instance, experimental work has shown that three- to six-month-old infants readily and successfully imitate the absolute pitch of vocalizations (Kessen et al., 1979) and produce more fully resonant sounds in vocal turn taking with caregivers when these adults are responding contingently to this feature (Bloom et al., 1987; see also Goldstein et al., 2003). When exposed to a combination of faces and voices with either matched or mismatched articulation, preverbal infants have also been shown to preferentially imitate vowels associated with the congruent pairings (Legerstee, 1990; see also Kuhl and Meltzoff, 1996).

Successful learning continues to keep the infant engaged with its environment throughout the early babbling stage, as demonstrated in experiments involving naturalistic play sessions. In these studies, caregivers are found to be more responsive to speech-like vocalizations that include fully resonant vowels and consonant–vowel-like transitions than they are to sounds that lack these features (Keller and Schölmerich, 1987; Hsu and Fogel, 2003; Gros-Louis et al., 2006). When engaged in face-to-face interactions, mothers and infants take turns vocalizing (Anderson et al., 1977; Papoušek et al., 1985), with both parties playing active roles in coordinating this joint behavior (Jaffe et al., 2001). As part of that process, mothers playing with their two- to five-month-old infants often match the infants' vocalizations and do so accurately (Papoušek and Papoušek, 1989; Papoušek, 1991). However, outcomes are very different when caregivers do not routinely produce speech or show well-coordinated responses to infant vocalizations. Hearing infants of nonspeaking deaf parents, for example, produce vocalizations that are acoustically disorganized and are delayed in using spoken language relative to infants with hearing parents (Petitto et al., 2004).

The Origins of Canonical Babbling

Taken together, the evidence indicates that infants have a rich history of speech-relevant experience well before they begin to produce speech-like, prelinguistic sounds. As noted earlier, however, work on early language development often takes canonical babbling as the starting point (e.g., MacNeilage, 1998). Further, some approaches characterize canonical babbling as being preprogrammed, preordained, or a maturational outcome. Deacon (1997), for example, points to myelination of cortical neurons as the critical factor in the emergence of canonical babbling. While these sorts of maturational events are likely both necessary and important, we suggest that the evident impact of auditory input and social engagement on vocalization indicates that they cannot in and of themselves be sufficient explanations for the emergence of canonical babbling.

Reflections on the Hypothesis

The babbling-scaffold hypothesis differs most fundamentally from traditional views of infant vocal development in imputing a central role for learning virtually from the moment of birth. On the production side, the infant is argued to first acquire volitional control of the larynx, while also more gradually gaining proficiency in supralaryngeal articulation. Self-stimulation through vocalization is proposed to be important throughout this development, beginning with cortical mapping initially produced by nonlinguistic vocalization but then continuing as early babbling gives rise to canonical babbling and then full-blown speech. Finally, social learning is seen as being critical at every stage of the process. In this view, even the earliest vocal behavior is importantly subject to contingency learning, with nonlinguistic and prelinguistic vocal behavior both promoting and being influenced by social interaction with caregivers. Although discussion is necessarily brief, we now examine some of the assumptions and implications of our approach in the larger context of development, the effects of hearing impairment, and vocal ontogeny in humans and primates.

Self-Stimulation in Development

One premise of the babbling-scaffold hypothesis is that the infant's own behavior plays an important role throughout its vocal development, specifically including a facilitating effect of innately grounded, nonlinguistic sounds on cortical mapping and volitional control of the larynx. Although this argument may seem unusual, it is in fact common for behavior or activity generated by one part of the immature brain to be critical for some other behavior or brain area. In comparative psychology, such occurrences are well-known and have played a central role in nature–nurture debates (Ho, 1998).

In a classic study of chick development, for instance, Kuo (1966) showed that the embryonic heartbeat plays an important role in stimulating and entraining raising and lowering of the head, opening and closing of the beak, and later swallowing of amniotic fluid. While still in the egg, in other words, activity of the chick's heart stimulates development of coordinated movements in quite different systems. At the purely neural level, many instances of spontaneous correlated activity in one part of the nervous system providing critical experience to neural circuits in other regions in advance of external stimulation have recently been uncovered (Feller, 1999; Wong, 1999).

There are fewer demonstrations of this kind specifically in the vocal domain, but some telling examples are available. For instance, work by Gottlieb (1963) has demonstrated that a newly hatched but socially isolated wood duckling can recognize conspecific vocalizations expressly because it has heard its own calls while in the egg. The influence that vocal self-stimulation can have across brain systems has been shown even more dramatically by Cheng (1992). In this work, the "nest" coos of female ring doves were found to stimulate hormone release in the vocalizer herself, thereby playing a functional role in advancing the courtship and mating process with a male. While the brain systems are different in the two cases, this last example in particular illustrates the kind of self-stimulation we are suggesting to be important in the early stages of human vocal development.

Hearing Impairment and Canalization

In proposing that experience plays a critical role throughout infant vocal development, the babbling-scaffold hypothesis must also account for the seeming imperturbability of this process up to and including the onset of canonical babbling. Here, the important observation is that canonical babbling emerges predictably in the face of a variety of possible risk factors, including deprivation due to socioeconomic circumstances, hearing multiple languages in early infancy, and mental retardation (Oller, 2000). Although the reliable emergence of canonical babbling in spite of such handicaps has been taken as evidence of biological depth and canalization, it need not be. Finding that a trait emerges in both normative and nonnormative circumstances does not in and of itself imply that is it guided by biology, as it could be that critical experiential factors remain operative in each instance.

In the case of vocal ontogeny, we suggest that none of the important factors—the presence of a primate-like, nonlinguistic repertoire, development of nascent cortical mapping through proprioceptive and auditory feedback, and contingent caregiver responses to vocalization—are likely to be importantly affected by socioeconomic circumstances or simultaneous exposure to multiple languages. The learning processes proposed to be at work during the nonlinguistic and early prelinguistic stages are also expected to be robust to many mental retardation effects.

However, attaining full-fledged canonical babbling and later speech is a long and challenging process. The infant must go far beyond the rudimentary volitional control of laryngeal and supralaryngeal production that we have focused on, including integration of auditory, motor, and social input that is both qualitatively and quantitatively much more complex. While we thus expect mentally retarded children to routinely acquire volitional vocal control and rudimentary speech, mature language requires that they both be able to process speech effectively and be sensitive to a variety of kinds of feedback to their sounds. If either aspect is impaired, language development is likely to be requisitely incomplete.

The babbling-scaffold approach is arguably also consistent with finding a strong correlation between the age at which hearing-impaired infants receive effective hearing aids and observed delays in the onset of canonical babbling. From this perspective, significant hearing impairment must have immediate and detrimental effects on vocal development—for instance, in hindering both self-stimulation by the infant's nonlinguistic vocalizations and early learning about caregiver responses. Consistent with Eilers and Oller's (1994) report, the hypothesis therefore predicts that even hearing impairment that is corrected very early will delay vocal development at later stages, particularly the acquisition of volitional control required for canonical babbling. While this prediction may seem obvious, it does not follow from the alternative perspective of viewing nonlinguistic vocalization, early babbling, and canonical babbling as being separable developmental events. The same rationale applies to finding that the amount of delay in canonical babbling is proportional to the age at which a hearing-impaired infant receives aid, although that expectation is likely compatible with a variety of theories of babbling. A more specific prediction is that at the time of receiving a hearing aid, the infant's progress in volitional control of laryngeal production, instrumental use of nonlinguistic vocalization, and transition from nonlinguistic to prelinguistic sounds should be correlated with the amount of onset delay observed later for canonical babbling.

Vocal Development in Humans and Primates

A central theme of the chapter has been that vocal development and vocal behavior are fundamentally different in humans and primates, yet also importantly similar. For primates, we have emphasized the separation between vocal production and reception and traced it to a lack of direct cortical connections to brain stem neurons innervating peripheral vocal anatomy. Consistent with this observation, projects attempting to teach apes to produce spoken language have met with little success despite heroic training efforts (Hayes, 1951; Kellogg and Kellogg, 1933). In contrast, bonobos, chimpanzees, orangutans, and gorillas have all been able to acquire significant language-related skills when the medium has involved manually controlled actions such as gesturing or selecting among visual symbols. The key difference is most likely that these animals have full volitional control over these kinds of motor actions, but not over vocal production.

It follows that the great divide between vocal communication in humans and primates does not merely reflect differences in relative cognitive abilities. In fact, primates have been found to best demonstrate language-relevant skill when the tasks involve the reception rather than the production side of the equation. Monkeys, for instance, have been found to be sensitive to grammar-like statistical regularities in auditory input (Hauser et al., 2001; Fitch and Hauser, 2004), and apes exposed to spoken language have acquired significant comprehensive vocabularies, even in the absence of explicit training (Brakke and Savage-Rumbaugh, 1995; Lyn and Savage-Rumbaugh, 2000; Williams et al., 1997).

We interpret such evidence to suggest that at least some primate species are probably capable of forming the sensorimotor maps, semantic representations, and illocutionary motivations involved in human language. In fact, linked cortical mapping of proprioceptive and auditory information from hearing their own calls may be routine in primates, at least at a rudimentary level. If so, however, neither that mapping nor their demonstrated processing capabilities can take them into the realm of speech-like communication in the absence of more direct cortical control of vocal production. We have argued that corticobulbar connections are critical at every stage of vocal-production learning in human infants, and that without them, primates have no possibility of even beginning such a process. Instead, animals whose communicative production is mediated solely or primarily through affective and limbic mechanisms will forever remain inflexible and context bound. While we do not claim that the simple appearance of expanded corticobulbar connections in a primate would inexorably lead to language-like vocal communication, we do argue that it would be a critical step. In humans, we see these pathways as the linchpin of a developmental process that may not have required any other special abilities or evolutionary adaptations. Rather, we suggest that the developmental process contributes much of the uniqueness of spoken language, proceeding as a cascade in which each new capability and its associated behaviors opens the door to additional and qualitatively more sophisticated communicative skills and interactions.

Conclusions: Development of Contextual Flexibility in Vocal Communication

The ubiquity of spoken language in humans can sometimes lessen the appreciation of its sophistication and complexity, and the extended period of development that is involved in learning to use it. Thus, while the rapid pace of speech ontogeny can suggest innateness, the learning required for becoming a fully competent speaker of a language extends well into middle childhood or beyond (Ferguson et al., 1992; Kuhl, 2007). In this chapter, we have argued for a critical role of learning from the very earliest stages of prelinguistic production as well, by linking the emergence of volitionally controlled prelinguistic sounds to the innate and primate-like nonlinguistic vocalizations that precede them. While the connections are indirect, we are thereby arguing that continuity does exist between vocal communication in primates and spoken language in humans. However, the two are nonetheless qualitatively discontinuous, with the expanded corticobulbar connections present in humans opening the door for speech-related sensorimotor and instrumental learning processes that primates might be capable of, but have no opportunity to realize.

The view that human vocal development is a process that begins virtually at birth has guided proposals about self-stimulation through nonlinguistic vocalization, as well as gradual acquisition of volitional control over first laryngeal and then supralaryngeal production. Some predictions follow, including that even very early correction of hearing impairment should noticeably speed vocal development and that the child's progress in instrumental and volitional control of vocalization should be correlated with the amount of delay observed in canonical babbling. Another expectation that flows from the proposed connection between nonlinguistic and prelinguistic vocalizations in early infancy is that greater variety in the earliest nonlinguistic vocalizations in a hearing infant should produce earlier emergence of volitional vocal abilities as a result of forming better sensorimotor cortical mapping. Finally, although hearing-impaired infants produce the same kinds of nonlinguistic vocalizations as their normally hearing peers, we expect that more detailed comparisons will reveal that they are nonetheless being hindered in the instrumental use of sounds and in the transition from nonlinguistic to volitional control of vocal production. If so, these are impediments occurring long before the well-documented canonical-babbling delays occurring in impaired infants and are likely to be contributing to those effects.

Contextual Flexibility

In its simplest form, the babbling-scaffold hypothesis argues that the unique flexibility of spoken language in humans may ultimately be traceable to combining an evolutionary innovation (increased corticobulbar connections) with an evolutionary legacy (innate vocal production but more sophisticated auditory learning). We have throughout put cortical control over vocal production at the center, more or less equating it with volitional control. That characterization is, of course, oversimplified, particularly in light of evidence that many brain areas are involved in language-related processing, both in production and in comprehension. We can nonetheless assert that the absence of direct cortical control over peripheral vocal anatomy that characterizes many mammals other than humans powerfully restricts the flexibility that can be achieved in their vocal signaling. There is perhaps some irony in noting that these are asymmetrical constraints, with the potential for flexibility and sophistication in signal reception and processing far outstripping possible vocal expressiveness. Bringing the cortex into play in vocal production may thus have been quite a small evolutionary change occurring in some early human ancestor, while the symmetry that was thereby created between producing and responding to vocal signals was likely a key factor in unlocking the vast potential of contextually flexible communication.

Acknowledgments

This work was supported in part by National Institute of Mental Health Prime Award 1 R01 MH65317-01A2, Subaward 8402-15235-X, and in part by the Center for Behavioral Neuroscience, STC Program of the National Science Foundation under Agreement No. IBN-9876754. Many thanks to Kurt Hammerschmidt, Chuck Snowdon, and Kim Oller for their helpful comments on earlier versions of this chapter.

References

Anderson BJ, Vietze P, Dokecki PR. 1977. Reciprocity in vocal interactions of mothers and infants. *Child Development* 48: 1676–81.

Bachorowski J-A, Owren MJ. In press. Vocal expressions of emotion. In *The Handbook of Emotion*, 3rd ed., ed. M Lewis, JM Haviland-Jones, LF Barrett. New York: Guilford Press.

Beaumont SL, Bloom K. 1993. Adults' attributions of intentionality to vocalizing infants. *First Language* 13: 235–47.

Bell SM, Ainsworth MDS. 1972. Infant crying and maternal responsiveness. *Child Development* 43: 1171–90.

Bloom K, D'Odorico L, Beaumont S. 1993. Adult preferences for syllabic vocalizations: Generalizations to parity and native language. *Infant Behavior and Development* 16: 109–20.

Bloom K, Lo E. 1990. Adult perceptions of vocalizing infants. *Infant Behavior and Development* 13: 209–19.

Bloom K, Russell A, Wassenberg K. 1987. Turn taking affects the quality of infant vocalizations. *Journal of Child Language* 14: 211–27.

Brakke K, Savage-Rumbaugh ES. 1995. The development of language skills in bonobo and chimpanzee—I. Comprehension. *Language and Communication* 15: 121–48.

Brosch M, Scheich H. 2003. Neural representation of sound patterns in the auditory cortex of monkeys. In *Primate Audition: Ethology and Neurobiology*, ed. A Ghazanfar, pp. 151–75. Boca Raton, FL: CRC Press.

Brown CH, Alipour F, Berry DA, Montequin D. 2003. Laryngeal biomechanics and vocal communication in the squirrel monkey (*Saimiri boliviensis*). *Journal of the Acoustical Society of America* 113: 2114–26.

Cheney DL, Seyfarth RM. 1990. *How Monkeys See the World*. Chicago: University of Chicago Press.

Cheng M-F. 1992. For whom does the female dove coo? A case for the role of self-stimulation. *Animal Behaviour* 43: 1035–44.

Cheung SW, Nagarajan SS, Schreiner CE, Bedenbaugh PH, Wong A. 2005. Plasticity in primary auditory cortex of monkeys with altered vocal production. *Journal of Neuroscience* 25: 2490–503.

Deacon TW. 1997. *The Symbolic Species*. New York: Norton.

Dennett D. 1983. Intentional systems in cognitive ethology: The "Panglossian" paradigm defended. *Behavioral and Brain Sciences* 6: 343–55.

D'Odorico L. 1984. Non-segmental features in prelinguistic communications: An analysis of some types of infant cry and non-cry vocalizations. *Journal of Child Language* 11: 17–27.

Doupe AJ, Kuhl PK. 1999. Birdsong and human speech: Common themes and mechanisms. *Annual Review of Neuroscience* 22: 567–631.

Eilers RE, Oller DK. 1994. Infant vocalizations and the early diagnosis of severe hearing impairment. *Journal of Pediatrics* 124: 199–203.

Elowson AM, Snowdon CT, Lazaro-Perea C. 1998. "Babbling" and social context in infant monkeys: Parallel to human infants. *Trends in Cognitive Sciences* 2: 35–43.

Feller MB. 1999. Spontaneous correlated activity in developing neural circuits. *Neuron* 22: 653–6.

Fernald A. 1992. Maternal vocalizations to infants as biologically relevant signals: An evolutionary perspective. In *The Adapted Mind: Evolutionary Psychology and the Generation of Culture*, ed. H Barkow, L Cosmides, J Tooby, pp. 391–428. Oxford: Oxford University Press.

Ferguson CA, Menn L, Stoel-Gammon C, eds. 1992. *Phonological Development: Models, Research, Implications.* Timonium, MD: York Press.

Fitch WT, Hauser MD. 2004. Computational constraints on syntactic processing in a nonhuman primate. *Science* 303: 377–80.

Flax J, Lahey M, Harris K, Boothroyd A. 1991. Relations between prosodic variables and communicative functions. *Journal of Child Language* 18: 3–20.

Fry DB. 1966. The development of the phonological system in the normal and deaf child. In *The Genesis of Language*, ed. F Smith, GA Miller, pp. 187–206. Cambridge, MA: MIT Press.

Furrow D. 1984. Young children's use of prosody. *Journal of Child Language* 11: 201–13.

Goldman HI. 2001. Parental reports of 'MAMA' sounds in infants: An exploratory study. *Journal of Child Language* 28: 497–506.

Goldstein MH, King AP, West MJ. 1998. Consistent responses of human mothers to prelinguistic infants: The effect of prelinguistic repertoire size. *Journal of Comparative Psychology* 113: 52–8.

Goldstein MH, King AP, West MJ. 2003. Social interaction shapes babbling: Testing parallels between birdsong and speech. *Proceedings of the National Academy of Sciences* 100: 8030–5.

Gottlieb G. 1963. A naturalistic study of imprinting in wood ducklings (*Aix sponsa*). *Journal of Comparative and Physiological Psychology* 56: 86–91.

Green JA, Gustafson GE. 1997. Perspectives on an ecological approach to social communicative development in infancy. In *Evolving Explanations of Development: Ecological Explanations to Organism–Environment Systems*, ed. C Dent-Read, P Zukow-Goldring, pp. 515–46. Arlington, VA: American Psychological Association.

Green S. 1975. Variation of vocal pattern with social situation in the Japanese macaque (*Macaca fuscata*): A field study. In *Primate Behavior, Vol. 4*, ed. LA Rosenblum, pp. 1–102. New York: Academic Press.

Gros-Louis J, West MJ, Goldstein MH, King AP. 2006. Mothers provide differential feedback to infants' prelinguistic sounds. *International Journal of Behavioral Development* 30: 509–16.

Gustafson GE, Green JA (1991). Developmental coordination of cry sounds with visual regard, and gestures. *Infant Behavior and Development* 14: 51–7.

Hammerschmidt K, Freudenstein T, Jürgens U. 2001. Vocal development in squirrel monkeys. *Behaviour* 138: 1179–204.

Hauser MD. 1996. *The Evolution of Communication*. Cambridge MA: MIT Press.

Hauser MD, Evans CS, Marler P. 1993. The role of articulation in the production of rhesus monkey (*Macaca mulatta*) vocalizations. *Animal Behaviour* 45: 423–33.

Hauser MD, Marler P. 1993. Food-associated calls in rhesus macaques (*Macaca mulatta*): I. Socioecological factors. *Behavioral Ecology* 4: 194–205.

Hauser MD, Newport EL, Aslin RN. 2001. Segmentation of the speech stream in a nonhuman primate: Statistical learning in cotton-top tamarins. *Cognition* 78: B53–B64.

Hayes C. 1951. *The Ape in Our House*. New York: Harper.

Ho M-W. 1998. Evolution. In *Comparative Psychology: A Handbook*, ed. G Greenberg, MM Haraway, pp. 107–19. New York: Garland.

Hsu H, Fogel A. 2003. Stability and transitions in mother–infant face-to-face communication during the first 6 months: A microhistorical approach. *Developmental Psychology* 39: 1060–82.

Jaffe J, Beebe B, Feldstein S, Crown CL, Jasnow MD. 2001. Rhythms of dialogue in infancy. *Monographs of the Society for Research in Child Development* 66: Serial No. 265.

Jürgens U. 2002. Neural pathways underlying vocal control. *Neuroscience and Biobehavioral Review* 26: 235–8.

Kaas JH, Hackett TA, Tramo MJ. 1999. Auditory processing in primate cerebral cortex. *Current Opinion in Neurobiology* 9: 164–70.

Keller H, Schölmerich A. 1987. Infant vocalizations and parental reactions during the first 4 months of life. *Developmental Psychology* 23: 62–7.

Kellogg WN, Kellogg LA. 1933. *The Ape and the Child: A Study of Environmental Influence upon Early Behavior*. New York: McGraw-Hill.

Kent RD, Vorperian HK. 2006. In the mouths of babes: Anatomic, motor, and sensory foundations of speech development in children. In *Language Disorders from a Developmental Perspective*, ed. R Paul, pp. 55–82. Mahwah, NJ: Lawrence Erlbaum Associates.

Kessen W, Levine J, Wendrich KA. 1979. The imitation of pitch in infants. *Infant Behavior and Development* 2: 93–9.

Knörnschild M, Behr O, von Helversen O. 2006. Babbling behavior in the sac-winged bat (*Saccopteryx bilineata*). *Naturwissenschaften* 93: 451–4.

Koopmans-van Beinum FJ, Clement CJ, van den Dikkenberg-Pot, I. 2001. Babbling and the lack of auditory speech perception: A matter of coordination? *Developmental Science* 4: 61–70.

Kuhl PK. 2007. Is speech 'gated' by the social brain? *Developmental Science* 10: 110–20.

Kuhl PK, Meltzoff AN. 1992. The bimodal perception of speech in infancy. *Science* 218: 1138–41.

Kuhl PK, Meltzoff AN. 1996. Infant vocalizations in response to speech: Vocal imitation and developmental change. *Journal of the Acoustical Society of America* 100: 2425–38.

Kuo ZY. 1966. *The Dynamics of Behavior Development: An Epigenetic View*. New York: Random House.

Legerstee M. 1990. Infants use multimodal information to imitate speech sounds. *Infant Behavior and Development* 13: 343–54.

Lester BM, Boukydis CFZ. 1992. Infantile colic: Acoustic cry characteristics, maternal perception of cry and temperament. *Infant Behavior and Development* 15: 15–26.

Lieberman P. 1975. *On the Origins of Language*. New York: Macmillan.

Lieberman P. 2002. The neural bases of language. *Yearbook of Physical Anthropology* 45: 36–62.

Locke J. 2001. First communion: The emergence of vocal relationships. *Social Development* 10: 294–308.

Lyn H, Savage-Rumbaugh ES. 2000. Observational word learning in two bonobos (*Pan paniscus*): Ostensive and non-ostensive contexts. *Language and Communication* 20: 255–73.

Lynch MP, Oller DK, Steffens M. 1989. Development of speech-like vocalizations in a child with congenital absence of cochleas: The case of total deafness. *Applied Psycholinguistics* 10: 315–33.

MacNeilage PF. 1998. The frame/content theory of evolution of speech production. *Behavioral and Brain Sciences* 21: 499–546.

Masataka N. 1992. Attempts by animal caretakers to condition Japanese macaque vocalizations result inadvertently in individual-specific calls. In *Topics in Primatology, Vol. 1: Human Origins*, ed. T Nishida, WC McGrew, P Marler, M Pickford, FBM de Waal, pp. 271–8. Tokyo: University of Tokyo.

Masataka N. 1993. Relation between pitch contour of prelinguistic vocalizations and communicative functions in Japanese infants. *Infant Behavior and Development* 16: 397–401.

Moody DB. 1995. Classification and categorization procedures. In *BioMethods, Vol. 6: Methods in Comparative Acoustics*, ed. GM Klump, RJ Dooling, RR Fay, WC Stebbins, pp. 293–305. Basel, Switzerland: Birkhauser Verlag.

Myers RE. 1976. Comparative neurology of vocalization and speech: Proof of a dichotomy. *Annals of the New York Academy of Sciences* 280: 745–57.

Newman JD. 2003. Auditory communication and cortical auditory mechanisms in the squirrel monkey: Past and present. In *Primate Audition: Ethology and Neurobiology*, ed. A Ghazanfar, pp. 227–46. Boca Raton, FL: CRC Press.

Niemiec AJ, Moody DB. 1995. Constant stimulus and tracking procedures for measuring sensitivity. In *BioMethods, Vol. 6: Methods in Comparative Acoustics*, ed. GM Klump, RJ Dooling, RR Fay, WC Stebbins, pp. 65–78. Basel, Switzerland: Birkhauser Verlag.

Oller DK. 1980. The emergence of speech sounds in infancy. In *Child Phonology: Vol. 1. Production*, ed. G Yeni-Komshian, J Kavanagh, C Ferguson, pp. 93–112. New York: Academic Press.

Oller DK. 1986. Metaphonology and infant vocalizations. In *Precursors of Early Speech*, ed. B Lindblom, Z Zetterstrom, pp. 21–35. New York: Stockton Press.

Oller DK. 2000. *The Emergence of the Speech Capacity.* Mahwah, NJ: Lawrence Erlbaum Associates.

Oller DK, Eilers RE. 1988. The role of audition in infant babbling. *Child Development* 59: 441–9.

Owren MJ. 2003. Vocal production and perception in nonhuman primates provide clues about early hominids and speech evolution. *ATR Symposium HIS Series* 1: 1–19.

Owren MJ, Dieter JA, Seyfarth RM, Cheney DL. 1992a. Evidence of limited modification in the vocalizations of cross-fostered rhesus (*Macaca mulatta*) and Japanese (*M. fuscata*) macaques. In *Topics in Primatology, Vol. 1: Human Origins*, ed. T Nishida, WC McGrew, P Marler, M Pickford, F de Waal, pp. 257–70. Tokyo: University of Tokyo.

Owren MJ, Dieter JA, Seyfarth RM, Cheney DL. 1992b. 'Food' calls produced by adult female rhesus (*Macaca mulatta*) and Japanese (*M. fuscata*) macaques, their normally-raised offspring, and offspring cross-fostered between species. *Behaviour* 120: 218–31.

Owren MJ, Dieter JA, Seyfarth RM, Cheney DL. 1993. Vocalizations of rhesus (*Macaca mulatta*) and Japanese (*M. fuscata*) macaques cross-fostered between species show evidence of only limited modification. *Developmental Psychobiology* 26: 389–406.

Owren MJ, Rendall D. 1997. An affect-conditioning model of nonhuman primate vocal signaling. In *Perspectives in Ethology, Vol. 12: Communication*, ed. DH Owings, MD Beecher, NS Thompson, pp. 299–346. New York: Plenum Press.

Papoušek H, Papoušek M. 1984. Qualitative transitions during the first trimester of human postpartum life. In *Continuity of Neural Function Prenatal to Postnatal Life*, ed. HFR Prechtl, pp. 220–44. London: Spastics International Medica.

Papoušek M. 1989. Determinants of responsiveness to infant vocal expression of emotional state. *Infant Behavior and Development* 12: 507–24.

Papoušek M. 1991. Early ontogeny of vocal communication in parent–infant interactions. In *Nonverbal Vocal Communication: Comparative and Developmental Approaches*, ed. H Papoušek, U Jürgens, M Papoušek, pp. 230–61. Cambridge, England: Cambridge University Press.

Papoušek M, Papoušek H. 1989. Forms and functions of vocal matching in interactions between mothers and their precanonical infants. *First Language* 9: 137–58.

Papoušek M, Papoušek H, Bornstein MH. 1985. The naturalistic vocal environment of young infants. In *Social Perception in Infants*, ed. TM Field, N Fox, pp. 269–98. Norwood, NJ: Ablex.

Patel ER, Owren MJ. 2007. Acoustic and behavioral analyses of gecker distress vocalizations in young rhesus macaques (*Macaca mulatta*). *Journal of the Acoustical Society of America* 121: 575–85.

Petitto LA, Holowka S, Sergio LE, Levy B, Ostry DJ. 2004. Baby hands that move to the rhythm of language: Hearing babies acquiring sign language babble silently on the hands. *Cognition* 93: 43–73.

Pierce JD. 1985. A review of attempts to condition operantly alloprimate vocalizations. *Primates* 26: 202–13.

Ploog D. 1988. Neurobiology and pathology of subhuman vocal communication and human speech. In *Primate Vocal Communication*, ed. D Todt, P Goedeking, D Symmes, pp. 195–212. Berlin: Springer-Verlag.

Poulson CL. 1983. Differential reinforcement of other-than-vocalization as a control procedure in the conditioning of infant vocalization rate. *Journal of Experimental Child Psychology* 36: 471–89.

Pulvermüller F. 1999. Words in the brain's language. *Behavioral and Brain Sciences* 22: 253–336.

Pulvermüller F, Preissl H. 1991. A cell assembly model of language. *Computation in Neural Systems* 2: 455–68.

Rendall D. 2003. Acoustic correlates of caller identity and affect intensity in the vowel-like grunt vocalizations of baboons. *Journal of the Acoustical Society of America* 113: 3390–402.

Rendall D. 2004. 'Recognizing' kin: Mechanisms, media, minds, modules and muddles. In *Kinship and Behaviour in Primates*, ed. B Chapais, C Berman, pp. 295–316. Oxford: Oxford University Press.

Rheingold HL, Gerwitz JL, Ross HW. 1959. Social conditioning of vocalizations in the infant. *Journal of Comparative and Physiological Psychology* 52: 68–73.

Riede TR, Zuberbühler K. 2003. Pulse register phonation in Diana monkeys alarm calls. *Journal of the Acoustical Society of America* 113: 2919–26.

Robb MP. 2003. Bifurcations and chaos in the cries of full-term and preterm infants. *Folia Phoniatrica et Logopaedica* 55: 233–40.

Routh DK. 1969. Conditioning of vocal response differentiation in infants. *Developmental Psychology* 1: 219–26.

Scheiner E, Hammerschmidt K, Jürgens U, Zwirner P. 2002. Acoustic analyses of developmental changes and emotional expression in the preverbal vocalizations of infants. *Journal of Voice* 16: 509–29.

Scheiner E, Hammerschmidt K, Jürgens U, Zwirner P. 2004. The influence of hearing impairment on preverbal emotional vocalizations of infants. *Folia Phoniatrica et Logopaedica* 56: 27–40.

Seyfarth RM, Cheney DL, Marler P. 1980. Vervet monkey alarm calls: Semantic communication in a free-ranging primate. *Animal Behaviour* 28: 1070–94.

Sinnott J. 1995. Methods to assess the processing of speech sounds by animals. In *BioMethods, Vol. 6: Methods in Comparative Acoustics*, ed. GM Klump, RJ Dooling, RR Fay, WC Stebbins, pp. 281–92. Basel, Switzerland: Birkhauser Verlag.

Sugiura H. 1998. Matching of acoustic features during the vocal exchange of coo calls by Japanese macaques. *Animal Behaviour* 55: 673–87.

Sutton D, Larson C, Lindeman RC. 1974. Neocortical and limbic lesion effects on primate phonation. *Brain Research* 71: 61–75.

Talmadge-Riggs G, Winter P, Ploog D, Mayer W. 1972. Effects of deafening on the vocal behavior of the squirrel monkey. *Folia Primatologica* 17: 404–20.

Trachy RE, Sutton D, Lindeman RC. 1981. Primate phonation: Anterior cingulate lesion effects on response rate and acoustical structure. *American Journal of Primatology* 1: 43–55.

Vogt BA, Barbas H. 1986. Structure and connections of the cingulated vocalization region in the rhesus monkey. In *The Physiological Control of Mammalian Vocalization*, ed. JD Newman, pp. 203–25. New York: Plenum Press.

Weisberg P. 1963. Social and nonsocial conditioning of infant vocalizations. *Child Development* 34: 377–88.

Westermann G, Miranda E. 2004. A new model of sensorimotor coupling in the development of speech. *Brain and Language* 89: 393–400.

Williams SL, Brakke KE, Savage-Rumbaugh ES. 1997. Comprehension skills of language-competent and nonlanguage-competent apes. *Language and Communication* 17: 304–17.

Winter P, Handley P, Ploog D, Schott D. 1973. Ontogeny of squirrel monkey calls under normal conditions and under acoustic isolation. *Behaviour* 47: 230–9.

Wolff PH. 1969. The natural history of crying and other vocalizations in early infancy. In *Determinants of Infant Behavior, Vol. 4*, ed. BM Foss, pp. 113–38. London: Methuen.

Wong RO. 1999. Retinal waves and visual system development. *Annual Review of Neuroscience* 22: 29–47.

Wood RM, Gustafson GE. 2001. Infant crying and adults' anticipated caregiving responses: Acoustic and contextual influences. *Child Development* 72: 1287–1300.

9 Cognitive Precursors to Language

Brian MacWhinney

Introduction

Language is a unique hallmark of the human species. Although many species can communicate in limited ways about things that are physically present, only humans can construct a full narrative characterization of events occurring outside the here and now. By using language for social coordination, humans have achieved a remarkable level of control over their environment (Sterelny, this volume). However, given the demonstrable adaptive advantages provided by language, why have other species not developed communicative systems of similar range and power? One possibility is that the unique recursive structures of human language arose through a singular event in the recent evolution of our species (Bickerton, 1990; Hauser et al., 2002; Mithen, 1996). According to this view, this recent event has not been replicated in other species because it depends on certain preexisting conditions that were unique to the human species. But, then, what are these unique preexisting conditions, and why did they appear only in hominids?

The thesis developed in this chapter is that human language depends on a quartet of characteristics found in combination only in hominids. This quartet of human characteristics worked together to constitute a unique ecological niche. This unique niche then produced ongoing evolutionary pressures (Geary, 2005) for increasingly complexity in hominid social semiotics. The four characteristics of this niche are bipedalism, manual dexterity, neoteny, and social bonding. This quartet of characteristics specified a niche that supported continually richer and richer communicative patterns. According to this account, language is not an accidental mutation but rather a natural expression of what it means to be a bipedal creature that relies on tools, social chatter, and the communal support of neotenous offspring. Without the copresence of each of these four features, the coevolutionary pressure (Deacon, 1997; Givón, 1998) toward greater semiotic complexity would have been absent.

A Quartet of Characteristics

Bipedalism

Within this quartet of essential human characteristics, the one that stands out most conspicuously in the fossil record is bipedalism. By 4 million years ago (mya), the hips, femurs, and tibia of australopithecines had evolved to resemble those of modern man (McHenry, 1986). The presence of hominid footprints in a layer of hardened ash from about 3.5 mya indicates that, by this time, australopithecines walked much like modern man (Tattersall, 1993), although their stature was shorter. This emergence of bipedalism was not a sudden event. Before 4 mya, hominids, such as *Ardipithecus ramidus*, inhabited both arboreal and terrestrial environments simultaneously. Moreover, within both arboreal and terrestrial habitats, apes and hominids deploy a variety of methods for walking and climbing (Stanford, 2006). Thus, there was no single moment when our ancestors abandoned the trees altogether. In fact, modern man can still climb trees when necessary.

Bipedalism brought with it a series of evolutionary costs. It placed increased mechanical pressure on the neck, the spine, and all the joints of the legs. The reliance on the hind limbs for walking made them less functional for climbing. The lungs had to adapt to support the breathing needed for running. Perhaps the most important of these costs involved the narrowing of the hips that makes parturition more difficult (Hockett and Ascher, 1964). As the infant's head grows in circumference, this problem becomes even worse. The evolutionary answer to this dilemma was to have the baby emerge earlier, before the skull could reach an unmanageable diameter. One result of this hastening of the time of birth is that basic sensorimotor systems such as vision and motor coordination are not consolidated until the eighth week of infancy (Johnson and Morton, 1991). This tendency toward premature parturition forced humans to provide heightened social support for increasingly dependent infants.

Manual Control

Despite these costs, bipedalism brought with it several advantages. With the arms free, we could use our hands for new purposes (Coppens, 1995). The hands remained useful for climbing trees and hanging from branches, but now they had a wider range of functions, including combat, manual communication, and tool production. The first evidence of toolmaking comes from about 2.4 mya with the appearance of Oldowan tools from the Gona and Omo Basins in Ethiopia. The tools found there include a variety of choppers, scrapers, bone points, and diggers (Potts, 1988). Although the shaping of these tools did not require the level of planning and design required by the tools of the Upper Paleolithic, their production still requires a high level of manual dexterity and control. This dexterity was supported by the development of an increasingly refined precision grip in *Homo habilis* and then *Homo erectus* (Jolly and Plog, 1982).

Neoteny

Neoteny involves the retention of juvenile, or even infantile, characteristics in adults. For example, infants have full cheeks, soft chins, and thin hair. When we see similar facial characteristics in an adult, we say that they have a "baby face." In human brain growth, we see evidence for neoteny in terms of the maintenance of infantile levels of cortical plasticity well into adolescence and early adulthood. In stricter terms (Rice, 1997), what we see in human brain development is neoteny combined with a secondary lengthening of the overall development process. However, during the first years of life, the primary change in developmental synchrony involves simple neoteny, including the delay of aspects of the last months of gestation into the first months of infancy.

Neoteny has wide-ranging consequences for cognitive and social development. Students of human evolution (Bjorklund and Pellegrini, 2002; Gould, 1977; Montagu, 1955) have often emphasized the extent to which neoteny can be used to explain diverse aspects of human cognitive flexibility. However, the most obvious immediate effect of neoteny is that the human infant remains dependent on parents for food, care, and guidance well into middle childhood. In this regard, the human infant is altricial, relying on support from adults rather than raw instinct. During infancy, the baby is dependent for many months on consistent care from the mother. Once established, this reliance on parental support can then extend further and further across the life span. The exact shape of this dependence varies markedly from societies like the Ik of Kenya (Turnbull, 1972) that cease support of children by age five to modern urban societies that maintain support for offspring well into the college years.

This extended period of human neoteny provides rich support for teaching children manual, cognitive, and social skills (Vygotsky, 1934) and for maintaining cortical plasticity (Elman et al., 1996; Julész and Kovacs, 1995) even into adolescence. In other species with protracted childhoods, such as elephants, we see a similar rich potential for the development of flexible cultural, cognitive, and communicative structures (Lee, 1986). Neoteny also opens up the door to further selectional pressures. If the parents fail to provide support or if older children interfere with that support, children may die. When resources are limited or when the child appears to be developmentally abnormal, the child may be abandoned. If the population is migratory, these pressures can be intensified. Although pressures of this type are reduced in modern society, they were certainly operative across earlier stages in evolution.

Social Bonding

Within this quartet of defining human characteristics, perhaps the most ancient feature is the hominid emphasis on tight social organization. In this regard, humans are very much like highly social bonobo chimpanzees (de Waal, 1995; de Waal and Lanting, 1997). Dunbar (1997) has argued that the maintenance of large social groups requires a high ratio of brain weight to body weight. However, what is crucial here is not simply the brain–body ratio, but also the ability to maintain cortical plasticity well into childhood (Finlay, 2005). In effect, neuronal neoteny works to promote social bonding. Primate groups rely on a variety of mechanisms to maintain and shape social bonding, including grooming, preening, touching, chattering, copulating, playing, and fighting. All of these devices are involved in face-to-face communication. However, two of these mechanisms have undergone further elaboration in humans. These are the mechanisms of face-to-face eye contact and vocal chatter.

When chimpanzees assume a sitting position, they can easily maintain continual face-to-face eye contact with the others in their group (Stanford, 2003). In the sitting position, they do not have to bend their neck up to maintain eye contact. However, if the chimpanzee leaves the sitting position, eye contact is broken. In humans, the movement to the upright posture provided fuller support for maintaining ongoing eye contact. This, along with the freeing of the hands, provided a shared visual space for the elaboration of gestural signals. As many have argued (Hewes, 1973), it is likely that hominids went through a period of relying heavily on gestural communication. It would be difficult to imagine that this did not happen, given the fact that chimpanzees make such extensive use of facial and body gesture (Call, this volume) and the fact that they are good at learning new signs (Savage-Rumbaugh, 2000). With our eyes locked even more continuously into contact, we are better able to track facial movements expressing joy, acceptance, fear, surprise, and anger. If our gaze extends a bit more broadly, we are also able to track gestures of the head, torso, and hands.

The second form of increased support for communication and social bonding involves the increasing reliance on chatter. Dunbar (1997) argues that this reliance on chatter and gossip was crucial in permitting hominids to maintain maximum group size. In this way, the forces of social bonding directly supported the emergence of increasingly precise vocal communication.

Cognitive Consequences

Each of these four characteristics appears to some degree in chimpanzees, gorillas, gibbons, and orangutans. Chimpanzees also display some neoteny, probably for similar purposes. The orangutan also has a precision grip and a thumb not unlike the human thumb. Young chimpanzees often assume a bipedal stance. Gibbons, bonobos, and baboons all maintain rich systems for supporting social bonding. However, within this shared framework of a basic primate emphasis on social bonding and neoteny, the human commitment to bipedalism and its consequences for parturition (Hockett and Ascher, 1964) pushed our ancestors into an even deeper reliance on neoteny and opened up promising new avenues for use of the hands.

None of these adaptations, by itself, provides direct support for the evolution of language. Rather, this quartet of characteristics led to the emergence of a new social and cognitive platform that could later support the evolution of protolanguage. Donald (1998) has referred to this new platform as the "executive suite" and held that it was the basis for a "mimetic revolution" (Donald, 1991) that occurred during the evolution of *Homo erectus*. In the following sections, we will examine how the ongoing quartet of human characteristics shaped these further aspects of human cognition. We will later see that each of these additional cognition adaptations plays a role in shaping human language and facilitating language evolution. Each of these developments involved a continual refinement of the "executive suite" across the four million years of hominid bipedalism.

Consequences of Bipedalism

According to my reasoning, our ancestors' commitment to bipedalism opened up an evolutionary pressure for refinement of the cognitive systems used to represent and navigate through space. As hominids began to rely less and less on trees for refuge, they began to range over a wider territory, while still retaining information about trees and hiding places. This meant that they needed to develop improved means of representing spaces and distances. All species must have some way of representing their territory. However, hominids faced the task of representing a large, often changing, territory in which they were both the hunters and the hunted. To do this, they needed to develop extended methods for spatial encoding. Holloway (1995) has presented evidence from endocasts (plaster casts of the interiors of skulls) indicating that there was a major reorganization of parietal cortex after about 3 mya. This reorganization involved the reduction of primary visual striate cortex and the enlargement of extrastriate parietal cortex, angular gyrus, and supramarginal gyrus. Much of the evidence for Holloway's analysis comes from traces of the changing positions of the lunate sulcus and the intraparietal sulcus over time. According to Holloway, the areas that were expanded during these changes in the parietal cortex support three basic cognitive functions:

1. Processing in the dorsal (parietal) stream of the visual field is important for representing actions of the other in terms of one's own body image (Ramachandran and Hubbard, 2001).

2. The association areas of parietal cortex maintain a map of the environment for navigation in the new bipedal mode (Wilkins and Wakefield, 1995).

3. The supramarginal gyrus is involved in face perception. Expansion of this area would facilitate the development of social patterns and memory for social relations. However, there is evidence that the neural adaptations for face recognition have general consequences for object recognition and categorization that will eventually impact linguistic categorization.

The first two functions emerge from the adoption of bipedal gait. The third function is linked to the promotion of social bonding.

Because the move to a terrestrial environment was gradual (Corballis, 1999), hominids needed to maintain the use of the hands in both the arboreal and terrestrial environments. The arboreal environment favors the development of a specific type of motor imagery. Povinelli and Cant (1995) have noted that increases in body weight for larger apes such as orangutans make it important to be able to plan motions through the trees. To do this, the animal needs a map of the self as it executes possible motor actions. Assuming that hominids or their ancestors were similar to orangutans in arboreal navigation, the reflexes of this penchant for postural adaptation may still be evident in the human enjoyment of dance, exercise, and sport. The pressures in the arboreal environment that had favored some limited form of brain lateralization were then carried over to the terrestrial environment (McManus, 1999). This ability to shift quickly between alternative environments required neural support for competing postural and perceptual systems.

Consequences of Manual Dexterity

Apes have good control of reaching and basic object manipulation (Ingmanson, 1996). However, with both hands free, hominids were able to explore still further applications. Monkeys and other primates have "mirror" neurons in premotor cortex that respond with equal force when an action such as "grabbing" is carried out either by the self or by the other, including a human. This mechanism provides a way of equating actions performed by the self with actions or postures performed by the other. This system provided support for the learning of tool use by the increasingly neotenous hominid children. Young hominids could learn to use branches and clubs by imitating their elders. They could acquire the ability to chip one stone against another to form primitive hand axes. The adaptive value of tracking and emulating patterns of tool usage is clear.

The ability to imitate a series of actions requires construction of stored mental images of specific motor actions and postures. To plan the actions involved in chipping an axe, we must be able to call up an image of the desired product, and we must be able to sequence a long series of specific motions that are needed to locate good stones and devise methods for chipping edges. In this regard, the ability to construct a planned sequence of actions appears to be a unique property of hominids, as opposed to monkeys and apes. Students of primate tool use (Anderson, 1996; Visalberghi and Limongelli, 1996) have shown that chimpanzees and capuchin monkeys can use tools in a productive and exploratory way. However, they do not appear to make planful use of mental imagery to limit their search through possible methods of tool use. Instead, they apply all directly perceptible methods in hopes that one may succeed.

Consequences of Neoteny

As children's brains became more plastic, they became increasingly responsive to parental teaching. This led, in turn, to the expansion and consolidation of group norms. In accord with the operation of the Baldwin Effect (Baldwin, 1897), children who were able to pick up social norms were more reproductively successful. These good learners could then breed new generations of good learners. In effect, this gave rise to an evolutionary arms race that favored the growth of a larger, more plastic brain (Geary, 2005). Much of the support for good learning came from overall brain size and plasticity, but there was also expansion of more specific mechanisms such as statistical learning (Aslin et al., 1999), motor control through practice (Oller, 2000), and imitation (Tomasello, 1999) that provide further support for smooth learning of social norms, including language.

Consequences of Social Bonding

Increased reliance on vocalization in face-to-face interaction produced three major consequences. The first was an increased reliance on vocalization to mark group membership. For many species, vocalizations provide an effective method of signaling group membership. Songbirds, parrots, and hummingbirds have developed methods for learning songs (Konishi, 1995) to mark both individuals and membership in a territorial group. At about two mya, hominids moved from being the hunted to being the hunters, thereby providing a high-protein diet that could fuel the metabolic requirements of the growing brain.

The shift to a diet based on hunting then brought pressure to mark group control of hunting territories. One way of achieving this is through the development of a group-based vocal dialect. Monkeys do not appear to construct local dialects or otherwise structure their call system through learning (Seyfarth and Cheney, 1999). Yerkes and Learned (1925) and others have tried to condition chimpanzee vocalizations in the laboratory and have failed. Neotenous human infants, on the other hand, rely on highly plastic cortical mechanisms to control vocalization (Oller, 2000). This allows them to pick up the sound patterns of their community through mere exposure. As a result, each hominid group can build a local vocal accent that is passed on to the next generation.

The second major consequence of the rise of face-to-face interaction was the strengthening of bonding relationships. Communication has a clear role in relations between the sexes. By bringing these expressions under cognitive control, it is possible to fine-tune both courtship and sexual deception (Buss, 1999; Miller, 2001). The same gestures and vocalizations involved in sexual interactions are also operative between mothers and their infants (Harlow, 1958). Parents and their babies engage in reciprocal (Trevarthen, 1984) flirting and play, much like adults who are in love. These interactions serve three functions. First, they can convince the mother to promote the child's physical well-being. Second, they can cement the infant's secure attachment to the mother, thereby promoting a variety of other developments. Third, they can serve to acculturate the child into the conversational norms of the adult group. The learning of these conversational functions occurs smoothly because the child is locked into face-to-face vocalization (Locke, 1995). In considering the role of face-to-face vocalization in supporting prosocial relations in hominid groups, we must not forget the potentially asocial, divisive role played by aggressive males (Anders, 1994; Goodall, 1979), as well as the compensatory use of face-to-face communication to dampen male aggression.

The third consequence of increased face-to-face interaction was the growth of mechanisms for social perspective taking. Building on the basic primate ability to mirror the actions of conspecifics (Rizzolatti et al., 1996) and relying on new methods for gestural communication, humans became increasingly adept at sharing aspects of their thoughts and tracking the thoughts of others. Without language, this perspective taking was confined to the here and now, but it nonetheless provided a central cognitive support for the eventual emergence of grammatical structure (MacWhinney, 2005b).

Support for Vocalization

Each of these adaptive contexts provided pressures toward increasing flexibility in vocal communication. However, to achieve this flexibility required a fundamental restructuring of vocal control. One form of restructuring was neuronal. In macaques (Jürgens, 1979), control of the vocal system relies on the periaqueductal gray matter (PAG) of the lower midbrain. Additional midbrain regions can stimulate the PAG, but the neocortex does not control or initiate primate vocalizations. Primate vocalization relies on direct connections to midbrain motivational areas (Pandya et al., 1988). Human language continues to rely on this underlying limbic architecture to provide emotional coloring to vocalization. The linkage of the vocal system to limbic mechanisms provides grounding in terms of arousal (brain stem and amygdala), motivation (basal ganglion), patterning (striatal-thalamic circuits), and memory (limbic circuits; Tucker, 2002). Humans also retain some direct links between audition and these limbic circuits as evidenced in the directness of our responses to sounds such as infant cries or the growls of predators (and see Hammerschmidt and Fischer, this volume).

In humans, this midbrain system has been supplemented by a cortical system. Electrical stimulation of both the supplemental motor area and the anterior cingulate of the frontal cortex can reliably produce vocalization. Ploog (1992) has shown that humans have more direct pyramidal connections between motor cortex and the speech and vocalization areas of the brain stem than do monkeys. Certain areas of the limbic system, such as the anterior thalamic limbic nuclei, have grown disproportionately large in humans. These nuclei serve the supplementary motor area and premotor and orbital frontal cortex. The expansion of these structures points to increased limbic input to the cortex, as well as input from the cortex to the limbic structures. Tucker (2002) shows that the basic adaptation here involved the absorption of the primate external striatum by the neocortex (Nauta and Karten, 1970).

The shift to cortical control of vocalization relies on adaptation of preexisting cortical pathways for orofacial gestural control. MacNeilage (1998) has argued that the primate gesture of lip smacking is the source of the core consonant–vowel (CV) syllabic structure of human language. The CV syllable has similarities in motoric structure to lip smacking. Moreover, it is produced in an area of inferior frontal cortex close to that used for lip smacking and other vocal gestures. In addition, humans have developed additional cortical control of phonation. Because apes have so little control over the opening and closing of their larynx, they will drown if placed in water. Hominids may have developed cortical control over laryngeal function through some early exposure to an aquatic environment (Morgan, 1997), control of breathing during running, or early attempts at singing. Infants achieve control over basic phonation by three months (Oller, 2000), whereas control of CV structure is not solidified until several months later (Oller, 2000). It is possible that this sequence follows a natural logic that governs both ontogeny and phylogeny.

The Platform

By the time of *Homo erectus*, humans had evolved a set of abilities that provided a crucial platform for the evolution of language (Donald, 1998). Many of the crucial aspects of human language, including those involved centrally in grammar, recursion, and productivity, depend directly on components of this platform. The components of this platform included the following:

1. An increased ability to represent space hierarchically and recursively.
2. A cognitive mechanism for remembering faces and fine within-category visual distinctions.
3. A well-developed body map, suitable for projection to the actions of others.
4. An ability to construct and rehearse plans based on the tracking of the actions and perspectives of others.
5. Emotional and attentional commitment to face-to-face interactions with attending gestures.
6. Increased learning abilities in human infants with special focus on statistical learning, imitation, and motor practice.
7. Cortical control of vocalization and an ability to acquire group vocal patterns.

None of these abilities involve a reliance on language narrowly defined. In the terms of Hauser, Chomsky, and Fitch (2002), these adaptations would all be considered to be part of the faculty of language as broadly defined (FLB). However, to dismiss these preconditions as nondecisive for the emergence of language would be to miss the point. Core features of language such as recursion and perspective marking (which invokes the concept of c-command in modern linguistic theory) rely on these specific cognitive preconditions. The attempt to separate FLB from the faculty of language narrowly defined (or FLN) glosses over the ways in which essential linguistic features depend on cognitive preconditions established millions of years ago.

Mimesis

Donald (1991) argues that *Homo erectus* had achieved a basic level of mimetic communication. He believes that it was this achievement that allowed the species to occupy all of Eurasia and Africa. Mimetic communications involve the depiction of an object or activity by using some characteristic to depict the whole. For example, running can be signaled mimetically by taking a few steps of running. An object can be signaled mimetically by pointing at it or sketching its shape. Mimesis in the vocal mode is more likely to describe sounds associated with emotional attitudes and the sounds of actions, animals, or objects. In general, mimesis achieves reference through partonymy, or mention of a part to express the whole. Givón (2002) has suggested that, during mimesis, the gestural system "trained" the vocal system. This training would have involved the coordination of timing, content, and even hierarchical order between the two systems. According to McNeill (1985), language and gesture are parallel expressive modalities that emerge together in real time from "growing points" arising within an embodied mental model. McNeill's account is particularly compatible with the idea that the links between language and gesture are evolutionarily quite old and well established.

It is important to distinguish the vocal imitation involved in the learning of bird song or babbling patterns from the online conceptual imitation involved in mimesis. In the learning of bird song, the introduction of variability facilitates the beginnings of communicative flexibility (Hausberger et al., this volume). However, mimesis takes this flexibility to a totally new conceptual level. It combines the flexibility of vocal and gestural control with the preexisting interest in detecting and expressing intentions. Although mimetic expressions were likely to be clumsy and incomplete, they provided an initial method for achieving a coconstruction of narrative events and intentions. In this way, mimesis provided support for the eventual expression of perspective taking through grammatical markings and syntax. It seems unlikely that the storage and retrieval of conventionalized mimetic sequences could be achieved simply by linking up older areas or by reusing earlier connections. Instead, additional computational space was needed to store the multitude of new visual and auditory images. In addition, evolution favored the growth of neural systems in frontal cortex for storing and switching between perspectives (Decety et al., 2002). Because mimesis arose in a haphazard way from earlier pieces of lip smacking, pointing, gesture, and rhythm, it was impossible to formulate a socially systematized coding method for the storage of mimetic communications. Instead, patterns and forms had to be learned and stored as holistic unanalyzed sequences. For example, when we chop wood, there is a complete interpenetration of muscle actions, visual experiences, hand positions, and sounds. We can think of this as a single merged form such as I-pull-hands-back-lift-axe-drop-split-chips-wood-cut. Mimetic forms have this same unanalyzed quality. This lack of analysis is not the result of chunking or automatization, since the gestalt is not constructed by a system of combinatorial semantics. Instead, each chunk is a raw, unanalyzed whole that is fully grounded on direct action and perception (Gibson, 1977). Because they are fully grounded, productions of these mimetic gestalts are easy to decipher. However, mimetic gestalts provide little support for cognitive organization.

The Lexicon

The expansion of the brain in *Homo erectus* was not enough by itself to trigger the emergence of material culture. Instead, humans needed some way to systematize the growth in vocal and gestural mimetic processes that had occurred during the Pliocene. The core of the new system involved the introduction of a set of phonological contrasts (Hockett and Altmann, 1973) that could build a productive lexicon. By coding words into a compact set of contrastive features, early *Homo sapiens* were able to conventionalize, learn, store, and retrieve a limitless set of names for things. To achieve accurate articulation of these contrasts, a further set of adaptations was needed for the serial ordering of actions and the precise articulation of sounds. These adaptations included reduction in size of the canines, adaptation of the arytenoids, bending of the vocal tract (Lieberman, 1975), and shaping of the musculature of the tongue. Each of these modifications led to a separate and meaningful increment in our ability to articulate clearly a full inventory of phonetic contrasts.

Word learning depends on several of the cognitive preconditions discussed earlier. The infant must detect statistical regularities that will determine word forms. By listening carefully to the speech of caretakers, the infant can tune in to the shape of words. Then, by following the cues of gaze direction and pointing, they can pick up names for things (MacWhinney, 1998). Infants can then learn to produce these sounds themselves by matching their vocalizations to those they hear during babbling (Westermann, this volume). Gupta and MacWhinney (1997) show how consolidation of word shapes depends on the system for phonological rehearsal (Cowan, 1992; Gathercole and Baddeley, 1993; Nairne, 1990). The final storage of lexical forms is achieved in auditory cortex, deep in the temporal lobe (König et al., 2005). Broca's area and motor cortex are responsible for shaping the motoric form of words (Blumstein, 2001). This means that vocal rehearsal must involve coordinations between three very separate brain areas—premotor cortex, motor cortex, and auditory cortex. Li, Farkas, and MacWhinney (2004) have shown how these patterns of connectivity can be modeled through self-organizing neural networks that represent local maps in auditory, articulatory, and semantic space.

The neuronal systems for encoding these connections provided further neuronal support for the evolution of spoken language. It is difficult to determine the age of these neuronal developments. However, evidence regarding the growth of peripheral support for speech suggests that these developments were under way by at least 300,000 years ago (MacLarnon and Hewitt, 1999). The use of phonological rehearsal to consolidate memories for lexical forms is one of the first steps toward the creation of linguistic productivity (Hockett and Altmann, 1973). This system involves a level of recursive combination within individual vocal gestures. However, except in marginal cases of phonetic symbolism (Paget, 1930), the individual components of words are not linked to components of meaning. Instead, it is the word as a whole that is linked to meaning as a whole (Li et al., 2004).

Recursion and Item-Based Patterns

In the human child, the joining of words into productive constructions relies on a system of item-based patterns, proposed first by MacWhinney (1982) and further investigated by Tomasello, Lieven, and Pine (Pine and Lieven, 1993; Tomasello, 2000). Given the centrality of item-based learning for child language, it is reasonable to suppose that the ability to combine words into item-based patterns constituted a major evolutionary achievement for modern humans. This achievement rests heavily on the consolidation of the ability to learn individual words. Once the word learning ability is in place, the movement to combinations of words involves the introduction of a method of linking words to predications.

The basic mechanisms supporting predication were already existent in nonhuman primates (McGonigle and Chalmers, 1992, 2002, 2006). A chimpanzee understands that a banana is yellow or that a river is too wide to cross by jumping. The real challenge involved consolidation of a neural mechanism for expressing predications through the auditory–vocal channel.

In order to express predications, human language relies on a system of item-based patterns. These patterns associate a particular predicate such as "red" or "hit" with its argument. For example, "red" is associated with the object that it describes, which occurs in a following slot. Similarly, "hit" takes two arguments. The first argument is a "hitter," who appears in the slot before the verb, and the second argument is the "thing being hit," which appears in the slot after the verb.

Children spend much of the time between about sixteen and thirty months of age learning these item-based patterns (MacWhinney, 1987). They pick up these patterns by processing simple inputs provided by parents. Consider the case of a child who has learned the word "doggie." The parent then says "big doggie." The child notices the size of the dog and assumes that "big" refers to size. This learning episode allows the child to acquire "big" as a lexical operator that takes as its argument an entity like "dog." The value of the argument that fills this slot is initially set to "dog," but this value can be generalized on the basis of additional input. Crucially, the child views the operator "big" as describing the size of its head argument. Later, when the child also hears "big cookie," the semantic range of the head ("cookie" in this case) is generalized to inanimate objects. Eventually, the item-based pattern "big + X" is linked up with other patterns such as "nice + X" to yield a generalized pattern that is no longer based on single items but operates across groups of lexical items that share a core feature. This new structure is called a "feature-based pattern" (MacWhinney, 1982). In this account, learning begins with words, advances through item-based patterns, and then is extended through feature-based patterns.

By themselves, item-based and feature-based patterns provide predication, but not recursion. In order to generate language recursively, item-based patterns must produce intermediary structures called "clusters" (Hudson, 1984; MacWhinney, 1987) or phrases. Consider the sentence "Your big dog chased my frightened cat." The phrase "your big dog" is produced by the recursive clustering of two item-based patterns. First, "big" is attached to "dog" to produce "big dog." Then "your" is attached to the cluster of "big dog." This whole cluster then functions as the first argument of the verb "chases," which takes "my frightened cat" as its second argument. This repeated deployment of item-based patterns requires an additional processing mechanism that can store a complex item such as "your big dog" in working memory for subsequent combination with the predicate "chased." The ability to store such memories can rely in part on existing action planning structures (Greenfield, 1991), but a system that uses item-based constructions to form combinations in real time would require these systems to perform at a higher level than previously and with a greater focus on manipulating lexical objects.

This account emphasizes the unique role of item-based patterns in the recursive construction of sentences. In this view, sentences do not emerge from abstract derivational processing across multiple constraint-based modules (Chomsky, 1986; Chomsky and Lasnik, 1993) but rather through online incremental processing of simple links between lexical items. In the terms of current dialogs in linguistics and psycholinguistics, this emergentist position (MacWhinney, 2005a; O'Grady, 2005) contrasts with minimalist formalism (Chomsky, 1995). However, both emergentism and formalism agree on the centrality of recursion in generating linguistic productivity and form. Minimalism singles out recursion as the pivotal development leading to the sudden emergence of language. In the emergentist account, on the other hand, recursion arises gradually through the implementation of memory structures based on preexisting motoric planning abilities and methods for spatial hierarchicalization that go back millions of years. Moreover, as we will see, the emergentist view treats recursion not as the final step toward the evolution of language but as a method for facilitating the further introduction of perspective taking into grammar.

Recently, researchers (Mithen, 1996) have attempted to link the appearance of recursion to some sudden recent evolutionary event. One account focuses on the claim that 70,000 years ago there was an evolutionary bottleneck that brought the number of our direct ancestors down to perhaps 10,000 individuals worldwide (Stringer and McKie, 1996). This ancestral population coexisted with other hominids, such as the Neanderthals, who are not our direct ancestors. Eswaran, Harpending, and Rogers (2005) argue that this evolutionary bottleneck arose as successful new generations were being produced by a slow wave of expansion from Africa. This wave brought with it the seeds of modern creativity that had begun to emerge earlier in Africa (McBrearty and Brooks, 2000). When we think in terms of a population of 10,000 individuals from whom we all may have descended, it is important to remember that it was not the case that human populations worldwide had declined to this low level. Rather, the bottleneck reflects the fact that individuals at the front of this wave were far more successful reproductively than those not in the crest of the wave.

Along the front of this wave, modern humans assimilated only marginally with the archaic populations they replaced. However, there were enough interactions to leave traces in nuclear DNA. The current distribution of populations suggests that the genetics of this new group could be characterized by the presence of a group of at least eight genes, including the *FOXP2* gene, linked to some articulatory and motor difficulties (Enard et al., 2002). The complexity of this wave model and the polygenetic determination involved in the determination of language-related features suggests that it would be best to view this wave of creative individuals as possessing a wide range of traits on both linguistic and social dimensions.

Perspective Taking

Mithen (1996) has argued that the increased artistic production in the Upper Paleolithic resulted from new methods for linking ideas between otherwise modularized cognitive functions. Mithen's analysis, while overstated in various ways (McBrearty and Brooks, 2000), still seems to capture a core aspect of recent human evolution. The idea that language serves as a method for integrating thought has appealed to a very diverse set of scholars from Plato and Vygotsky to Dennett (1996) and Carruthers (2002). However, it would be a mistake to imagine that the low-level mechanism of recursive combination was sufficient to trigger the creative productions found in Europe after 30,000 years ago or earlier in Africa. Instead, it seems much more likely that item-based patterns and recursion served to construct a new linguistic platform that could then support ongoing advances in social cognition, perspective taking, and child rearing.

The solidification of language as a means of expressing creative and productive thought required the social construction of a set of grammatical devices for marking perspective taking. These devices used the system of item-based patterns, along with grammatical affixes to mark the flow of perspective through agents, positions, and actions. These markers are encoded and positioned at the level of the item-based pattern. However, they work together in sentences to trigger perspective taking and shifting across recursive systems of predicates. Together, this system of markers allows us to construct embodied representations of the infinite array of meaningful combinations that can be encoded by sentences and combinations of sentences.

Let us consider some examples of the application of perspective shifting to linguistic structure. (For a fuller recent account, see MacWhinney, 2005b) For our first example, let us return to the earlier sample sentence (1):

(1) Your big dog chased my frightened cat.

Here, we begin the process of interpretation with the word "your." However, because this word is marked as possessive, we delay commitment to a point of view until hearing the subject "dog." We then interpret "dog" as the perspective from which to understand the sentence and construct a secondary perspective in which "you" are the owner of this dog. Then, we hear "chased" and link the dog to the chasing and open up a slot for something being chased. This slot is then filled by "cat." Although primary perspective remains on "dog" throughout this sentence, an additional secondary, social perspective is opened for attaching "my" and "frightened" as descriptors of "the cat."

In general, perspective switching in English leads us to select the first noun as the primary perspective (Gernsbacher, 1990) and to shift away from this primary perspective only when strong syntactic cues signal a shift. Strong cues of this type appear in passives like (2) and clefts like (3):

(2) My frightened cat was chased by your big dog.

(3) It was my frightened cat that your big dog chased.

More subtle effects of perspective shifting and marking can be detected throughout grammar. Consider an ambiguous sentence such as (4):

(4) Tim saw the Grand Canyon flying to New York.

Here, the syntax licenses a competition between "Tim" and "Grand Canyon" as the attachment sites for "flying to New York." Although we may attempt both attachments, only one makes good sense. Sentences like (5) through (8) show how the intervention of a perspective shift can block reflexivization:

(5) Jessie stole a picture of herself.

(6) *Jessie stole a picture of her.

(7) Jessie stole me a picture of herself.

(8) Jessie stole me a picture of her.

Reflexivization is not possible in (6). Formal linguistic theory explains this by holding that "picture nouns" block the c-command relation that supports the clause-mate constraint. Perspective theory interprets this instead in terms of the need for an intervening perspective shift as in (8) to license the move away from ego perspective required for reflexives. Sentence (9) illustrates how ongoing construction of a spatial scene can lead to alternative perspectival constructions.

(9) The adults in the picture are facing away from us, with the children.

This brief excursion into the theory of item-based patterns and the marking of perspective has provided only a limited sample of the phenomena involved, intended to illustrate the basic shape of these processes. The thesis is that the humans at the front of the wave of expansion from Africa (Eswaran et al., 2005) were in possession of the postulated newly evolved ability to make recursive application of lexical structures. The cognitive ability to generate linguistic recursion must necessarily have been accompanied by a heightened ability for recursive manipulation of images and plans.

By linking this new ability to socially devised methods for marking perspective shifting in language, these new groups were able to raise the use of human language to new, more powerful levels.

As they competed with less articulate groups, more articulate groups were able to strengthen group solidarity and purpose by constructing unique cultural histories. They did this by creating totemistic myths (Freud, 1913) that formalized kinship loyalties in the terms of hunter–gatherer shamanism. The creation of these myths depended on new methods for describing the not-here and not-now in terms of coherent ontological narratives (Frazer, 1890). By articulating their visions of these myths and the spirit world, shamans and priests were able to achieve additional control (Geary, 2005) and status in their alliance with chieftains. These myths solidified the bonds between group members and their families, committing them to sacrifices in the name of the group. They also allowed the group to maintain contacts of exogamy and trade with related groups near the frontier of expansion. Although these cultural and linguistic innovations were concentrated among the population most involved in the dynamic expansion across Eurasia, they eventually diffused back to Africa itself. As this diffusion progressed into the Neolithic, it involved the spread of cultural innovations rather than additional mutations. Throughout these developments, increasingly complex recursive sentence patterns were developed to express the details of perspective taking and perspective shifting needed for increasingly complex myths, narrations, and speeches. These new "memes" (Blackmore, 1999) then spread to become nearly universal components of human society, fueling all of the developments of modern civilization.

Summary

This chapter has suggested that human language evolved gradually within a unique context that included four preconditions: bipedalism, manual dexterity, neoteny, and social bonding. The evolution of cognitive control of vocalization and gesture received support from parent–child interaction, as well as other social bonding mechanisms, and between-group identification. Once cortical control was established, *Homo erectus* was able to organize mimetic sequences for communication regarding intentions. However, these sequences lacked stable reference and were eventually supplemented by fixed items from a vocal lexicon. Then, using item-based patterns that relate arguments to their heads, speakers at the crest of a new wave of linguistic creativity constructed methods for recursive combination of words into phrases and sentences. At this point, vocal methods permitted a full mapping of embodied human perspectives through grammatical devices marking perspective shift. Together, these features of cortical control, lexical mapping, recursion, and perspective marking constitute the cognitive underpinnings for human language.

References

Anders T. 1994. *The Origins of Evil: An Inquiry into the Ultimate Origins of Human Suffering*. Chicago: Open Court.

Anderson J. 1996. Chimpanzees and capuchin monkeys: Comparative cognition. In *Reaching into Thought: The Minds of the Great Apes*, ed. AE Russon, KA Bard, ST Parker, pp. 23–56. Cambridge, England: Cambridge University Press.

Aslin RN, Saffran JR, Newport EL. 1999. Statistical learning in linguistic and nonlinguistic domains. In *The Emergence of Language*, ed. B MacWhinney, pp. 359–80. Mahwah, NJ: Lawrence Erlbaum Associates.

Baldwin J. 1897. Organic selection. *Science* 5: 634–6.

Bickerton D. 1990. *Language and Species*. Chicago: University of Chicago Press.

Bjorklund D, Pellegrini A. 2002. *The Origins of Human Nature: Evolutionary Developmental Psychology*. Washington, DC: American Psychological Association.

Blackmore S. 1999. *The Meme Machine*. Oxford: Oxford University Press.

Blumstein S. 2001. Deficits of speech production and speech perception in aphasia. In *Handbook of Neuropsychology, Vol. 3*, ed. R Berndt, 95–113. Amsterdam: Elsevier.

Buss D. 1999. *Evolutionary Psychology: The New Science of Mind*. Needham Heights, MA: Allyn and Bacon.

Carruthers P. 2002. The cognitive functions of language. *Behavioral and Brain Sciences* 25: 657–74.

Chomsky N. 1986. *Barriers*. Cambridge, MA: MIT Press.

Chomsky N. 1995. *The Minimalist Program*. Cambridge, MA: MIT Press.

Chomsky N, Lasnik H. 1993. The theory of principles and parameters. In *Syntax: An International Handbook of Contemporary Research*, ed. J Jacobs, pp. 1–32. Berlin: Walter de Gruyter.

Coppens Y. 1995. Brain, locomotion, diet, and culture: How a primate, by chance, became a man. In *Origins of the Human Brain*, ed. J-P Changeux, J Chavaillon, pp. 4–12. Oxford: Clarendon Press.

Corballis MC. 1999. Phylogeny from apes to humans. In *The Descent of Mind: Psychological Perspectives on Hominid Evolution*, ed. MC Corballis, S Lea, pp. 40–70. Oxford: Oxford University Press.

Cowan N. 1992. Verbal memory span and the timing of spoken recall. *Journal of Memory and Language* 31: 668–84.

de Waal F. 1995. Bonobo sex and society. *Scientific American* 264: 82–8.

de Waal F, Lanting F. 1997. *Bonobo: The Forgotten Ape*. New York: Wiley.

Deacon T. 1997. *The Symbolic Species: The Co-evolution of Language and the Brain*. New York: Norton.

Decety J, Chaminade T, Grezes J, Meltzoff AN. 2002. A PET Exploration of the neural mechanisms involved in reciprocal imitation. *Neuroimage* 15: 265–72.

Dennett D. 1996. *Kinds of Minds*. New York: Basic Books.

Donald M. 1991. *Origins of the Modern Mind*. Cambridge, MA: Harvard University Press.

Donald M. 1998. Mimesis and the executive suite: Missing links in language evolution. In *Approaches to the Evolution of Language*, ed. JR Hurford, MG Studdert-Kennedy, C Knight, 44–67. New York: Cambridge University Press.

Dunbar R. 1997. *Grooming, Gossip, and the Evolution of Language*. Cambridge, MA: Harvard University Press.

Elman JL, Bates E, Johnson M, Karmiloff-Smith A, Parisi D, Plunkett K. 1996. *Rethinking Innateness*. Cambridge, MA: MIT Press.

Enard W, Przeworski M, Fisher S, Lai C, Wiebe V, et al. 2002. Molecular evolution of FOXP2, a gene involved in speech and language. *Nature* 418: 869–72.

Eswaran V, Harpending H, Rogers A. 2005. Genomics refutes an exclusively African origin of humans. *Journal of Human Evolution* 49: 1–18.

Finlay B. 2005. Rethinking developmental neurobiology. In *Beyond Nature–Nurture: Essays in Honor of Elizabeth Bates*, ed. M Tomasello, DI Slobin, pp. 195–218. Mahwah, NJ: Lawrence Erlbaum Associates.

Frazer J. 1890. *The Golden Bough*. Oxford: Oxford University Press.

Freud S. 1913. *Totem and Taboo: Some Points of Agreement between the Mental Lives of Savages and Neurotics*. New York: Penguin.

Gathercole V, Baddeley A. 1993. *Working Memory and Language*. Hillsdale, NJ: Lawrence Erlbaum Associates.

Geary D. 2005. *The Origin of Mind: Evolution of Brain, Cognition, and General Intelligence*. Washington, DC: American Psychological Association.

Gernsbacher MA. 1990. *Language Comprehension as Structure Building*. Hillsdale, NJ: Lawrence Erlbaum Associates.

Gibson JJ. 1977. The theory of affordances. In *Perceiving, Acting, and Knowing: Toward an Ecological Psychology*, ed. RE Shaw, J Bransford, pp. 67–82. Hillsdale, NJ: Lawrence Erlbaum Associates.

Givón T. 1998. On the co-evolution of language, mind and brain. *Evolution of Communication* 2: 45–116.

Givón T. 2002. The visual information-processing system as an evolutionary precursor of human language. In *The Evolution of Language out of Pre-language*, ed. T Givón, B Malle, pp. 3–51. Amsterdam: Benjamins.

Goodall J. 1979. Life and death at Gombe. *National Geographic* 155: 592–620.

Gould SJ. 1977. *Ontogeny and Phylogeny*. Cambridge, MA: Harvard University Press.

Greenfield P. 1991. Language, tools and brain: The ontogeny and phylogeny of herarchically organized sequential behavior. *Behavioral and Brain Sciences* 14: 531–95.

Gupta P, MacWhinney B. 1997. Vocabulary acquisition and verbal short-term memory: Computational and neural bases. *Brain and Language* 59: 267–333.

Harlow H. 1958. The nature of love. *American Psychologist* 13: 573–685.

Hauser M, Chomsky N, Fitch T. 2002. The faculty of language: What is it, who has it, and how did it evolve? *Science* 298: 1569–79.

Hewes GW. 1973. Primate communication and the gestural origin of language. *Current Anthropology* 14, 5–24.

Hockett CF, Altmann SA. 1973. A note on design features. In *Animal Communication*, ed. TA Sebeok, pp. 61–72. Bloomington, IA: Indiana University Press.

Hockett CF, Ascher R. 1964. The human revolution. *Current Anthropology* 5: 135–67.

Holloway R. 1995. Toward a synthetic theory of human brain evolution. In *Origins of the Human Brain*, ed. J-P Changeux, J Chavaillon, pp. 42–60. Oxford: Clarendon Press.

Hudson R. 1984. *Word Grammar*. Oxford: Blackwell.

Ingmanson EJ. 1996. Tool-using behavior in wild *Pan paniscus*: Social and ecological considerations. In *Reaching into Thought: The Minds of the Great Apes*, ed. AE Russon, KA Bard, ST Parker, pp. 190–210. New York: Cambridge University Press.

Johnson M, Morton J. 1991. *Biology and Cognitive Development: The Case of Face Recognition*. Oxford: Basil Blackwell.

Jolly CJ, Plog F. 1982. *Physical Anthropology and Archeology*. New York: Alfred Knopf.

Julész B, Kovacs I, eds. 1995. *Maturational Windows and Adult Cortical Plasticity*. New York: Addison-Wesley.

Jürgens U. 1979. Neural control of vocalization in nonhuman primates. In *Neurobiology of Social Communication in Primates*, ed. HD Steklis, MJ Raleigh, pp. 82–98. New York: Academic Press.

König R, Heil P, Budinger E, Scheich H. 2005. *The Auditory Cortex: A Synthesis of Human and Animal Research*. Mahwah, NJ: Lawrence Erlbaum Associates.

Konishi M. 1995. A sensitive period for birdsong learning. In *Maturational Windows and Adult Cortical Plasticity*, ed. B Julesz, I Kovacs, pp. 87–92. New York: Addison-Wesley.

Lee P. 1986. Early social development among African elephant calves. *National Geographic Research Journal* 2: 22–34.

Li P, Farkas I, MacWhinney B. 2004. Early lexical development in a self-organizing neural network. *Neural Networks* 17: 1345–62.

Lieberman P. 1975. *On the Origins of Language: An Introduction to the Evolution of Human Speech.* New York: Macmillan.

Locke J. 1995. Development of the capacity for spoken language. In *The Handbook of Child Language*, ed. P Fletcher, B MacWhinney, pp. 278–302. Oxford: Basil Blackwell.

MacLarnon A, Hewitt G. 1999. The evolution of human speech. *American Journal of Physical Anthropology* 109: 341–63.

MacNeilage P. 1998. The frame/content theory of evolution of speech production. *Behavioral and Brain Sciences* 21: 499–546.

MacWhinney B. 1982. Basic syntactic processes. In *Language Acquisition, Vol. 1: Syntax and Semantics*, ed. S Kuczaj, pp. 73–136. Hillsdale, NJ: Lawrence Erlbaum Associates.

MacWhinney B. 1987. The competition model. In *Mechanisms of Language Acquisition*, ed. B MacWhinney, pp. 249–308. Hillsdale, NJ: Lawrence Erlbaum Associates.

MacWhinney B. 1998. Models of the emergence of language. *Annual Review of Psychology* 49: 199–227.

MacWhinney B. 2005a. Item-based constructions and the logical problem. *ACL* 2005: 46–54.

MacWhinney B. 2005b. The emergence of grammar from perspective. In *The Grounding of Cognition: The Role of Perception and Action in Memory, Language, and Thinking*, ed. D Pecher, RA Zwaan, pp. 198–223. Mahwah, NJ: Lawrence Erlbaum Associates.

McBrearty S, Brooks A. 2000. The revolution that wasn't: A new interpretation of the origin of modern human behavior. *Journal of Human Evolution* 39: 453–563.

McGonigle BO, Chalmers M. 1992. Monkeys are rational! *Quarterly Journal of Experimental Psychology* 45B: 189–228.

McGonigle BO, Chalmers M. 2002. The growth of cognitive structure in monkeys and men. In *Animal Cognition and Sequential Behavior*, ed. S Fountain, M Bunsey, J Danks, M McBeath, 264–313. Boston: Kluwer.

McGonigle BO, Chalmers M. 2006. Ordering and executive functioning as a window on the evolution and development of cognitive systems. *International Journal of Comparative Psychology.*

McHenry HM. 1986. The first bipeds: A comparison of the *A. afarensis* and *A. africanus* postcranium and implications for the evolution of bipedalism. *Journal of Human Evolution* 15: 177–91.

McManus IC. 1999. Handedness, cerebral lateralization, and the evolution of language. In *The Descent of Mind: Psychological Perspectives on Hominid Evolution*, ed. MC Corballis, S Lea, pp. 194–217. Oxford: Oxford University Press.

McNeill D. 1985. So you think gestures are nonverbal? *Psychological Review* 92: 350–71.

Miller G. 2001. *The Mating Mind: How Sexual Choice Shaped the Evolution of Human Nature.* New York: Anchor.

Mithen S. 1996. *The Prehistory of the Mind: The Cognitive Origins of Art, Religion, and Science.* London: Thames and Hudson.

Montagu A. 1955. Time, morphology, and neoteny in the evolution of man. *American Anthropologist* 57: 27–35.

Morgan E. 1997. *The Aquatic Ape Hypothesis.* London: Souvenir Press.

Nairne JS. 1990. A feature model of immediate memory. *Memory and Cognition* 18: 251–69.

Nauta WJH, Karten HJ. 1970. A general profile of the vertebrate brain, with sidelights on the ancestry of cerebral cortex. In *The Neurosciences*, ed. GC Quarton, T Melnechuck, G Adelman, pp. 7–26. New York: Rockefeller University Press.

O'Grady W. 2005. *Syntactic Carpentry.* Mahwah, NJ: Lawrence Erlbaum Associates.

Oller DK. 2000. *The Emergence of the Speech Capacity*. Mahwah, NJ: Lawrence Erlbaum Associates.

Paget R. 1930. *Human Speech*. New York: Harcourt Brace.

Pandya DP, Seltzer B, Barbas H. 1988. Input-out organization of the primate cerebral cortex. In *Comparative Primate Biology: Neurosciences*, ed. H Steklis, J Irwin, pp. 38–80. New York: Liss.

Pine J, Lieven E. 1993. Reanalysing rote-learned phrases: Individual differences in the transition to multiword speech. *Journal of Child Language* 20: 551–71.

Ploog DW. 1992. Neuroethological perspectives on the human brain: From the expression of emotions to intentional signing and speech. In *So Human a Brain: Knowledge and Values in the Neurosciences*, ed. A Harrington, pp. 3–13. Boston: Birkhauser.

Potts R. 1988. *Early Hominid Activities at Olduvai Gorge*. New York: Aldine de Gruyter.

Povinelli DJ, Cant JGH. 1995. Arboreal clambering and the evolution of self-conception. *Quarterly Journal of Biology* 70: 393–421.

Ramachandran VS, Hubbard EM. 2001. Synaesthesia: A window into perception, thought and language. *Journal of Consciousness Studies* 8: 3–34.

Rice S. 1997. The analysis of ontogenetic trajectories: When a change in size or shape is not heterochrony. *Proceedings of the National Academy of Sciences* 94: 907–12.

Rizzolatti G, Fadiga L, Gallese V, Fogassi L. 1996. Premotor cortex and the recognition of motor actions. *Cognitive Brain Research* 3: 131–41.

Savage-Rumbaugh S. 2000. Linguistic, cultural and cognitive capacities of bonobos (*Pan paniscus*). *Culture and Psychology* 6: 131–53.

Seyfarth R, Cheney D. 1999. Production, usage, and response in nonhuman primate vocal development. In *Neural Mechanisms of Communication*, ed. M Hauser, M Konishi, pp. 57–83. Cambridge, MA: MIT Press.

Stanford CB. 2003. *Upright: The Evolutionary Key to Becoming Human*. New York: Houghton Mifflin.

Stanford CB. 2006. Arboreal bipedalism in wild chimpanzees: Implications for the evolution of hominid posture and locomotion. *American Journal of Physical Anthropology* 129: 225–31.

Stringer C, McKie R. 1996. *African Exodus*. London: Pimlico.

Tattersall I. 1993. *The Human Odyssey: Four Million Years of Human Evolution*. New York: Prentice-Hall.

Tomasello M. 1999. *The Cultural Origins of Human Communication*. New York: Cambridge University Press.

Tomasello M. 2000. The item-based nature of children's early syntactic development. *Trends in Cognitive Sciences* 4: 156–63.

Trevarthen C. 1984. Biodynamic structures, cognitive correlates of motive sets and the development of motives in infants. In *Cognition and Motor Processes*, ed. W Prinz, AF Sanders, pp. 327–50. Berlin: Springer.

Tucker D. 2002. Embodied meaning. In *The Evolution of Language out of Pre-language*, ed. T Givón, B Malle, pp. 51–82. Amsterdam: Benjamins.

Turnbull C. 1972. *The Mountain People*. New York: Simon and Schuster.

Visalberghi E, Limongelli L. 1996. Acting and undertanding: Tool use revisited through the minds of capuchin monkeys. In *Reaching into Thought: The Minds of the Great Apes*, ed. AE Russon, KA Bard, ST Parker, pp. 57–79. Cambridge, England: Cambridge University Press.

Vygotsky L. 1934. *Thought and Language*. Cambridge, MA: MIT Press.

Wilkins WK, Wakefield J. 1995. Brain evolution and neurolinguistic preconditions. *Behavioral and Brain Sciences* 18: 161–226.

Yerkes RM, Learned BW. 1925. *Chimpanzee Intelligence and Its Vocal Expressions*. Baltimore: Williams and Wilkins.

10 Language and Niche Construction

Kim Sterelny

Organism and Environment

In conventional evolutionary thought, selection is seen as shaping an evolving lineage to fit its environment, as a key is shaped to fit a lock. For example, the Australian interior is challenging for plant life. It is very dry. It is subject to fire and other disturbances. Soils are infertile. Plants respond. They evolve disturbance-tolerant (especially fire-tolerant) life cycles. They evolve physical adaptations to reduce water loss, and deep root systems to extract the moisture that is available. They are metabolically thrifty and guard their tissues with toxic chemicals. Eucalypts, banksias, and acacias fit an Australian world by adapting in response to the browning of Australia. The relationship between environment and lineage is asymmetrical: Environmental change has caused lineage change, but not vice versa.

However, though lineages sometimes adapt to an environment, sometimes they adapt their environment instead. Think not of banksias but termites. Termites live in massive structures that equilibrate temperature and humidity. They resist physical disturbance and exclude all but the most specialized of predators. Termites inhabit their own miniuniverses. There is a fit between termites and their world not because termites have unidirectionally adjusted to their world but because they have in part adjusted their world to fit them (Turner, 2000). As Richard Lewontin argued, organisms are active; they partially construct their own niches.[1] Richard Dawkins has developed a similar line of the argument from the perspective of gene selection theory, showing that the adaptive effects of genes are often expressed outside the organism within which they are located (Dawkins, 1982a). His interactionist perspective on the relationship between lineage and environment is particularly apt when considering the evolution of language. Language is a human product, so it would be odd to think of its evolution in an externalist way—to think of language as an autonomous feature of the environment to which humans have adapted. Moreover, we are niche constructors par excellence. We modify our physical and biological environment: The selective pressures on humans have altered radically as a consequence of the inventions of tools, weapons, shelters, clothes, boats, cooking. However, like other organisms, we also modify our informational environment. We are informavores, and anciently so: Our techniques for extracting resources from our environment, and protecting ourselves from its dangers, depend on access to, and use of, impressive amounts of information (Kaplan, Hill, et al., 2000). Thus, humans are active epistemic agents—informational engineers as well as active physical agents (Clark, 1997). We intervene in the world to improve our access to information and to improve our ability to use information. Think of an act as simple and as normal as marking a trail while out in the bush. This intervention enormously simplifies coordination problems (making it easier for others to follow you) and navigation problems (making it easier to find your way back).

In this chapter I aim to show the relevance of nice construction to the evolution of language. The invention and elaboration of language is itself a crucial alteration of the informational world in which humans live. It is a crucial example of information engineering. Language makes social learning vastly more accurate and powerful, and so it enhances our capacity for the intergenerational transfer of skills and information and thus improves our access to physical and biological resources (Bickerton, 2005). But it also transforms the selective forces acting on social interactions. For instance, language enhances selection for cooperation by magnifying the importance of reputation. Once language has been invented, gossip becomes a powerful mechanism of reward and punishment (Dunbar, 1996; Wilson, Wilcznski, et al., 2000). The role of language in facilitating the transmission of information within and across generations, and in amplifying selection for communication, sets up a coevolutionary loop between language and the rest of culture: Each allows the other to become more elaborate. The elaboration of human cultural and social worlds intensified selection for languages of greater precision and expressive power. As languages of greater power and accuracy became available, that resource enabled human social worlds to become more elaborate.

One aspect of the coevolutionary elaboration of human culture and the cognitive tools that sustain it is cost management, and that will be the focus of this chapter. Humans manage their lifeways with the aid of cognitive technologies of many kinds. These include language, depictive representation, the use of templates, the cognitive division of labor, and the organization of work spaces (Clark, 2002). However, while these technologies enhance the cognitive power of human wetware, they have cognitive costs (Sterelny, 2004, 2006a). These can be contained by appropriate modifications of the tools and the environment in which these tools are used. I shall argue that niche construction is part of the mix of meta-adaptations through which we contain and manage the cognitive costs of using language.

I begin with the problem of honesty. Sharing information is a special case of cooperation, and it inherits the usual problems of explaining the evolution and stability of cooperative social lives. However, language-like systems are especially puzzling instances of cooperation. In the "Honest Signaling" section, I explain that prima facie problem and begin its solution, appealing both to population structure and to epistemic action. Even at early stages of language evolution, hominids (I argue) were preadapted to assess signal reliability, and those preadaptations enabled them to contain deception costs. Next, in the section "The Grammaticalization of Protolanguage," I turn to the evolution of a fully grammaticalized language and argue that quarantining the costs of deception was one of the factors driving grammaticalization. Finally, in the "Symbols, Signals, and Information Load" section, I distinguish between signaling systems and symbol-using systems. In signaling systems, signal transmission covaries with a specific, ecologically salient feature of the environment (external or internal emotional). They are systems of natural signs. Learning the system as a whole may be cognitively challenging.[2] However, the basic signal–world relationship is covariation. Detecting covariation does not, in general, require special cognitive machinery: Associative mechanisms in rats, pigeons, and people detect covariation. But covariation is not the key to understanding the word–world relationship: Words do not covary in space and time with their targets, and hence symbol use does have special cognitive costs, costs that are contained, I shall argue, by the division of linguistic labor.

Honest Signaling

In their classic paper on the evolution of signaling, Krebs and Dawkins made vivid the evolutionary problem of the evolution of honest signals (Krebs and Dawkins, 1984). In a competitive world, how could an agent be advantaged by honestly signaling to another about risks, opportunities, or resources? It is true that in one respect, deception is self-limiting. If deception were pervasive, other agents would cease to respond to linguistic signals, and there would be no point in sending them. The payoff for deceptive signaling is apt to decline as its proportion of total traffic increases. Therefore, we do not need any special explanation of the fact that *so long as agents continue to talk*, most talk will be honest. But why do they continue to talk? We need an explanation of the survival and elaboration of mostly honest signaling in the face of temptations to defect and free ride. Such an explanation cannot rely on mechanisms of surveillance and enforcement, since these presuppose a rich linguistic and social environment. They presuppose an environment in which norms can be expressed, taught, and enforced. Policing mechanisms cannot explain how a simple protolanguage used for honest signaling became established as part of our ancestors' lives, for they presuppose something richer than such a protolanguage. Moreover, though the proportion of deceptive signals may be low, the potential cost to an individual of a lie can be very high. Lies can kill. Audiences cannot afford to ignore the threat of deception, even if it is rare.

Honest signaling is an instance of the general evolutionary problem of cooperation; cooperation, in turn, can be selected by synergy: Two agents can generate more return to each acting together than acting alone. Brian Skyrms' *The Stag Hunt* is a game-theoretic exploration of such synergies. Two hunters acting together can capture a stag, whereas each hunting individually can take only a hare apiece. Cooperation is favored because a stag is worth more than twice a hare (Skyrms, 2003). The power of coordination builds a temptation to cooperate even for the prudent, selfish agent.

A mere exchange of information may not generate significant synergies. If I tell you the location of a fruiting tree today, and you do the same for me tomorrow, no synergy has been generated. Indeed, given that it is always rational to discount future benefits (you might be killed before you have a chance to reciprocate), I am behind the game. However, agents talk in the course of coordinating and planning their actions, thus improving the synergistic payoffs of other joint activities. Moreover, cooperation is important in reducing variance. Even skilled foragers have unlucky days, so it can pay to share, even if sharing does not increase your average return. But sharing insures against those days. In a world where signals from the environment are noisy and equivocal, communication might likewise have risk reduction functions. You judge that the river is safe to cross. But even for the experienced agent, such judgments are uncertain. Perhaps sampling the opinions of others can reduce this uncertainty. Christian List has explored this idea using the Condorcet jury theorem. Suppose each member in a group has a better than .5 but worse than 1.0 chance of determining whether this river is indeed safe to cross. As the number of agents in the group goes up, the chance of a correct majority vote rises rapidly. Information pooling increases the reliability of judgment in a noisy and uncertain world (List, 2004). Risk reduction might help underpin the evolution of cooperative signaling.

The benefits of coordination, planning, and information pooling might explain why I should listen. But why should I signal? Despite the benefits of synergies, free riding is a potential barrier to the evolution of any type of cooperative social interaction. Moreover, there are specific barriers to the evolution of cooperative communication. Signaling creates a public good. Information leakage imposes a serious cost, for honest talk generates benefits from which third parties cannot be easily excluded. Eavesdropping imposes a tax on honest talk. This tax is likely to be high in the initial stages of language evolution. The more effortful we suppose protolinguistic communication to have been (and hence slower, repeated, and with redundancy) and the more dependent on pragmatic scaffolding for its interpretation (contextual cues, ostension, and the like), the harder it would be to exclude bystanders. Whispering conspiratorially in the dark requires advanced language skills, for it requires interpretation to be decoupled from context.[3] Moreover, defection through failure to reciprocate is hard to detect. How will other agents discover that you knew yesterday of some resource or danger that they discover tomorrow? Active deception will sometimes be unmasked, but failures to signal are a more cryptic form of free riding.

In short, there does seem to be good reason to believe that honest communication generates synergies. It is a direct aid to coordination in cooperative economic activities: The stag hunters who talk and plan will probably do better than those that remain mute. And information pooling increases the reliability of individual judgment in informationally noisy environments. But those benefits can be eroded by free riders, so how can those costs be contained? Population structure helps block the invasion of deceptive strategies. Game theoretic models show that in a mixed population, in which all agents interact with all agents, free riding prevents the stabilization of honest, low-cost signaling systems. However, if the population is structured into small knots of agents who interact primarily with one another, cooperative strategies can be stable (Grassly, Haeseler, et al., 2000). This general idea comes in a number of different forms. Thus, Tecumseh Fitch has suggested that protolanguage begins within the family, between a mother and her offspring. Family members have overlapping evolutionary interests, so predominantly honest signaling within the family would be no surprise (Fitch, 2004). I have suggested a related idea, arguing that conflicts of interest within human groups are partially suppressed by selection for cooperation on groups. This allows cooperative signaling to evolve in tandem with other forms of cooperation (Sterelny, 2003).

These ideas both link honest signaling to the minimization of competition between signaling agents, but I prefer my form of this idea to that of Fitch. For one thing, I do not think Fitch has a plausible model of how language leaves the family. He relies on the benefits of reciprocation to amplify the circle of conversational exchange. However, there are well-known problems in scaling up reciprocation from two-player to many-player contexts. Withdrawing cooperation is a very blunt instrument for policing cooperation in *n*-player interactions, for it penalizes cooperators as much as free riders (Sripada, 2005). Moreover, there is a gap in Fitch's story. His model seems to predict the evolution of protolanguage in chimpanzees, for they too are characterized by long-lasting associations between a mother and her offspring. Among common chimpanzees, a mother and her offspring often spend years foraging together, endlessly in one another's company. Why, then, is there no chimpanzee protolanguage?[4] The overall point, though, is that the costs of deception can be reduced by population structure. In part, population structure is independent of hominin agency: It is a response to the local geography and ecology. However, in other respects it is an effect of human agency: The extent to which the populations within a metapopulation differ from one another will depend on their customs of trade, intermarriage, and migration. Some patterns of group–group interaction will tend to damp down between-group differences; others will accentuate them.

The reduction of competition between conversational partners provides part of the explanation of the establishment of honest and cooperative signaling. Equally relevant were the preexisting capacities for epistemic action. Agents intervene in their physical environment, and they intervene in their informational environment too. Agents act epistemically to improve access to the information they need, engaging in both long-term informational engineering and short-term fixes. So consider, again, a leopard signal in the absence of a leopard. Will such signals tend to undermine the whole practice of signaling? Hardly: After all, no detection process is perfect. Leopard signals will never have been perfectly correlated with their adaptive targets. Protolanguage was not "born honest," beginning with signals that perfectly covaried with their targets and then becoming less reliable only with the invention of deception. Hominins did not need to wait for The Fall to first hear "leopards" without leopards. We evolved in environments in which predators, competitors, and prey hid, camouflaged themselves, and disguised their intentions. They degraded the informational environment. In such environments, despite an agent's best efforts, pickup of information can never be perfect, and the accuracy of information pickup constrains signal reliability.

Intelligent agents—and protolanguage-using hominins were intelligent—were likely to be aware of imperfect correlations between signal and target and to have established strategies to limit the costs of such failures. An agent hearing "leopard" has options in addition to those of ignoring the call or engaging in leopard flight. The audience can become more alert. In addition, they can probe the environment: They can actively scan, move to a better vantage point, and suppress extraneous noise. They can monitor others, including nonhuman others—a region of bush might have gone suspiciously quiet. If protolanguage is sufficiently rich, they can interrogate the signaler—asking exactly what the signaler saw; how well; how far away, perhaps while monitoring the state of the signaler. Responses to specific signals are not all-or-nothing, and audiences have tools to assess the reliability of specific signals. Those tools are needed because agents signal in difficult informational environments. Signal–target failures are bound to occur in such environments, and hence signal reliability assessment is needed independent of, and prior to, the threat of deception.

As we have seen, intelligent hominids are likely to have had capacities that preadapted them to the problem of deception. In assessing the reliability of signals, deception does not seem to be strikingly different from mere error. However, as we shall see in more detail in the "Symbols, Signals, and Information Load" section, language is not a signaling system. Much of the literature on the evolution of protolanguage treats it as a kind of super-vervetese: a system for communicating about the here and now. Signs are arbitrary, but they stand for objects in the signaler's local environment. Perhaps protolanguage began as such a system, but language is independent of the local environment. We can talk about the elsewhere and the elsewhen and about the possible, the impossible, and the imaginary as well as the actual. Communication systems that are decoupled from the current environment pose new reliability-checking problems. Changing positions for a better look is not much help in reliability-testing reports about subjects displaced in time, let alone displacements from the actual.

Even so, audiences are not helpless in the face of this expansion of the expressive power of language. Dan Sperber has argued that metalinguistic capacities are adaptations that limit the dangers of deceptions that are beyond the scope of direct checking. Metarepresentation is a tracking and calibration device. Tagging the source of information is essential if we are to keep track of the trustworthy. Thus, it is important to be able to represent the fact that it is Dave who said that the caves were empty and the creek was full. Moreover, we have a folk logic: We can explicitly represent the truth and falsity of statements, the soundness or invalidity of inferences. These are *folk tools*. For the first 2,000 years of its history, the discipline of logic may have done little more than systematize and organize folk judgments of validity and invalidity. Folk logic is a toolkit for assessing the reliability of what we are told when we are in no position to check the truth of assertions directly. It gives us techniques of indirect assessment. It is possible to reason without representing reasoning; it is possible to believe truly without the concept of truth. However, we could not assess and interrogate the assertions of others without metarepresentational tools in our language. These are tools that enable audiences to vet signals; they are tools for epistemic action (Cosmides and Tooby, 2000; Sperber, 2001; Sterelny, 2006b).

In short, humans have adapted their linguistic tools and their social environment to reduce the costs of unreliable signals while exploiting the benefits of honest communication. Restricting the size of conversational circles, exploiting our capacities for epistemic action, and constructing a metalinguistic apparatus to vet signals and signalers all help in keeping communication mostly honest and fairly reliable. We have engineered language and the environment in which we use it in response to the challenges of assessing reliability. I shall further suggest that fundamental change in the organization of language itself—the grammaticalization of protolanguage—is in part driven by vetting issues. Grammaticalization allows agents to quarantine the problem of deception, and hence to reduce the costs of using language, by using automated, encapsulated mechanisms to process those aspects of language for which no threat of deception arises.

The Grammaticalization of Protolanguage

I have just suggested that there are structures in language itself that function to reduce the costs of deception: our metarepresentational apparatus. I think that an even more fundamental feature of language, its regularized syntax, exists in part to reduce costs of deception. The usual supposition is that the evolution of language involves a trajectory from a system without a regularized morphology and syntax to a phonologically, morphologically, and syntactically patterned system. In the earlier, protolanguage-like stages of this trajectory, interpreting conversation would have depended very heavily on pragmatics: on context, gesture, and background knowledge. At this stage, utterances are short, consisting of just a few word-like elements but without fixed order, syntactic elements, or indications of mood (Bickerton, 1990; Jackendoff, 1999). As hominin communication evolved, it acquired patterned ways of indicating number, tense, aspect, subject and object, and illocutionary force. Context became less important. How is this trajectory to be explained? The evolution of a generative phonology and lexicon is, I suspect, a response to the increasing unpredictability of human environments. But syntax, I shall argue, helps quarantine the costs of assessing others' reliability. I will develop these ideas by contrasting them with an alternative.

One idea is that grammaticalization is driven by learning costs.[5] Rather than supposing that the human mind is adapted to the structure of language (by being provided with information about that structure innately), Simon Kirby, in association with Henry Brighton and Kenny Smith, suggests that language is adapted to human psychology. In particular, they argue that the fundamental organizational feature of language—its compositionality—is a consequence of selection for learning (Smith et al., 2003; Brighton et al., 2005). Holistic communication systems are ones in which every signal must be learned independent of every other signal. They model the transitions from these to compositional systems, that is, systems in which the significance of a signal depends on the units from which the signal is constructed together with its structure. In many of their simulations there is no evolution away from holistic systems. However, when this transition occurs, it is driven (they argue) by a learning bottleneck, for any human language must be learned by the $N + 1$ generation from the sample provided by generation N. As a consequence, languages are under cultural selection for learnability, since the $N + 1$ generation must reconstruct language from this limited input. There is a sample size filter through which language passes, and they argue that this explains why languages are compositional and recursive.

The general point that languages are under selection for learnability, and that features of language may be explained by structures of the language-using community rather than the structure of individual minds, is surely well-taken. However, Kirby's iterated learning model is misleadingly idealized. For one thing, protolanguage—if it is anything like the systems Bickerton or Jackendoff have in mind—is not a holistic system, for in holistic systems similarity in meaning is not correlated with similarity in signal. A holistic system maps signals onto the world, but knowledge of one signal–world correlation tells you nothing about the next. A signal meaning that "Bill is large" is no more likely to be similar to a signal meaning that "Bill is small" than one randomly chosen. No chunk about Bill appears in both signals. Distance in signal space is not correlated at all with distance in semantic space.

This holistic pattern does not characterize protolanguage, especially if iconic representation—utterances linked with gesture and mime—was important in the early stages of protolanguage. The difficulty of interpreting protolanguage is generated by ambiguity, not holism. Signal interpretation depends on context, so one signal type can have many meanings. And if word order is not fixed, different signal types can have the same meaning. However, though protolanguage utterances do not have regularized linguistic meaning, expressions (subparts of utterances) in protolanguage are *portable*. An expression that picks out Bill in one utterance can be reused to pick him out again in another. Similarity in signal space *is* correlated with similarity in semantic space, but the signal–world map is many-to-many. Even granted the inevitable simplifications of model building, the iterated learning model does not characterize the initial state insightfully.

Learnability doubtless plays some role, but compositional signal systems are mechanisms of phenotypic plasticity. They enable agents to produce novel signals in response to novel environments. Kirby, Brighton, and Smith model language change in a range of different environments. However, within each run of a model, the environment is fixed. I suspect that this idealization may explain the fact that in their models recursive organizations do not typically evolve from holistic systems. No system with a fixed menu of signals is adequate in changing or unpredictable environments. Yet human environments are variable: The array of artifacts, resources, dangers, individuals, and social relations important to an agent cannot be predicted in advance. These features of human environments change both within and across generations. Even in stable environments, the expressive power of such systems would be limiting. Agents with a fixed menu of distinct signals will have to leave times, dates, places, and perhaps even agents unspecified, thus throwing an enormous burden onto the pragmatics of interpretation. The problem of limited expressive power is yet more serious if the environment is changeable. A compositional system gives agents the ability to signal about many new states of affairs for free. If you have the capacity to signal that a tiger is at the lake, and to signal that a bear is in the cave, you get for free the capacity to signal that a bear is at the lake.

In the light of these considerations of novelty, Jackendoff is right to emphasize the importance of compositional phonology, for it allows speakers to generate an indefinite number of well-defined word blanks that can be recruited as terms for novel phenomena (Jackendoff, 1999). Protolanguages, as Jackendoff and Bickerton conceive them, contrast with holistic systems in being open-ended. Open-ended systems are needed by agents whose communicative needs cannot be predicted in advance, because their world cannot be predicted in advance. And ancient humans lived in an increasingly unpredictable environment, both because they themselves became increasingly important drivers of change, and because the climate became increasingly unstable (Potts, 1996, 1998). Selection for phenotypic plasticity selects for a readily expandable system, perhaps especially for a generative, indefinitely extendable lexicon. A regular phonology is important as well, for it imposes discreteness on the menu of basic referential signals. Without compensating adaptations, as the size of the lexicon increases, so too will the rate of misunderstanding: There will be an error threshold that constrains the size of the lexicon (Grassly et al., 2000). If there is selection for an increasing lexicon, there will also be selection for mechanisms making signals more obviously distinct from one another.

But what of syntax? One possibility is that compositionality quarantines the honesty problem and sets up an adaptive division of labor between consciously supervised and automatic language-processing cognitive systems. Let me begin with a crucial distinction: that between understanding and acceptance. Conflicts of interest and, hence, the temptation to deceive arise over acceptance of the sincerity of the speaker, but not over understanding of the speaker. All parties to a conversation need to have the cognitive resources to ask themselves, "Why is he saying that to me?" We need to be able to identify the pragmatic point of conversation in wholly cooperative interchanges, so that honest uses of language fulfill their coordinating role. Even more obviously, we need to have tools for evaluating others' purposes in conversation, to guard against unfriendly manipulation of our thoughts and deeds. Conflicts of interest make the assessment of another's motives and reliability important. The independence of language from local context makes it difficult. Thus, assessing the reliability and point of an utterance cannot safely be left to an encapsulated, automatic mechanism. We need the capacity to reflectively evaluate the expressed views of others to reduce the error costs of learning from others while keeping its benefits. This assessment can depend on any information to which an agent has access. Unmasking an utterance as deceptive (or merely unreliable) can turn on an apparently recondite detail. Much detective fiction turns on this: the subtle seeding of the plot with a tiny unmasking detail, which leads the protagonist, and can lead the reader, to the unmasking inference.

There is less need for reflective evaluation in identifying the features of an utterance that are internal to language: its lexical and phrasal organization. These features of utterances involve no temptation to deceive. There is no deception problem with syntax: Whatever the ultimate motives of people's talking—whether they are honest cooperators or defectors—it is in their interests to make the organization of their utterance—the lexical items in use: its clausal structure, its subject/predicate organization, and its tense/aspect organization—as transparent as possible to the target of their speech. Whatever your aims in speech, those aims will miscarry if you do not secure minimal understanding of what you say. Likewise, it is in the interests of conversational targets to understand what is said, whether or not they accept what is said to them. The expressed views of others are data, whether or not those views are true and whether or not those views are sincere. From this perspective, it is no surprise that while we have a quite rich and detailed folk logic, we have a much less elaborate folk syntax or folk phonology. If dialects really track in-group and out-group distinctions (Nettle and Dunbar, 1997), it might be important to notice that another agent sounds foreign, but there is no special need to be able to explicitly represent the aspects of speech that make other agents sound unfamiliar. And while the folk are quite good at noticing dialectal differences, those without any formal training in linguistics are pretty hopeless at identifying the source of those differences.

The form of language is relevant to its persuasive force and to its memorability, so there has long been reason to develop an awareness of how we say what is said. Thus, we are not completely without the capacity to represent form. Even so, form and organization generate no temptation to deceive. All parties in a linguistic transaction share an interest in making it as easy as possible to identify the lexical items and the way they have been structured into a sentence. Thus, it is adaptive both to individual language users and their speech community to automatize phonology, morphology, and syntax. We should expect speakers and targets to coevolve in ways that reduce the burdens of language processing. Ruth Millikan has written of language as a system of natural telepathy. In the normal case, when a speaker expresses the thought that there is a tiger in the bottom wood, the belief there is a tiger in the bottom wood appears in the audience's mind without that audience's representing the informational channel through which that information has been pumped from the speaker's mind to the target's mind (Millikan, 1998).

Human life is not cooperative enough for natural telepathy to be a good general model of linguistic interaction. However, a cut-down version of Millikan's idea is a good model: When a speaker asks a question, her audience takes her to have asked a question without the audience's having to attend to the informational channel that carries that structural information. For phonology, morphology, and syntax, Millikan's "natural telepathy" metaphor is apt. Language has been engineered to make these aspects of language transparent. They can then be subcontracted to automatic systems, saving the scarce resources of attention, metarepresentation, and top-down monitoring for domains where those resources are needed: for a decision on acceptance. Once conversational interaction was no longer limited to families, or to small groups of reliably cooperating and repeatedly interacting agents, the need for reliability assessment would increase. Yet this assessment is cognitively demanding. While we have these capacities, despite the importance of reliability assessment, we use them sparingly. The modularization of phonology and syntax might have been crucial to freeing the cognitive resources needed to engage in the attention-rich, informationally demanding, metalinguistic scrutiny of what is said to us.

The evolutionary transition from a basic protolanguage to full human language involved a multitude of changes to phonology, morphology, and syntax. There is no reason to suppose that a single selective force was responsible for all these changes. Kirby and his colleagues are surely right at some grain of analysis: Any change to the system that made it more patterned and more regular would have eased some learning problems, as samples of linguistic experience become better guides to the whole. Compositionality and portability allow adaptive linguistic response to environmental uncertainty. However, I also think it likely that as language became more complex, there was selection to automatize the processing of those aspects of language that could, safely, be automatized. Protolanguage, without a well-defined syntax, lacks a clear distinction between what is said and what a speaker is up to in saying what was said. Redesigning the system—regularizing it—introduces this distinction and thus allows much language processing to be automatized. Automatizing language will free attentional and memory resources, and those resources are always scarce. Freeing them will have many advantages. One, I have argued, is more effective counter-deception strategy.

Symbols, Signals, and Information Load

Formal models of the evolution of language often conceive language as a system of arbitrary, low-cost, referential signs. This is a reasonable first approximation, but it misrepresents the cognitive challenge of acquiring a lexicon. Referential signaling systems among animals are poor models of word use in human protolanguages, even early protolanguages. Referential calls are semantically very different from lexical items. The vervetese leopard call is not well-translated as "leopard." The leopard call signals a state of affairs: Roughly "there is a leopard nearby." Such signals are natural signs of the phenomena they indicate. Many shorebirds and waterbirds feed together in rather exposed places, and in rather dense flocks, and they use the alarm and escape behavior of other birds as a signal to take wing. They recruit the ecological activities of other birds as signals. There is a rough but useful covariation between the alarm flights of (say) pied stilts and danger, and many birds on an estuary will exploit that natural connection between one very easily observed phenomena—the cloud of stilts panicking into the air—and another, much less easily observed one (a stalking feral cat; Danchin, 2004). Likewise, vervets treat the leopard calls of other vervets as a natural sign of leopard presence. Natural signs are just covariations between one kind of event and another. Learning to use others' calls as natural signs is not especially challenging. Associative mechanisms can detect and exploit such contingent connections.

Words are not natural signs of their referents. In conversations we talk of the benefits for which we hope, and the potential dangers we wish to avoid. We talk of other times and places and describe the possible as well as the actual. We tell stories. For that reason, it would be quite hopeless to regard the English term "leopard" as a natural sign of leopards. Moreover, words are often not learned by association with their referents. Ostensive definition plays a role in language acquisition, but much of language is acquired via other representations: through depiction and description. Thus, associative mechanisms that are powerful enough to detect the covariation between the vervet leopard call and leopards would not suffice to map "tiger" onto tigers or "Charles Darwin" onto Charles Darwin. Few "tiger" tokens are copresent with tigers, and these days no "Charles Darwin" tokens appear with Charles Darwin.

This negative point is universally accepted. Words are not natural signs of their referents. They are symbols of their referents (Deacon, 1997). There is, however, no consensus whatever on the nature of lexical symbols: on what an agent has to understand to use "leopard" as a symbol of leopards. However, though there is no consensus on the nature of a linguistic symbol, on every view, using a symbol is informationally demanding. My approach will be as follows. I shall give *one account* of the informational demands of symbol use, then show how humans exploit their social environment to minimize those demands. I shall argue that the large and varied vocabularies we all command depend on a socially organized division of linguistic labor (Putnam, 1975). Importantly, this view of the effects of social organization on the informational demands of language does not depend on the specific account of the symbol–reference relationship I discuss. That account merely illustrates the way we collectively manage the problem of information load.

How is it that my tokens of "leopard" are grounded on leopards? On one family of views, "leopard" is a symbol of leopards (rather than, say, panthers) because competent users of English associate that term with something like an identifying criterion of leopards. The use of "leopard" as a symbol of leopards depends on a speaker's having the capacity to identify leopards as leopards. The use of language is information rich. Though referential expressions do not need to be used in the presence of their referents, their use as symbols depends on the presence of an informational image of leopards. No leopard has to be present, but identifying information does have to be present. Likewise, while Charles Darwin does not have to be present for me to talk about him, identifying information about him is present: I know Darwin to be the codiscoverer of the principle of natural selection, the author of *The Origin of Species* and *The Descent of Man*, and I use "Charles Darwin" intending thereby to refer to that very person. The use of linguistic symbols is anchored in descriptive information available to the parties of a conversation about the targets of those symbols.

There is an obvious objection to this view of linguistic symbols: Ignorant and mistaken speakers seem to be able to talk about leopards. Not everyone who can use the term "leopard" can identify leopards; some of those who talk about the Darwins of history are confused or ignorant about what they did. This is an important objection, but, equally, there is an important response. Within the community, there are leopard experts. They do have identifying knowledge of leopards. They use the term "leopard" as a symbol of leopards, for they use it with the intention of referring to that great cat that satisfies the following description: the largest spotted cat in Africa and Asia, solitary in habits, adult length (including tail) between 2 and 2.5 meters (Sunquist and Sunquist, 2002). In the same community, there will be speakers whose grip on the distinction between leopards, jaguars, and snow leopards is hazy, but no matter. They use the term "leopard" intending to refer to that great cat to which mammologists refer in using the term "leopard." That is a unique description of leopards: That description is true of one and only one great cat. The ignorant do have identifying knowledge of leopards, but it is second-hand knowledge; it is the knowledge that others know about leopards, together with a way of specifying those others (Jackson and Braddon-Mitchell, 1996; Jackson, 1997). The problem of reference—the symbol grounding problem—is solved collectively rather than individually.

The division of linguistic labor goes beyond our practice of learning from others. People routinely use linguistic tags to pick out items in the world they have never seen and about which they know little. Sometimes, of course, they learn enough to become identifiers in their own right: They learn the distinguishing features of leopards or the distinctive ideas of Darwin. However, the point is that the use of linguistic symbols does not depend on the cultural transmission of identifying information about the targets of those symbols: It is still *Darwin* that the hopelessly confused creationist mischaracterizes, even though he or she has a garbled view of Darwin's life and works. The creationist's capacity to deploy a symbol for Darwin despite failing to grasp Darwinian ideas depends on a division of linguistic labor within communities. Individuals have differential access to aspects of their environment and differential knowledge about that environment. Those with little access to leopards defer linguistically to those with lots: to the experience and information of those whose worlds are leopard rich. That experience and information grounds "leopard" on leopards for all of us. Collectively, but only collectively, we know what we are talking about.

The idea that a division of linguistic labor is crucial to our ability to use a large lexicon is not tied to any specific account of the relationship between symbol and target in the mouths of experts. So-called "causal theorists" of reference think that the relation between a linguistic symbol and its target is constituted by an appropriate causal chain between language and the world. My "Charles Darwin" tokens refer to Charles Darwin, because there is an appropriate causal chain between my tokens of that name and the man. On this view, I need have no identifying knowledge of Darwin at all—not even a second-hand description like: "By 'Darwin' I intend to refer to the person historians of biology call 'Darwin.'" The relevant fact is that my disposition to use that term is caused by speakers whose disposition to use that term was caused by speakers whose disposition to use that term was caused by speakers whose disposition to use that term ... was caused by speakers who knew and named Darwin. Thus, causal theorists are also committed to the view that there is an informationally privileged subset of "Darwin" users. Those who grounded the name on the individual knew him. They had to know what they were doing and to whom they were referring when Charles Robert Darwin acquired his names (Devitt and Sterelny, 1999). On this view, too, the division of linguistic labor is important, and those who ground lexical symbols on their targets carry an informational burden for the rest of us. This favor is reciprocated, for most of us play some role in grounding the language of our community, if not for kind terms like "leopard," at least with proper names. We are all experts on our own inner circle.

In recent years, Dan Dennett and Andy Clark have pointed to the importance of cognitive tools in explaining human cognitive competence and its evolution (Clark, 1999; Dennett, 2000; Clark, 2001, 2002). Undoubtedly, the most important of these tools is language. We make tools for thinking, and these tools enormously enhance our cognitive powers. However, while these tools empower us, they are themselves cognitively demanding, as all of us who have stared despairingly at a computer manual know. Language magnifies our cognitive powers, but at the same time learning and using a language impose cognitive costs. In this chapter, I have discussed a few aspects of those costs and of how we have rejigged our social and linguistic environment to reduce those costs. In particular, I have discussed quality control and its costs. Quality control is necessary to keep the benefits of learning from others while reducing its error costs. But it is itself potentially very expensive, and I have discussed ways in which those costs were managed in the transitions from protolanguage to language.[6]

Notes

1. See Lewontin (1982, 1985, 2000) and more recently, and in great empirical detail, Odling-Smee, Laland, et al. (2003).

2. This is especially the case if there is a large menu of distinct signals. Moreover, error complicates learning: No signal will covary perfectly with its target state in the world (see Grassly, Haeseler, et al., 2000).

3. (Krakauer, 2001) suggests that the cultural evolution of arbitrary signals patches the public information problem, for only members of the local community will understand those symbols: Public information will not flow beyond the boundaries of the local group.

4. There is a plausible ecological hypothesis explaining selection for increased cooperation between early hominin adults. As Africa became hotter and drier, our ancestors found themselves livings in woodlands and savannah. This change in habitat intensified selection for male cooperation in defense (Dunbar, 2001) and for reproductive cooperation among females (Key and Aiello, 1999). If that is right, we can explain why in our lineage alone selection on groups for cooperation partially suppressed selection on individuals for free riding.

5. Or, analogously, it is driven by error costs in transmission: For this idea, see Nowak and Krakauer (1999).

6. Thanks to the editors, David Krakauer, and an anonymous referee for very helpful feedback on an earlier version of this chapter, and thanks also to the workshop participants for their comments and questions.

References

Bickerton D. 1990. *Language and Species*. Chicago: Chicago University Press.

Bickerton D. 2005. *The Origin of Language in Niche Construction*. http://www.derekbickerton.com/blog/_archives/2005/3/28/486319.html.

Brighton H, Kirby S, et al. 2005. Cultural selection for learnability: Three principles underlying the view that language adapts to be learnable. In *Language Origins: Perspectives on Evolution*, ed. M Tallerman, pp. 291–309. Oxford: Oxford University Press.

Clark A. 1997. *Being There: Putting Brain, Body, and World Together Again*. Cambridge, MA: MIT Press.

Clark A. 1999. An embodied cognitive science? *Trends in Cognitive Sciences* 3: 345–50.

Clark A. 2001. Reasons, robots and the extended mind. *Mind and Language* 16: 121–45.

Clark A. 2002. *Mindware: An Introduction to the Philosophy of Cognitive Science*. Oxford: Oxford University Press.

Cosmides L, Tooby J. 2000. Consider the sources: The evolution of adaptation for decoupling and metarepresentation. In *Metarepresentation: A Multidisciplinary Perspective*, ed. D Sperber, pp. 53–116. Oxford, Oxford University Press.

Danchin E, et al. 2004. Public information: From nosy neighbors to cultural evolution. *Science* 305: 487–91.

Dawkins R. 1982. *The Extended Phenotype*. Oxford: Oxford University Press.

Deacon T. 1997. *The Symbolic Species: The Co-evolution of Language and the Brain*. New York: Norton.

Dennett DC. 2000. Making tools for thinking. In *Metarepresentation: A Multidisciplinary Perspective*, ed. D Sperber, pp. 17–29. Oxford: Oxford University Press.

Devitt M, Sterelny K. 1999. *Language and Reality: An Introduction to Philosophy of Language*. Oxford: Blackwell.

Dunbar R. 1996. *Grooming, Gossip and the Evolution of Language*. London: Faber and Faber.

Dunbar R. 2001. Brains on two legs: Group size and the evolution of intelligence. In *Tree of Origin*, ed. F. de Waal, pp. 173–92. Cambridge, MA: Harvard University Press.

Fitch T. 2004. Evolving honest communication systems: Kin selection and "mother tongues." In *Evolution of Communication Systems: A Comparative Approach*, ed. DK Oller, U Griebel, pp. 275–96. Cambridge, MA: MIT Press.

Grassly N, Haeseler A, et al. 2000. Error, population structure and the origin of diverse sign systems. *Journal of Theoretical Biology* 206: 369–78.

Jackendoff R. 1999. Possible stages in the evolution of the language capacity. *Trends in Cognitive Sciences* 3: 272–9.

Jackson F. 1997. *From Metaphysics to Ethics*. Oxford: Oxford University Press.

Jackson F and Braddon-Mitchell, D. 1996. *The Philosophy of Mind and Cognition*. Oxford: Blackwell.

Kaplan H, Hill K, et al. 2000. A theory of human life history evolution: Diet, intelligence and longevity. *Evolutionary Anthropology* 9: 156–85.

Key C, Aiello L. 1999. The evolution of social organization. In *The Evolution of Culture: An Interdisciplinary View*, ed. R Dunbar, C Knight, C Power, pp. 15–33. Edinburgh: Edinburgh University Press.

Krakauer DC. 2001. Selective imitation for a private sign system. *Journal of Theoretical Biology* 213: 145–57.

Krebs J, Dawkins R. 1984. Animal signals, mind-reading and manipulation. In *Behavioural Ecology: An Evolutionary Approach*, ed. JR Krebs, NB Davies, pp. 380–402. Oxford: Blackwell Scientific.

Lewontin RC. 1982. Organism and environment. In *Learning, Development and Culture*, ed. HC Plotkin, pp. 151–70. New York: Wiley.

Lewontin RC. 1985. The organism as subject and object of evolution. In *The Dialectical Biologist*, ed. RC Lewontin, R Levins, 85–106. Cambridge, MA: Harvard University Press.

Lewontin RC. 2000. *The Triple Helix*. Cambridge, MA: Harvard University Press.

List C. 2004. Democracy in animal groups: A political science perspective. *Trends in Ecology and Evolution* 19: 168–9.

Millikan R. 1998. Language conventions made simple. *Journal of Philosophy* 94: 161–80.

Nettle D, Dunbar R. 1997. Social markers and the evolution of reciprocal exchange. *Current Anthropology* 38: 93–9.

Nowak M, Krakauer DC. 1999. The evolution of language. *Proceedings of the National Academy of Sciences* 96: 8028–33.

Odling-Smee J, Laland K, et al. 2003. *Niche Construction: The Neglected Process in Evolution*. Princeton, NJ: Princeton University Press.

Potts R. 1996. *Humanity's Descent: The Consequences of Ecological Instability*. New York: Avon.

Potts R. 1998. Variability selection in hominid evolution. *Evolutionary Anthropology* 7: 81–96.

Putnam H. 1975. *Mind, Language and Reality: Philosophical Papers, Volume 2*. Cambridge, Cambridge University Press.

Skyrms B. 2003. *The Stag Hunt and the Evolution of Social Structure*. Cambridge, England: Cambridge University Press.

Smith K, Kirby S, et al. 2003. Iterated learning: A framework for the emergence of language. *Artificial Life* 9: 371–86.

Sperber D. 2001. An Evolutionary Perspective on Testimony and Argumentation. *Philosophical Topics* 29: 401–413.

Sripada CS. 2005. Punishment and the strategic structure of moral systems. *Biology and Philosophy* 20: 767–789.

Sterelny K. 2003. *Thought in a Hostile World*. New York: Blackwell.

Sterelny K. 2004. Externalism, epistemic artefacts and the extended mind. In *The Externalist Challenge: New Studies on Cognition and Intentionality*, ed. R Schantz, pp. 239–54. Berlin and New York: de Gruyter.

Sterelny K. 2006a. Cognitive load and human decision, or, three ways of rolling the rock up hill. In *The Innate Mind, Vol. 2: Culture and Cognition*, ed. P Carruthers, S Laurence, S Stich, pp. 218–33. Oxford: Oxford University Press.

Sterelny K. 2006b. Folk logic and animal rationality. In *Rational Animals?*, ed. S Hurley, M Matthew Nudds, pp. 291–312. Oxford: Oxford University Press.

Sunquist M, Sunquist F. 2002. *Wild Cats of the World*. Chicago: University of Chicago Press.

Turner JS. 2000. *The Extended Organism: The Physiology of Animal-Built Structures*. Cambridge, MA: Harvard University Press.

Wilson DS, Wilcznski C, et al. 2000. Gossip and other aspects of language as a group-level adaptations. In *The Evolution of Cognition*, ed. L Huber, C Heyes, pp. 347–66. Cambridge, MA: MIT Press.

IV UNDERPINNINGS OF COMMUNICATIVE CONTROL: FOUNDATIONS FOR FLEXIBLE COMMUNICATION

Part IV takes a lateral move. Instead of focusing directly on foundations for vocal communication, which is the primary mode of human language, it addresses two key domains of evolution that are related to vocal communication and that are assumed to have played important roles in language evolution. These domains are communication through gesture and the role of play in cognition and communication. The approach in both chapters is comparative.

Part IV begins with Call's intriguing summary of results on gestural communication in apes. The importance of the work can hardly be overemphasized because it is clear that joint attention, supported especially by pointing gestures, provides a crucial foundation for language development in humans. The results, largely from laboratories of Call and his colleagues, illustrate the creativity of gestural signaling as seen in naturalistic observation of apes. The discussion encourages further interest in manual dexterity and visuo–manual communication flexibility of apes but also draws attention to the fact that the numbers of communicative types that emerge in individual animals is quite small in comparison with the numbers of words and gestures young children acquire in the first two years of life. The chapter thus offers a crucially important empirical base for speculations about the role of manual dexterity in language evolution as proposed, for example, in the MacWhinney chapter.

Kuczaj and Makecha provide a theoretical perspective on the role of play in a variety of species, and how its evolution may have influenced the emergence of language. The chapter illuminates ways that flexibility of vocal control in humans may be in part a reflection of tendencies to play, tendencies that exist widely in mammals. Their review of research in cetaceans provides poignant examples of complex and cognitively elaborate play demanding extraordinary flexibility of action and intent. Further, the review addresses Kuczaj's research on vocal play in the crib by human infants, presenting dramatic illustrations of ways that play exemplifies the vocal exploratory inclination of children as well as the apparent drive to practice vocalization and linguistic structure as it emerges. The work suggests domains of action that may be the testing grounds yielding self-organization through exploratory action as posited, for example, in Owren and Goldstein and in Oller and Griebel and as modeled by Westermann in a later chapter.

11 How Apes Use Gestures: The Issue of Flexibility

Josep Call

Flexibility is one of the main features of advanced cognition. It is a feature that allows organisms to adapt to novel situations and produce efficient solutions to problems that they encounter, either when they are prevented from using their usual strategies or when those strategies do not elicit a desired outcome. Historically, this kind of flexibility has been investigated in problem-solving situations in which an individual cannot directly get access to a food reward but has to seek an indirect route to get the reward (Köhler, 1925; Yerkes, 1927). For instance, Köhler (1925) investigated the ability of chimpanzees to use various kinds of tools or to take spatial detours to get access to out-of-reach rewards. Since then, numerous studies have shown that many animals can solve these kinds of problems, and more importantly, they can flexibly use multiple strategies to achieve their goals (see Shettleworth, 1998). A good indicator of flexible problem solving resides in the ability to change strategies when required and to attempt multiple strategies if necessary, which on some occasions may even entail overcoming prepotent responses such as moving away from a desired food to take an indirect route to a goal.

Several authors have argued that the use of gestures is analogous to the use of tools in the kind of problem-solving situations described earlier (e.g., Bates, 1979; Gómez, 1990). Individuals are often faced with situations in which other individuals control access to certain resources and they have to use social strategies to get access to those resources. For instance, an infant may want to nurse from its mother or travel to a certain location. One can say that a gesture directed at manipulating an individual accomplishes the same function as manipulating a tool to obtain a reward that is out of reach.

From the point of view of this chapter, a key event in social interactions occurs when the recipient of such gestures does not comply with the actor's request. Can subjects flexibly deploy other gestures or combinations of gestures to obtain their goal? Traditionally, much of animal communication has been based on assigning one signal to one function, which makes it quite inflexible. Recent research, however, has shown that some aspects of primate vocal communication can be quite flexible (see Hammerschmidt and Fischer, this volume; Snowdon, this volume). Thus, several species can recognize and respond appropriately to a variety of calls, including those of other species. Additionally, they learn to use calls in a variety of contexts. However, nonhuman primates are unable to modify the structure of their calls in substantial ways to produce, for instance, non-species-specific calls. In this regard, the use of gestures, at least in apes, is much more flexible than vocalizations. There are three key aspects that make these gestures flexible. First, there is a means–ends dissociation between gestures and contexts (and functions) in which these gestures are applied. Second, apes adjust the sensory modality of their gestures to the attentional state of the recipient (e.g., tactile or acoustic components can be recruited through such actions as touching or clapping). Third, apes can produce novel gestures that are not part of their natural repertoire, with the most striking example being the adoption of some human gestures and signs. We examine each of these three aspects in turn.

Our analysis of gestures is based on a method developed by developmental psychologists to study the ontogeny of intentional communication in children (e.g., Bruner, 1981; Bates, 1979). Although the ethological literature contains a number of studies reporting the use of gestures in nonhuman animals, with very few exceptions (e.g., Plooij, 1984) their focus has not been on the intentional nature of gestures. Intentional gestures consist of body postures and movements that (1) are motorically ineffective, thus requiring the cooperation of the recipient to attain the goal, (2) are followed by a response waiting period on the part of the user, (3) are often accompanied by gaze alternation between the goal and the recipient, and (4) persist until the goal is reached. The gestural repertoires of various ape species range between twenty and forty gestures, excluding facial expressions. Ape gestures utilize three broad sensory modalities (visual, auditory, and tactile), depending on the dominant channel by which messages are conveyed.

The focus of this chapter will be on intentional gestures of the great apes, since these represent the best candidates if we wish to find evidence of flexible communication. This is not to say that great apes are the only species that display intentional communication. On the contrary, it is very likely that monkeys and other species are also capable of intentional communication, but until these data are available detailed comparisons between taxa must wait.

Means–Ends Dissociation

From an early age, human infants can perform simple actions to bring about certain outcomes (e.g., shake a rattle to produce a noise). Infants will repeat these actions until their goal is achieved, but they will not try new actions to produce the same

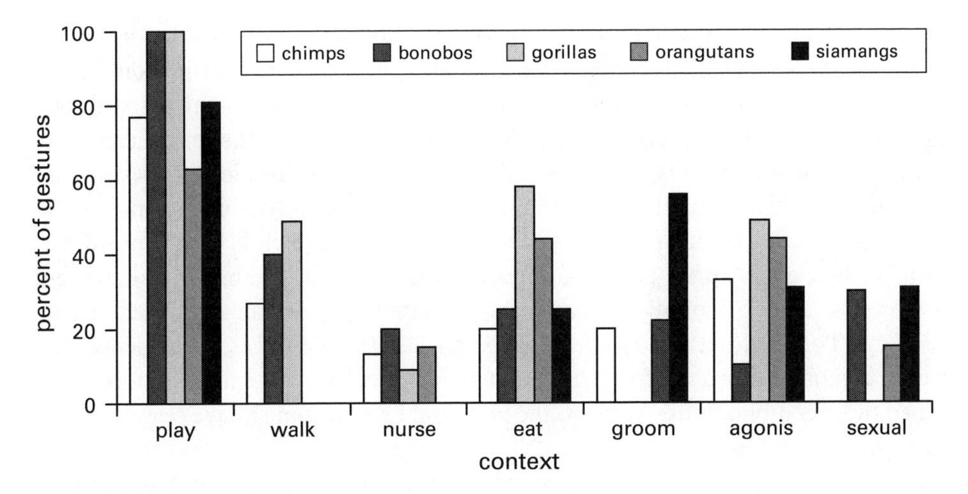

Figure 11.1
Percentage of gestures used in each context as a function of species (sources: Call and Tomasello, 2007a; Liebal, 2007a, b; Pika 2007a, b).

goal. One might say that the action (i.e., means) is not detached from the goal (i.e., end); instead, they are one and the same. One important developmental milestone occurs when infants learn to dissociate actions from goals and can try different means to achieve the same end (Piaget, 1952). This dissociation opens the door to further sophistication as observed when individuals coordinate multiple means to achieve a particular end.

Apes show a clear dissociation between means and ends. Multiple gestures can be used for a particular context (i.e., activity). Figure 11.1 presents the percentage of each type of gesture (out of the total repertoire) that was used in each of the contexts indicated in the figure in five ape species. The predominance of the play context can be explained by the fact that these studies were mostly based on youngsters. Nevertheless, as indicated in the figure, other contexts are also served by multiple gestures. Most of these gestures function as requests for various activities such as play, sex, grooming, body contact, food transfer, or change of posture or location.

Besides the use of multiple gestures to achieve a particular goal in a particular context, several studies have documented that approximately 20% to 30% of the total gestures produced by chimpanzees, orangutans, and siamangs appeared in combinations of two or more gestures. In most cases, gesture sequences are a consequence of repeating a single gesture in rapid succession, mostly because recipients did not respond (Leavens, Russell, and Hopkins, 2005; Liebal, Call, and Tomasello, 2004). Two-gesture combinations are by far the most frequent ones, accounting for about

70% of all the combinations observed (see Liebal, 2007a, b; Tomasello et al., 1994, for additional details). Persistence toward fulfilling a particular goal not only leads to the repetition of single gestures but also generates gesture combinations with two or more types of gestures. This is important because it shows that subjects can coordinate two or more gestures to achieve a particular goal. For instance, a weaning infant that wants his mother (who is lying prone on the ground) to change posture in order to gain access to the nipple may stomp the ground, stretch his arm toward her, touch her side, and slap the ground in rapid succession. Such combinations represent between 39% and 55% of all combinations generated by chimpanzees depending on the study (Tomasello et al., 1994; Liebal et al., 2004). There was no evidence that gesture combinations acquired a different meaning from the single gestures that formed those combinations. At most, the additional gestures may have served an emphatic function.

Another aspect of the dissociation between means and ends in gestural use is that a particular gesture can be used in multiple contexts and to serve multiple functions. Call and Tomasello (2007b) reported that between 40% and 80% of the gestures, depending on the ape species, were used in more than one context. Chimpanzees and bonobos showed the lowest percentage of multicontext gestures, whereas orangutans displayed the highest. For instance, the gesture "touch" is used to request that the recipient change her posture, travel with the individual to another location, or transfer food. Certain primate vocalizations such as baboon grunts and macaque coos also occur in a variety of contexts (Owren and Rendall, 2001).

Recruiting a gesture to fulfill multiple purposes raises the theoretical problem of how recipients of those gestures extract their intended meaning. On some occasions gestures appear together with other components of the communicative system such as facial expressions or vocalizations that also help the recipient to interpret the communicative message of the sender. Thus, a ground slap gesture can be interpreted in the context of play when it is accompanied by a play face or as an intimidation when accompanied by piloerection. In other occasions, however, those other communicative cues are not present, yet messages do get across. For instance, a reaching gesture can be used to beg for food or request that an infant approach her mother (see figure 11.2). Moreover, detailed analyses of the topography of gestures occurring in different contexts have found no evidence of systematic differences between them (e.g., Tomasello et al., 1997).

In summary, these data suggest that there is not a one-to-one correspondence between gestures and contexts of use. Multiple gestures are used in one context, while at the same time one gesture is used in multiple contexts. Some gestures are also used for multiple functions. One could say that many of the ape gestures constitute a "toolbox" that can be used flexibly to achieve particular goals.

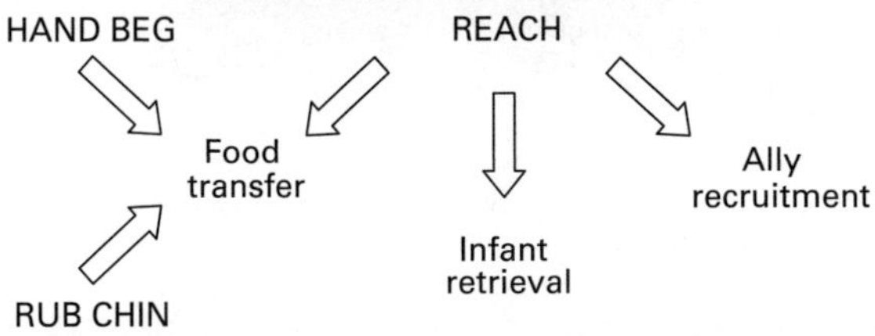

Figure 11.2
Example of means–ends dissociation. The gesture REACH can be used in three different contexts for different functions. Likewise, food transfer can be accomplished with three different gestures (REACH, RUB CHIN, and HAND BEG).

Audience Effects

Audience effects in the use of vocalizations have been documented in a variety of species (ground squirrels: Sherman, 1977; chickens: Evans and Marler, 1991). However, such differential vocal responses appear to be based on the mere presence or absence of certain individuals, not their attentional status. Unlike auditory signals, effectiveness of visual signals requires potential recipients both to be present and to attend to the signaler. This means that the use of visual as opposed to auditory signals requires a more refined tailoring to the attentional states of potential recipients.

Initial observations suggested that chimpanzees were sensitive to what others can see regarding their communicative signals. For instance, de Waal (1982) observed a chimpanzee covering his fear-grinning facial expression from opponents during agonistic interactions. Byrne and Whiten (1988, 1990) collated this and other observations from multiple researchers and suggested that some primates were capable of some levels of visual perspective taking. Although these records possess great heuristic value, they do not constitute systematic observations and, therefore, are vulnerable to multiple interpretations.

Tomasello et al. (1994) conducted systematic observations of gestural communication in young chimpanzees to assess whether they took into account the attentional

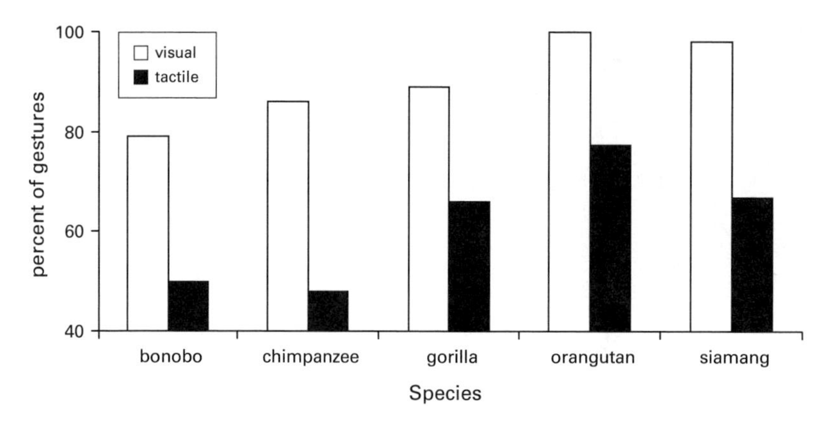

Figure 11.3
Percentage of visual and tactile gestures produced when the recipient was looking at the sender in five ape species (sources: Call and Tomasello, 2007; Liebal, 2007a, b; Pika 2007a, b).

orientation of the recipient when they used gestures. They found that chimpanzees preferentially used visual gestures when others were facing them (so that they could see them). In contrast, chimpanzees did not take into account body orientation when they used tactile gestures. Auditory gestures fell in between visual and tactile gestures, being used less specifically than visual gestures but more than tactile gestures. A possible explanation is that many auditory gestures such as slapping the ground with an open palm also have an important visual component. Since then, these results have been replicated in a different group of chimpanzees (Tomasello et al., 1997; Liebal et al., 2004) and extended to other ape species (Pika, 2007a, b; Liebal, 2007a, b). Figure 11.3 presents a summary of these results.

Experimental studies have also confirmed these findings. Call and Tomasello (1994) showed that orangutans gestured less frequently to someone who had his back turned to them compared to someone facing them. In fact, the back-turned condition produced the same amount of gestures as someone leaving the room. Povinelli and Eddy (1996) also found that chimpanzees spontaneously begged food from a human who was facing them significantly more often than from a human with the back turned. Kaminski, Call, and Tomasello (2004) replicated this result with orangutans, chimpanzees, and bonobos. In contrast, Povinelli and Eddy (1996) found that chimpanzees failed to spontaneously discriminate between a human (1) with her face visible and one with a bucket over her head, (2) with her eyes visible (and a blindfold over her mouth) and a human with her eyes covered by the blindfold, or (3) with her back turned but looking over her shoulder.

Some of these results, however, have been challenged. Kaminski et al. (2004) investigated the relative contribution of face and body orientation by presenting subjects with four conditions representing all possible combinations of body and face

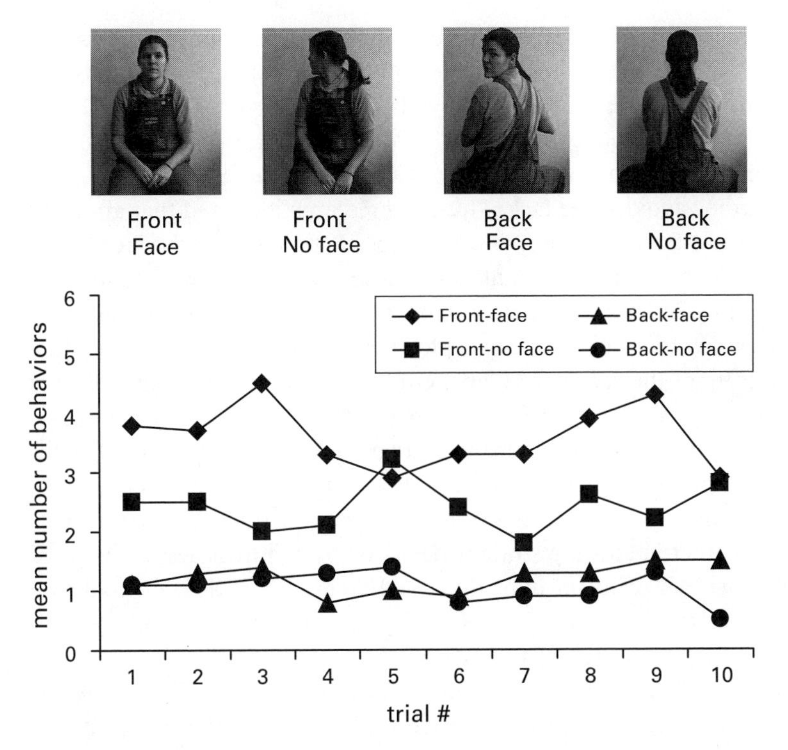

Figure 11.4
Mean number of behaviors produced by orangutans, chimpanzees, and bonobos to request food from the experimenter as a function of experimental condition and trial number (source: Kaminski et al., 2004).

orientation (figure 11.4). Thus, subjects saw a human experimenter with both her body and face oriented toward them, just the body but not the face, just the face but not the body, or neither the body nor the face. Results showed that apes were sensitive to face orientation when the body was oriented toward them but not when the body was facing away; this pattern remained unchanged during the course of the experiment (see figure 11.4). These results led us to hypothesize that ape observers extracted two types of information from the human experimenter. The face orientation informs them about whether someone can see them, while the body orientation informs them about whether the observer is disposed to give them food. Povinelli et al. (2003) also found that chimpanzees directed their begging gestures toward the location where a human experimenter's face was oriented. Gómez (1996) also found that chimpanzees—but only those chimpanzees that had been raised by humans—were sensitive to the face (and the eyes) of human recipients.

Other studies have reinforced the importance of the face as a stimulus to which subjects respond spontaneously in competitive situations. Chimpanzees and rhesus macaques preferentially steal food from a human competitor whose face is not directed toward them (Flombaum and Santos, 2005; Hare, Call, and Tomasello, 2006). One interesting outcome of these studies is that subjects seek or avoid the face depending on their communicative intent. When chimpanzees wanted to communicate with their partners, they actively sought the face, whereas when they tried to escape detection, they avoided the face. Thus, this research indicated that apes and monkeys are sensitive to the orientation of certain social stimuli (e.g., the face).

However, it is unclear whether individuals can also manipulate the attention of others when they want to communicate with them. One way to investigate this question consists of observing how chimpanzees deploy their gesture sequences depending on the attentional state of the recipient. Liebal et al. (2004) found that chimpanzees did not use gestures to call attention followed by gestures to request a goal (e.g., food, sex, or travel) when recipients were not attending. Instead, they used other ways to get the attention of distracted recipients. For instance, they used tactile gestures to convey their message. This strategy clearly sidestepped the problem of inattentive recipients. Another interesting strategy deployed by chimpanzees consisted of walking around a potential recipient until they were facing her and then producing their request by gesturing. It is therefore possible that subjects preferred to walk around to face the experimenter rather than use an attention getter.

Next, Liebal, Pika, Call, and Tomasello (2004) tested these two possibilities experimentally by presenting individuals with an inattentive human (with her back turned) and giving them the opportunity to either use an attention getter to call the human's attention or to walk around to face the human before begging for food from the human. Results showed that apes did not use auditory gestures to call the attention of the experimenter when she had her back turned. Instead all species moved around to face the experimenter when she took the food with her, thus corroborating the observational findings. This does not mean, however, that apes (or some apes species) are unable to use auditory gestures or vocalizations to call the attention of others. On the contrary, Hostetter, Hopkins, and Cantero (2001) found that chimpanzees uttered vocalizations faster and were more likely to produce vocalizations as their first communicative behavior when a human holding food was oriented away from them. In contrast, chimpanzees used manual gestures more frequently and faster when the human holding the food was facing the chimpanzee. The authors concluded from these findings that chimpanzees can distinguish a human's attentional state on the basis of the human's orientation and use this orientation as a prerequisite for successful visual communication. A critical difference between Hostetter et al.'s (2001) and Liebal, Pika, Call, and Tomasello's (2004) studies was that chimpanzees in the former could not position themselves in front of the experimenter.

In summary, chimpanzees adjust their gestures to the attentional state of the recipient. Not only do they use gestures whose sensory modality is appropriate to the state of the recipient but they also can alter the recipient's attentional state by placing themselves in front of her. When this is not an option, they may switch modalities by either touching the recipient or vocalizing, thus again showing the flexible deployment of various strategies to fulfill their goal. This evidence, taken together with other recent data on perspective taking, suggests that chimpanzees (see Tomasello and Call, 2006, for a recent review), at least, know what others can and cannot see.

Novelty

Creating or acquiring a gesture not previously available in the individual's repertoire and using it to solve a novel problem is perhaps one of the strongest indications of flexibility. The appearance of idiosyncratic gestures when interacting with conspecifics as well as the adoption of human gestures has been taken as strong evidence for flexible production.

Idiosyncratic Gestures

Several studies have documented the use of idiosyncratic gestures. First, Tomasello et al. (1994) compiled data from three study periods on the same group of chimpanzees and reported that 14% of the gestures were idiosyncratic, that is, unique to particular individuals. Moreover, 40% of the gestures were idiosyncratic during a particular given study period. That is, only one individual used a particular gesture during a particular observation period. Liebal (2007a, b) found that 10% and 20% of the gestures in orangutans and siamangs, respectively, were seemingly idiosyncratic. In contrast, no idiosyncratic gestures were observed in gorillas or bonobos (Pika, 2007a, b).

It is unlikely that these idiosyncratic gestures were the result of poor motor control because they did not appear to be random variations of other gestures. Variations on certain gestures were observed (e.g., Tomasello et al., 1997), but they were classified as variations, not separate gestures. Idiosyncratic gestures could not be easily classified under any of the other gestures, and they were aimed at obtaining a particular goal. For instance, an infant chimpanzee would rub his mother's chin to get her to drop onto his hand some of the food that she was chewing. Tomasello et al. (1994) hypothesized that these gestures became ritualized through repeated interactions between particular pairs of individuals to accomplish certain goals (e.g., food transfer)—a process called ontogenetic ritualization.

Novel Gestures

As noted previously, apes have a rich natural repertoire of gestural signals, some of which are idiosyncratic to particular individuals. In addition, apes can also acquire novel gestures that do not belong to their natural repertoire. Two of the most compelling cases that demonstrate this ability are the use of the pointing gesture and sign language.

Pointing

Using an extended index finger (or the whole hand) is not a gesture that apes use to direct the attention of their conspecifics. Some authors argue that the lack of pointing with an extended arm and finger are actually not surprising at all because chimpanzees point with their body orientation (Menzel, 1974). In other words, chimpanzees do not display pointing (with an extended arm) because they already have other ways to point. Moreover, de Waal and van Hooff (1981) observed that during recruitment episodes following an aggressive incident, chimpanzees use "side-directed" behavior to indicate enemies to their potential allies. They do this by using an extended arm toward the aggressors. Although this behavior could be construed as pointing at the enemies to inform potential allies, it is also true that chimpanzees use an extended arm toward their potential allies. Therefore, it is not clear whether pointing with the body or side-directed behavior constitute genuine cases of intentionally directing attention as opposed to behaviors that others can read, but that the sender does intend to send as referential acts.

Despite the lack of pointing gestures among apes, it is clear that they can spontaneously develop this gesture when they interact with humans. There are numerous examples of language-trained apes acquiring this gesture in the company of humans (Savage-Rumbaugh, Mcdonald, Sevcik, Hopkins, and Rubert, 1986; Krause and Fouts, 1997; Miles, 1990; Patterson, 1978). Most apes acquire the whole-hand point, but there is a minority of individuals that use the index-finger point. For instance, Krause and Fouts (1997) systematically studied the pointing of two language-trained chimpanzees and found that they were capable of pointing very accurately to various targets. In this study humans could reliably locate the position of rewards on the basis of the information provided by the chimpanzee gestures. In contrast, pointing behavior has not been documented either in the wild or in groups of captive chimpanzees interacting with other chimpanzees. Menzel (2005) reported that the language-trained chimpanzee Panzee pointed to various locations to inform a naive human about the presence of food. On some occasions she pointed to locations after delays of several hours and was also able to indicate the correct location on a TV monitor that humans could use to find the food.

Leavens and Hopkins (1998; see also Leavens, Hopkins, and Bard, 1996) have shown that even non-language-trained chimpanzees can develop pointing gestures to direct the attention of humans. These authors found that 40% of the chimpanzees they studied pointed with the whole hand (with 5% using an index finger extension) to indicate the location of food to humans. These pointing gestures were accompanied with gaze alternation between the human and the food, and the chimpanzees persisted in using this gesture until they received the food. These features suggest that pointing gestures were both intentional and referential. These gestures were only observed when chimpanzees interacted with humans, not with other chimpanzees. Since captive chimpanzees (both mother reared and human reared) interact with humans extensively, and humans are especially attuned to the pointing gesture, it is not surprising that chimpanzees can quickly capitalize on this and develop pointing gestures. Thus, the current evidence indicates that chimpanzees can develop the pointing gesture to direct the attention of humans to locations or objects that they desire regardless of their rearing history.

There are also some reports of pointing in monkeys. Hess, Novak, and Povinelli (1993) described a rhesus macaque that would indicate the location of food by extending her arm in the direction of the food. Mitchell and Anderson (1997) and Genty, Palmier, and Roeder (2004) also described capuchin monkeys and lemurs, respectively, using gestures to indicate the location of food, which on some occasions constituted cases of tactical deception, that is, trying to direct a human competitor away from a baited container. Despite this evidence, it is unclear whether monkey gestures qualify as pointing or show the degree of flexibility implied by the idea of pointing. One thing that we do not know is whether monkeys that point to indicate a container they want will also be able to point (as apes have been shown to do) in other situations, for instance, pointing to a tool that the experimenter needs in order provide the animal with a reward.

Another issue that skeptics raise against reported referential pointing in nonhuman primates is that apes point proficiently for others but they do not automatically comprehend others' pointing gestures. Such a difficulty with pointing comprehension, particularly distal pointing, in which the distance between the index finger and the referent are at least one meter apart, has been extensively documented both in monkeys and apes (see Call and Tomasello, 2005, for a review). Call and Tomasello (1994) found that while two orangutans pointed very accurately for a human to inform her about food location, they were much less proficient when they had to use a pointing human to locate a food reward.

These findings are not surprising in light of the well-documented dissociation between production and comprehension not just in pointing but also in the acquisition of communicative signals in a variety of species (Morrel-Samuels and Herman, 1993; Savage-Rumbaugh, 1993). Nevertheless, this is not to say that monkeys and apes are not capable of learning to use pointing gestures as a cue to locate food. On the contrary, there are several studies that have shown that enculturated apes or trained apes and monkeys can comprehend pointing (e.g., Itakura and Tanaka, 1998; Itakura and Anderson, 1996). Moreover, it appears that other species such as dogs, goats, and marine mammals are more proficient than apes at comprehending pointing. Hare,

Brown, Williamson, and Tomasello (2002) directly compared chimpanzees, wolves, and dogs and found that the dogs outperformed the other two species, which performed at chance levels. One possible explanation for this result is that dogs have been selected for following pointing. Indeed, puppies with little exposure to humans follow pointing to distal locations (Agnetta, Hare, and Tomasello, 2000). However, selection for point following cannot be the whole explanation because both domestic goats and domesticated foxes also follow pointing to some extent (Hare et al., 2005; Kaminski et al., 2005). An alternative hypothesis is that domestic species, not being afraid of humans, pay more attention to them and can then exploit pointing signals. This would also explain why other animals with which humans interact routinely can also comprehend pointing.

It is important to consider why these signals develop in some settings but not in others. Pointing is readily observed in the laboratory but not in the wild. One possibility is that chimpanzees already point in the wild, but they use different means (Menzel, 1974). However, it is difficult to know whether these other means constitute intentional communication or simple eavesdropping information acquired by the recipient from the individual's gaze direction or body orientation, even though the individual did not intend to communicate.

Another possibility is that chimpanzees do not point for others in the wild to request objects or actions. Pointing is a skill that only develops in the laboratory. Leavens, Hopkins, and Bard (2005) have argued that the captive setting creates a new set of challenges and opportunities different from those that apes encounter in the wild. The captive niche, combined with apes' adaptable behavior and cognition, is what generates new variants. According to Leavens et al. (2005), the combination of two factors (one physical and one social) in particular could explain the appearance of pointing. First, access to food or other rewards is restricted by impassable physical barriers. Thus, if a banana is sitting outside a cage beyond reach, individuals cannot get access to it. Second, apes experience food provisioning in their everyday lives and therefore can expect to receive food from humans.

Although Leavens et al. (2005) rightly considered the participation of both physical and social components in the emergence of pointing, their focus on the social component may have been too narrow. In particular, they alluded to the fact that captive chimpanzees experience "histories of daily provisioning by adult human caretakers," but this would not explain why apes point for other things besides food, including objects that they want to get, tools that the experimenter needs to use, or locations where they want to go. These activities appear to fall beyond the "provisioning experience" scope and suggest that apes may capitalize on a generalized human willingness to share food, help, and comply with others' requests; this is a willingness that underlies our system of cooperation and cultural learning (Tomasello et al., 2005). Apes may be particularly successful at tapping into our altruistic motivation precisely because they can use gestures such as pointing, an action with which humans identify and to which they respond very readily.

Sign Language

Apes can be trained to use some aspects of human sign language (e.g., Fouts, 1975; Gardner, Gardner, and van Cantfort, 1989; Miles, 1990; Patterson, 1978). Most studies have used American Sign Language or some version of it. The size of the sign repertoire that has been achieved with apes varies between 100 and 200 signs (Miles and Harper, 1994), which is well beyond the natural gestural repertoire of apes. Apes use signs to communicate with humans and also among each other. They even produce signs when they are looking at magazines on their own (Jensvold and Fouts, 1993). In some cases, these signs are iconic (i.e., their topography is reminiscent of the referent they represent), while in other cases they are truly symbolic. Signing apes can use signs to request things like food or play, to answer queries, or to ask questions. The use of this system of communication is important because it allows the apes to engage in displaced reference, that is, they can refer to specific actions or objects that are not currently perceptually available. Note that displaced reference is virtually nonexistent in their species-specific communicative systems.

A more controversial issue is whether apes that use sign language do so in the same way that humans do. Some of the signs that they use correspond to grammatical categories such as nouns, verbs, adverbs, pronouns, and adjectives. However, it is unclear whether they process these categories the way children do. Moreover, although they produce sign combinations, it is unclear that they have a notion of syntax. In fact, there is little evidence that these combinations follow syntactic rules (e.g., Terrace, Petitto, Sanders, and Bever, 1980).

In summary, apes create a number of idiosyncratic gestures in their natural repertoires to fulfill certain functions. Invariably, these gestures disappear within a few years once their function is no longer relevant for the animal. Additionally, chimpanzees can acquire gestures that are not in their natural repertoires to solve novel problems. Pointing, and especially sign language, constitute prime examples. The use of sign language is particularly interesting because it allows apes to designate specific actions or objects that are not currently available.

Gestures, Intentional Gestures, and Vocalizations

Flexibility is one of the key features of many ape gestures, and possibly of those of other primates, although much more research is necessary to confirm this point. There are three aspects that make ape gestures flexible. First, apes display a clear dissociation between gestures and contexts (as well as functions). Namely, they can use multiple gestures (and combinations of gestures) in pursuing a particular goal (i.e., to serve the same function). In addition, apes use particular gestures in multiple contexts. Second, apes adjust their gesture production to the attentional state of the recipient, not just their presence or absence. Thus, they use visual gestures preferentially when others are looking, whereas they use tactile gestures regardless of the attentional state of the recipient. Third, apes can acquire new gestures to solve novel problems as evidenced by the creation of idiosyncratic gestures and the adoption of human gestures and signs that allow apes to engage in displaced reference.

It is important to emphasize that the gestures treated in this chapter constitute only a subset of signals that apes use in visual communication. Not all visual signals are as flexible as those that have been described here. For instance, facial expressions are typically hardwired (some are present soon after birth), highly stereotyped in appearance, and linked with specific emotional states. It is also unclear how much control individuals have over their production/usage and whether they can create novel facial expressions. Nevertheless, there is some evidence suggesting that apes and monkeys learn to use facial expressions in appropriate contexts, although again it is unclear whether subjects would be able to produce facial expressions (e.g., fear) in the absence of the corresponding emotional state.

Another important comparison can be established between the gestures covered in this chapter and the vocalizations of nonhuman primates. Like gestures, vocalizations show an important degree of contextual flexibility when it comes to usage and comprehension (see Hammerschmidt and Fischer, this volume; Snowdon, this volume, for examples). Baboon grunts and macaque coos occur in a variety of contexts (Owren and Rendall, 2001), a finding that is similar to some of the findings on gestures described here. Cotton-top tamarins produced intense predator alarm calls upon seeing an experimenter wearing a veterinarian's attire, while a live boa constrictor (a natural predator for these monkeys) only produced mild arousal, at a level equivalent to seeing a laboratory rat (Snowdon, this volume; Hayes and Snowdon, 1990).

In terms of comprehension, several species of monkeys can comprehend the alarm calls of other species (Seyfarth and Cheney, 1990; Zuberbühler, 2000). Japanese and rhesus macaques can comprehend the calls of their cross-fostered offspring despite the interspecific difference in call production between mothers and offspring (Owren, Dieter, Seyfarth, and Cheney, 1992). Tamarins that heard (and observed) a conspecific alarm calling after tasting peppered tuna fish, which the listeners themselves had not tasted, subsequently reduced the consumption of that food item (Snowdon and Boe, 2003).

It is precisely in the area of call production that we find the largest difference in flexibility between gestures and vocalizations. Apes, and possibly other primates, can produce novel signals and adapt them to novel functions to an extent not observed in vocal communication. The development of pointing to request help from humans is the prime example, but research on sign language shows that this is not an isolated example. The acquisition and use of these signals is not trivial, as it allows individuals to communicate about specific objects and actions that may not be currently present to an extent that goes well beyond the capabilities offered by their natural system of communication.

Despite the flexibility of ape gestures, it is important to briefly point out some of their limitations. Two are particularly noticeable when compared to human gestures. First, novel gestures are not transmitted across generations in apes. Cultural variations in visual gestures are virtually nonexistent in chimpanzees or orangutans. Thus, although an individual may invent a new gesture to solve a problem, there is little evidence that this gesture is adopted by other individuals (see Tomasello et al., 1997, for some negative results). Second, ape gestures are invariably tied to an imperative format, not a declarative one (e.g., Rivas, 2005). Thus, unlike children, apes do not use gestures to direct the attention of others to objects or events *with the purpose* of sharing social attention and emotion about external entities.

Conclusion

One of the key features of ape gestures is flexibility. Several aspects, including a means–ends dissociation, a refined audience effect capability, and the creation of novel gestures, support this conclusion. Moreover, it is the combination of these three aspects that allows apes to produce flexible solutions to a variety of problems. Thus, faced with a recipient that could transfer food to them, apes evaluate the attentional state of the recipient and select gestures accordingly or change position to alter the attentional state and then use gestures, either single gestures or a combination of gestures. If apes are interacting with humans, apes will acquire particularly effective gestures such as pointing. Moreover, they will be able to use those gestures not only to request food but to indicate the location of a tool that the experimenter needs to get a reward.

For someone familiar with the behavior that apes can deploy in problem-solving situations in the physical domain (e.g., using a stick to get an out-of-reach banana), the previous description will not come as a surprise because there too, subjects routinely deploy multiple strategies flexibly to obtain particular goals. Whether flexibility in physical and social problems represents two separate domains or simply two sides of the same coin is a fascinating question that awaits further research.

References

Agnetta B, Hare B, Tomasello M. 2000. Cues to food location that domestic dogs (*Canis familiaris*) of different ages do and do not use. *Animal Cognition* 3: 107–12.

Bates E. 1979. *The Emergence of Symbols*. New York: Academic Press.

Bruner J. 1981. Intention in the structure of action and interaction. In *Advances in Infancy Research, Vol. 1*, ed. L Lipsett, pp. 41–56. Norwood, NJ: Ablex.

Byrne RW, Whiten A. 1988. Toward the next generation in data quality: A new survey of primate tactical deception. *Behavioral and Brain Sciences* 11: 267–71.

Byrne RW, Whiten A. 1990. Tactical deception in primates: The 1990 database. *Primate Report* 27: 1–101.

Call J, Tomasello M. 1994. Production and comprehension of referential pointing by orangutans (*Pongo pygmaeus*). *Journal of Comparative Psychology* 108: 307–17.

Call J, Tomasello M. 2005. What chimpanzees know about seeing revisited: An explanation of the third kind. In *Joint Attention: Communication and Other Minds*, ed. N Eilan, C Hoerl, T McCormack, J Roessler, pp. 45–64. Oxford: Oxford University Press.

Call J, Tomasello M. 2007a. The gestural communication of chimpanzees. In *The Gestural Communication of Apes and Monkeys*, ed. J Call, M Tomasello, pp. 17–39. Mahwah, NJ: Lawrence Erlbaum Associates.

Call J, Tomasello M. 2007b. *The Gestural Communication of Apes and Monkeys*. Mahwah, NJ: Lawrence Erlbaum Associates.

de Waal FBM. 1982. *Chimpanzee Politics*. London: Jonathan Cape.

de Waal FBM, van Hooff JARAM. 1981. Side-directed communication and agonistic interactions in chimpanzees. *Behaviour* 77: 164–98.

Evans CS, Marler P. 1991. On the use of visual images as social stimuli in birds: Audience effects on alarm calling. *Animal Behaviour* 41: 17–26.

Flombaum JI, Santos LR. 2005. Rhesus monkeys attribute perceptions to others. *Current Biology* 15: 447–52.

Fouts RS. 1975. Capacities for language in great apes. In *Socioecology and Psychology of Primates*, ed. RH Tuttle, pp. 371–90. The Hague: Mouton.

Gardner RA, Gardner BT, van Cantfort TE. 1989. *Teaching Sign Language to Chimpanzees*. Albany, NY: State University of New York Press.

Genty E, Palmier C, Roeder JJ. 2004. Learning to suppress responses to the larger of two rewards in two species of lemurs, *Eulemur fulvus* and *E. macaco*. *Animal Behaviour* 67: 925–32.

Gómez JC. 1990. The emergence of intentional communication as a problem-solving strategy in the gorilla. In *"Language" and Intelligence in Monkeys and Apes*, ed. ST Parker, KR Gibson, pp. 333–55. New York: Cambridge University Press.

Gómez JC. 1996. Non-human primate theories of (non-human primate) minds: Some issues concerning the origins of mind-reading. In *Theories of Theories of Mind*, ed. P Carruthers, PK Smith, pp. 330–43. Cambridge, England: Cambridge University Press.

Hare B, Brown M, Williamson C, Tomasello M. 2002. The domestication of social cognition in dogs. *Science* 298: 1634–6.

Hare BH, Call J, Tomasello M. 2006. Chimpanzees deceive a human competitor by hiding. *Cognition* 101: 495–514.

Hare BH, Plyyusnina I, Ignacio N, Schepina O, Stepika A, Wrangham R, Trut L. 2005. Social cognitive evolution in captive foxes is a correlated by-product of experimental domestication. *Current Biology* 15: 226–30.

Hayes SL, Snowdon CT. 1990. Predator recognition in cotton-top tamarins (*Saguinus oedipus*). *American Journal of Primatology* 20: 283–91.

Hess J, Novak MA, Povinelli DJ. 1993. 'Natural pointing' in a rhesus monkey, but no evidence of empathy. *Animal Behaviour* 46: 1023–5.

Hostetter AB, Hopkins WD, Cantero M. 2001. Differential use of vocal and gestural communication by chimpanzees (*Pan troglodytes*) in response to the attentional status of a human (*Homo sapiens*). *Journal of Comparative Psychology* 115: 337–43.

Itakura S, Anderson JR. 1996. Learning to use experimenter-given cues during an object-choice task by a capuchin monkey. *Current Psychology of Cognition* 15: 103–12.

Itakura S, Tanaka M. 1998. Use of experimenter-given cues during object-choice tasks by chimpanzees (*Pan troglodytes*), an orangutan (*Pongo pygmaeus*), and human infants (*Homo sapiens*). *Journal of Comparative Psychology* 112: 119–26.

Jensvold MLA, Fouts RS. 1993. Imaginary play in chimpanzees (*Pan troglodytes*). *Human Evolution* 8: 217–27.

Kaminski J, Call J, Tomasello M. 2004. Body orientation and face orientation: Two factors controlling apes' begging behavior from humans. *Animal Cognition* 7: 216–23.

Kaminski J, Riedel J, Call J, Tomasello M. 2005. Domestic goats (*Capra hircus*) follow gaze direction and use social cues in an object choice task. *Animal Behaviour* 69: 11–18.

Köhler W. 1925. *The Mentality of Apes*. New York: Vintage Books.

Krause MA, Fouts RS. 1997. Chimpanzee (*Pan troglodytes*) pointing: Hand shapes, accuracy, and the role of eye gaze. *Journal of Comparative Psychology* 111: 330–6.

Leavens DA, Hopkins WD. 1998. Intentional communication by chimpanzees: A cross-sectional study of the use of referential gestures. *Developmental Psychology* 34: 813–22.

Leavens DA, Hopkins WD, Bard KA. 1996. Indexical and referential pointing in chimpanzees (*Pan troglodytes*). *Journal of Comparative Psychology* 110: 346–53.

Leavens DA, Hopkins WD, Bard KA. 2005. Understanding the point of chimpanzee pointing. Epigenesis and ecological validity. *Current Directions in Psychological Science* 14: 185–9.

Leavens DA, Russell JL, Hopkins WD. 2005. Intentionality as measured in the persistence and elaboration of communication by chimpanzees (*Pan troglodytes*). *Child Development* 76: 291–306.

Liebal K. 2007a. The gestural communication of siamangs. In *The Gestural Communication of Apes and Monkeys*, ed. J Call, M Tomasello, pp. 131–58. Mahwah, NJ: Lawrence Erlbaum Associates.

Liebal K. 2007b. The gestural communication of orangutans. In *The Gestural Communication of Apes and Monkeys*, ed. J Call, M Tomasello, pp. 69–98. Mahwah, NJ: Lawrence Erlbaum Associates.

Liebal K, Call J, Tomasello M. 2004. The use of gesture sequences in chimpanzees. *American Journal of Primatology* 64: 377–96.

Liebal K, Pika S, Call J, Tomasello M. 2004. To move or not to move: How apes alter the attentional states of humans when begging for food. *Interaction Studies* 5: 199–219.

Menzel C. 2005. Progress in the study of chimpanzee recall and episodic memory. In *The Evolution of Consciousness in Animals and Humans*, ed. H Terrace, J Metcalfe, pp. 188–224. New York: Oxford University Press.

Menzel EW. 1974. A group of young chimpanzees in a one-acre field: Leadership and communication. In *Behavior of Nonhuman Primates*, ed. AM Schrier, F Stollnitz, pp. 83–153. New York: Academic Press.

Miles HLW. 1990. The cognitive foundations for reference in a signing orangutan. In *"Language" and Intelligence in Monkeys and Apes*, ed. ST Parker, KR Gibson, pp. 511–39. Cambridge, England: Cambridge University Press.

Miles HLW, Harper SE. 1994. "Ape language" studies and the study of human language origins. In *Hominid Culture in Primate Perspective*, ed. D Quiatt, J Itani, pp. 253–78. Niwot, CO: University Press of Colorado.

Mitchell RW, Anderson JR. 1997. Pointing, withholding information, and deception in capuchin monkeys (*Cebus apella*). *Journal of Comparative Psychology* 111: 351–61.

Morrel-Samuels P, Herman LM. 1993. Cognitive factors affecting comprehension of gesture language signs: A brief comparison of dolphins and humans. In *Language and Communication: Comparative Perspectives*, ed. HL Roitblat, LM Herman, PE Nachtigall, pp. 311–27. Hillsdale, NJ: Lawrence Erlbaum Associates.

Owren MJ, Dieter JA, Seyfarth RM, Cheney DL. 1992. Evidence of limited modification in the vocalizations of cross-fostered rhesus (*Macaca mulatta*) and Japanese (*M. fuscata*) macaques. In *Topics in Primatology: Human Origins*, ed. T Nishida, WC McGrew, P Marler, M Pickford, FMB de Waal, pp. 257–70. Tokyo: University of Tokyo Press.

Owren MJ, Rendall D. 2001. Sound on the rebound: Returning form and function to the forefront in understanding nonhuman *primate vocal signaling. Evolutionary Anthropology* 10: 58–71.

Patterson F. 1978. Linguistic capabilities of a lowland gorilla. In *Sign Language and Language Acquisition in Man and Ape*, ed. FCC Peng, pp. 161–201. Boulder, CO: Westview Press.

Piaget J. 1952. *The Origins of Intelligence in Children*. New York: Norton.

Pika S. 2007a. The gestural communication of bonobos. In *The Gestural Communication of Apes and Monkeys*, ed. J Call, M Tomasello, pp. 41–67. Hillsdale, NJ: Lawrence Erlbaum Associates.

Pika S. 2007b. The gestural communication of gorillas. In *The Gestural Communication of Apes and Monkeys*, ed. J Call, M Tomasello, pp. 99–130. Hillsdale, NJ: Lawrence Erlbaum Associates.

Plooij FX. 1984. *The Behavioral Development of Free-Living Chimpanzee Babies and Infants.* Norwood, NJ: Ablex.

Povinelli DJ, Eddy TJ. 1996. What young chimpanzees know about seeing. *Monographs of the Society for Research in Child Development* 61(3).

Povinelli DJ, Theall LA, Reaux JE, Dunphy-Lelii S. 2003. Chimpanzees spontaneously alter the location of their gestures to match the attentional orientation of others. *Animal Behaviour* 66: 71–9.

Rivas E. 2005. Recent use of signs by chimpanzees (*Pan troglodytes*) in interactions with humans. *Journal of Comparative Psychology* 119: 404–17.

Savage-Rumbaugh ES. 1993. Language learnability in man, ape, and dolphin. In *Language and Communication. Comparative Perspectives*, ed. HL Roitblat, LM Herman, PE Nachtigall, pp. 457–84. Hillsdale, NJ: Lawrence Erlbaum Associates.

Savage-Rumbaugh ES, Mcdonald K, Sevcik RA, Hopkins WD, Rubert E. 1986. Spontaneous symbol acquisition and communicative use by pygmy chimpanzees (*Pan paniscus*). *Journal of Experimental Psychology: General* 115: 211–5.

Seyfath RM, Cheney D. 1990. The assessment by vervet monkeys of their own and another species' alarm calls. *Animal Behaviour* 40: 754–64.

Sherman PW. 1977. Nepotism and the evolution of alarm calls. *Science* 197: 1246–53.

Shettleworth SJ. 1998. *Cognition, Evolution, and Behavior*. New York: Oxford University Press.

Snowdon CT, Boe CY. 2003. Social communication about unpalatable foods in tamarins (*Saguinus oedipus*). *Journal of Comparative Psychology* 117: 142–8.

Terrace HS, Petitto LA, Sanders RJ, Bever TG. 1980. On the grammatical capacity of apes. In *Children's Language*, ed. KE Nelson, pp. 371–495. New York: Gardner Press.

Tomasello M, Call J. 2006. Do chimpanzees know what others see—or only what they are looking at? In *Rational Animals*, ed. S Hurley, M Nudds, pp. 371–84. Oxford: Oxford University Press.

Tomasello M, Call J, Nagell K, Olguin R, Carpenter M. 1994. The learning and use of gestural signals by young chimpanzees: A trans-generational study. *Primates* 35: 137–54.

Tomasello M, Call J, Warren J, Frost GT, Carpenter M, Nagell K. 1997. The ontogeny of chimpanzee gestural signals: A comparison across groups and generations. *Evol. Comm.* 1: 223–59.

Tomasello M, Carpenter M, Call J, Behne T, Moll H. 2005. Understanding and sharing intentions: The origins of cultural cognition. *Behavioral and Brain Sciences* 28: 1–17.

Yerkes RM. 1927. The mind of a gorilla. *Gen. Psych. Monog.* 2: 1–193.

Zuberbühler K. 2000. Interspecific semantic communication in two forest monkeys. *Proceedings of the Royal Society of London B* 267: 713–8.

12 The Role of Play in the Evolution and Ontogeny of Contextually Flexible Communication

Stan Kuczaj and Radhika Makecha

Roots of Communicative Flexibility in Play

The significance of the human capacity for flexible behavior, cognition, and communication has long been debated by scholars, theologians, and politicians. Central to these debates is the recognition that humans are not automatons that invariably react in some set way to specific situations but can instead respond flexibly when they interact with their social and physical environments. Of course, humans are not always flexible, but the *capacity* for flexibility is certainly a crucial aspect of human nature. This capacity is reflected in our behavior, our thought, and our communication. However, there is little consensus concerning either the evolutionary origins of this capacity or the manner in which it emerges during ontogeny.

Reynolds (1976) suggested that the evolution of the human capacity for flexibility was influenced by four phenomena: (1) increasing delays in maturation, (2) increasing abilities to manipulate objects (and increasing interest in doing so), (3) increasing reliance on observational learning, and (4) increasing significance of play for social development. These phenomena are not mutually exclusive but are instead intertwined aspects of our evolutionary history. Increased delays in maturation facilitated the survival of large brained but physically inept newborns, infants, and toddlers, resulting in more significant costs for parents in the rearing of individual offspring. This likely contributed to the emergence of complex social systems that spread the costs of child rearing among the group (Bogin, 1990; Hrdy, 1999; Oller and Griebel, 2005). Such social systems further increased the likelihood that the children of group members would survive and at the same time provided more opportunities for observational learning.

In addition to the emergence of increasingly complex social systems, our ancestors developed an enhanced intrigue for objects, possibly inspired by the recognition that some objects could be used as tools, an interest that continues unabated to this day. Interest in objects resulted in significant accomplishments in object manipulation, including the manufacturing of tools. Fascination with objects may also have inspired our ancestors to attend to others' manipulation of objects, such observational learning enabling individuals to ascertain the significance of various objects as tools by watching others, thus reducing the reliance on trial-and-error learning. Of course, observational learning need not always involve objects, for it is possible to learn a variety of types of behavior via observation as well as by playing with other people.

Finally, and most importantly for this chapter, the emergence of play provided developmental mechanisms to facilitate the acquisition of flexible thinking (Kuczaj et al., 2006; Piaget, 1951; Špinka, Newberry, and Bekoff, 2001). Although each of the factors outlined by Reynolds was undoubtedly important in the evolution of the human capacity for flexibility, in the remainder of this chapter we will focus on the possible roles of play in the evolution and ontogeny of flexible communication systems.

What Is Play, and Why Is It Important?

Although we believe that play behaviors evolved to facilitate the survival and reproductive success of individual animals, play rarely has immediate survival-oriented or reproduction-oriented functions. Play has evolutionary consequences because play allows young animals to practice behaviors that facilitate survival and reproductive success in later life. When animals play, three basic types of behavior are involved: modification of a produced or observed behavior, imitation of another's behavior, and repetition of a behavior produced by the self (Kuczaj, 1998). Modification involves the transformation of a preceding activity (produced by the self or others). Imitation and repetition both involve the (sometimes partial) reproduction of a preceding behavior, the difference between the two being the source of the model behavior. In imitation, the model is provided by another. In repetition, the model is provided by the self. Play often involves both modifications and imitation/repetitions. These processes may be important for the ontogeny and evolution of flexible thought and flexible communication (Kuczaj, 1982, 1983, 1998; Kuczaj et al., 2006; Piaget, 1952; Špinka, Newberry, and Bekoff, 2001; Weir, 1962).

The notion that play is an important aspect of the ontogeny of flexible thought rests on the opportunities that play provides organisms to create novel experiences for themselves and their playmates. These novel experiences enhance individual behavioral variability, individual creativity, and individual innovations that may spread throughout the group (Kuczaj et al., 2006). Not all species engage in these sorts of play activities. Play that incorporates novel experiences is cognitively demanding and, consequently, may have evolved only in species that possessed the requisite cognitive abilities (Špinka, Newberry, and Bekoff, 2001). Of course, it is difficult to determine whether play resulted *from* the emergence of particular cognitive abilities or contributed *to* the emergence of such abilities. The most likely evolutionary scenario is one in which play and cognitive abilities facilitated the increasing complexity of one another.

At this point, it is worth noting that all play is not the same. For example, play is not always novel, for players often play familiar games and produce familiar behaviors during play. Behaviors may even be repeated over and over again during play, such repetitious play sometimes involving known behaviors but at other times consisting of behaviors the player is in the process of acquiring. The repetition of a behavior that is not yet in the organism's repertoire may help the player learn the behavior, while the repetition of a known behavior may help the player to consolidate and/or maintain a behavior it has learned. Thus, both familiar and novel play behaviors may facilitate the ontogeny of an organism's behavior repertoire (Baldwin and Baldwin, 1974; Kuczaj, 1983, 1998; Kuczaj et al., 2006; Piaget, 1952), perhaps even those involved in the emergence of flexible communication systems.

Solitary Play

In addition to ranging from familiar to novel, play also varies in terms of the extent to which it involves individuals other than the player. Some play is solitary and so only involves the player (and perhaps imaginary playmates). Children's and animals' interactions with their environment during solitary play may facilitate cognitive growth and cognitive flexibility in that the player has control over the play activity that is unimpeded by other players. In these cases, players often make their play more complex, which results in the outcomes becoming less predictable. For example, a child who has perfected her ability to toss a beanbag into a bucket from a distance of two feet might then try to toss the bag with her eyes closed, with her other hand, or from a greater distance. The outcome of these variations is that she is less successful at tossing the bag into the bucket, but she will nonetheless persist in making the task of tossing the beanbag into the bucket more difficult. Obviously, the goal of the young girl is not to simply toss the bag into the bucket. If so, she would not make the task more challenging. The challenge itself must be what is intrinsically reinforcing. Basically, the play process is often more important than the play product (Kuczaj et al., 2006). Given that play is often considered to consist of actions that serve no readily determined immediate survival or reproductive function, it stands to reason that play behaviors are more important than play products.

Many years ago, Piaget (1952) suggested that moderately discrepant events were critically important for cognitive development. Such events are both somewhat familiar and somewhat novel, hence the phrase "moderately discrepant." The young girl described above did not begin trying to toss a bag with her eyes closed, but instead did so after perfecting the toss with her eyes open. When she closed her eyes while trying to toss the bag into the bucket, doing so became both novel (eyes closed) and familiar (the same type of activity from the same distance).

Given that the play process seems to be more important than the play product and that players use play to create moderately discrepant events, solitary object play should be enhanced by novelty, complexity, and unpredictability. And such is the case for both humans and animals (Kuczaj and Trone, 2001; Piaget, 1952; Pellegrini and Bjorklund, 2004). For example, Kuczaj, Lacinak, Garver, and Scarpuzzi (1998) described a form of play produced by killer whales in which the whales used bits of fish to bait seagulls. Successful baiting resulted in the capture of a live seagull, which the whale then played with much as a cat plays with a mouse. Perfecting the technique of baiting and catching seagulls required considerable practice, during which time the young whales were rarely reinforced (except by the occasional near catch of a gull). Although young whales appeared to learn about gull baiting by observing other whales do so, each whale developed a slightly different technique, demonstrating that observational learning consisted of some form of goal enhancement (the goal being to catch a seagull by baiting it with fish) rather than an exact mimicry of observed behaviors. Play helped each whale perfect its own style of catching gulls, each individual style resulting from trial-and-error learning. And once a whale had learned to capture a gull, it could then obtain live toys with which to play.

As we observed the emergence of gull baiting among a group of killer whales, we believed that the ultimate goal of such behavior was the capture of a live gull. Such captures resulted in interesting play objects for the whales, but we soon learned that catching a gull was not the primary goal of gull baiting. While a whale was learning to catch gulls, the challenge of doing so seemed to maintain the behaviors involved in such attempts. When a whale had perfected its technique, it was rewarded by the opportunity to play with its captive. However, playing with a gull was evidently not sufficiently interesting to the killer whales to maintain the behavior for long. Certain whales became so proficient at gull baiting that they could seemingly catch gulls at will (one whale caught five gulls on five attempts within a sixty-minute period). These whales then did something that surprised us. They modified their gull baiting behaviors to make catching gulls more difficult. For example, one whale had perfected its technique of lurking below the surface as a piece of fish floated on the surface and catching the gull the instant that it touched the fish. This whale subsequently modified its behavior to try to catch the gulls approximately four to five feet *above* the surface of the water as the gull swooped down toward the fish. This change resulted in the whale catching far fewer gulls, but the whale nonetheless persisted at this new behavior. Clearly, the whale was motivated by the challenge rather than by the outcome itself. If catching gulls was all that was important, the whale should have maintained the behavior that resulted in more gull captures rather than modifying its behavior to make it more difficult to catch gulls.

What factors might have led to the evolution of the need to challenge oneself during play? We suspect that the factors that selected for the capacity for flexible problem solving were also relevant in the selection for the need to challenge oneself during play. For example, predators that were better able to adapt to changing ecologies and to the various defensive capabilities of prey may have been more likely to survive and reproduce than were individuals that possessed more rigid hunting techniques. Play provided a context in which to practice behavioral flexibility in relatively safe conditions, and so young animals that engaged in play behaviors that promoted flexible behavior and subsequent flexible hunting skills may have been more likely to survive and reproduce, the result being the evolution of a species in which play facilitated the development and maintenance of a capacity for flexible behavior. In general, we expect that the need to challenge oneself during play will be found in members of a species for which flexible behavior and problem solving are important, but not in members of a species that is more reliant on more rigid stereotypical behaviors.

Possible Role of Evolution of Play in the Evolution of Flexible Communication

Whales, like human children, modify their play behavior to produce moderately discrepant events. Such behavior characterizes the play of many species (Kuczaj et al., 2006; Pellegrini and Bjorklund, 2004), and such play may facilitate the ontogeny of contextually flexible thought. If contextually flexible thought is necessary for contextually flexible communication, then solitary play may indirectly contribute to the emergence of contextually flexible communication by virtue of play's effect on requisite cognitive abilities. For example, the explorative nature of play helps players learn to view objects and events in different ways and so facilitates the acquisition of many-to-one and one-to-many mappings, a key element of contextually flexible communication. Griebel and Oller (this volume) consider the functional decoupling of signal and function in evolution as a key step in the evolution of communicative flexibility, and it is possible that play was instrumental in this decoupling. Spontaneous variable vocalization may be the most primitive sign of decoupling in acoustic communication, and it has sometimes been characterized as playful vocalization in the literature on infant vocal development (e.g., see Oller, 2004, and later section in this chapter on sound play).

Social Play

Although many animals engage in forms of solitary play, social play seems very important for the young of social species. Playing in the presence of others is necessary but not sufficient to constitute social play. Parallel play occurs in a social context (others are present), but an individual's play does not involve others. It is essentially solitary play that occurs in the presence of others. Parallel play in human children decreases with increasing age. True social play involves at least two players who are interacting with one another. Such play increases in relative frequency for human children as they get older. Animals engage in both social and parallel play, but general developmental trends in these forms of play have been specified in only a few species (see Burghardt, 2005, for a general review). Nonetheless, animal social play involving peers is known to increase the complexity of play behavior (Kuczaj et al., 2006) and may also facilitate the imitation of novel behaviors (Reynolds, 1976; Miklosi, 1999; Kuczaj et al., 2006).

Does Social Play Influence Communicative Flexibility?

The significance of solitary play for contextually flexible communication may rest on solitary play's role in the ontogeny of flexible thought, but the significance of social play for the emergence of communication systems has been argued to be much more direct, as evidenced by the following quotation: "To the extent that communication systems are vital to the creation and maintenance of social systems, play appears to be the medium through which the young learn to use and understand these systems" (Lancy, 1980, p. 489). The significance of social play for communication was also emphasized by Sutton-Smith (1980), who argued that social play is a kind of communication. Garvey and Berndt (1977) pointed out that in order to play with others, children must communicate more than "this is play." Play partners must also discuss the play theme, each other's roles in the theme, the status of pretend objects, the play setting, and so on. The basic idea of all this is that social play is impossible without cooperation among the players, and that cooperation depends on communication. More specifically, successful social play involves turn taking, role playing, and appreciation of another's intentions. Language accompanying the social play of preschool children tends to be more complex than that which occurs in other contexts (Chance, 1979). This may result from preschool children's producing more sophisticated forms of language when they are comfortable. Play is typically a comfortable setting for young children, particularly when they are playing with friends, and so may result in more complex language. Regardless of the veracity of this claim, children's social play may help them acquire conversational competence and communicative flexibility.

Play Signals

Although animals clearly lack the language abilities of young human children, animal play may nonetheless facilitate the development and maintenance of social bonds and social interaction skills, including the ability to respond to others in a flexible manner (Baldwin and Baldwin, 1974; Bekoff, 1984; Byers, 1984; Colvin and Tissier, 1985; Fairbanks, 1993; Kuczaj et al., 2006; McCowan and Reiss, 1997;

Snowdon, Elowson, and Roush, 1997; but see Sharpe, 2005, for an opposing opinion). Creels and Sands (2003) suggested that the inability to understand a social situation is stressful and that animals with poor abilities to do so have higher chronic levels of stress. Play may help animals learn to assess social situations by learning to interpret communicative cues and thereby reduce the incidence of social stress. Play, then, may enhance the ontogeny of appropriate social behaviors. However, the extent to which animal play facilitates the development of communication skills is unknown.

Bateson (1955, 1972) was one of the first to suggest that play was important in the ontogeny and evolution of communication. He argued that play requires an ability to distinguish pretense and reality, *as well as an ability to communicate this distinction to others*. In order to engage in social play, play partners must share what Bateson termed "play frames," which are essentially shared meanings or concepts created during play. Bateson emphasized the importance of play "signals" in the establishment and maintenance of play frames. According to Bateson, play signals are different from those associated with more serious behaviors and serve to inform the recipient that "these actions in which we now engage do not denote what those actions for which they stand would denote" (1972, p. 180). The fact that play signals communicate that behaviors produced during play do not connote the same seriousness that they normally would in other contexts has led to the conclusion that play signals are metacommunicative (also see Bekoff, 1975, 1995). However, we suspect that play signals are not necessarily metacommunicative in the sense that the producers of such signals reflect on the meaning of the signals that they are producing. It seems more plausible that the motivation to play with others may result in the automatic activation of signals that indicate a readiness to play, and that such an automatic system emerged during the evolution of social play to decrease the likelihood that playing animals might misinterpret each other's intent.

As Bateson suspected, play signals are common among species that engage in social play. Members of social species typically engage in both solitary and social play, but play signals are more likely to accompany social play than solitary object play (Biben and Symmes, 1986). Some play signals are acoustic in nature. Play signal vocalizations have been reported in a variety of species, including chimpanzees (Matsusaka, 2004), cotton-top tamarins (Goedeking and Immelmann, 1986), bottlenose dolphins (Blomqvist, Mello, and Amundin, 2005), dwarf mongooses (Rasa, 1984), rhesus monkeys (Gard and Meier, 1977), and squirrel monkeys (Biben and Symmes, 1986). Other play signals involve other sorts of behaviors. For example, chimpanzees sometimes signal play via head tilts (Sade, 1973) and play faces (van Hoof, 1967). Play postures and/or play faces are used by a number of species, including dogs (Bekoff, 1995), spotted hyenas (Drea et al., 1996), and squirrel monkeys (Baldwin and Baldwin, 1974). We have also observed bottlenose dolphin calves positioning themselves in water perpendicular to a playmate and remaining stationary until the other animal came forward and pushed the calf sideways through the water. We believe that the stationary perpendicular posturing was a play signal indicating a willingness to be pushed. This behavior was only produced during play contexts and always resulted in one of two outcomes. Either it was ignored by the other dolphins, or the posing dolphin was pushed sideways along the surface of the water. If the posing dolphin was pushed, it sometimes posed again in order to solicit another push. On some occasions, two dolphins alternated between being the "pusher" and the "pushee," suggesting that they were engaging in a form of cooperative play, perhaps even a primitive form of turn taking.

To sum up, the communicative status of play signals is well established. Animals use play signals to indicate a willingness to play, to ascertain another's willingness to play, and to verify that ongoing activity is still play (Bekoff, 1975; Bekoff and Byers, 1981; Fagen, 1981; Feddersen-Petersen, 1991; Hailman, 1997; Hailman and Dzelzkalns, 1974; Loeven, 1993). Play signals also help in the interpretation of other signals (Hailman, 1977; Bekoff and Allen, 1998). For example, coyotes respond differently to threat gestures that are preceded by a play signal than to those that are not (Bekoff, 1975).

During social play, play signals may be produced by the player acting upon another or by the player being acted upon. Van Hoof and Preuschoft (2003) reported that playing chimpanzees taking the active role were more likely to display a "wide-open" mouth, while chimpanzees in the passive role were more likely to produce rapid "play chuckles." Matsusaka (2004) found that chimpanzee "play pants" were most likely to occur during social play, and that the recipient of the activity (e.g., the one being chased or tickled) was most likely to produce "play pants." These signals were more likely to be followed by bouts of "play aggression" than by other forms of behavior. Although certain forms of play signals are more likely to occur in certain contexts, it is important to remember that play signals are not automatically produced in response to others' behaviors but are instead "flexibly related to the occurrence of events in a play sequence" (Bekoff and Allen, 1998, p. 109).

For example, Bekoff (1977) examined the play bows produced by dogs during play bouts. The form of play bows was relatively constant throughout play bouts, but the duration of the play bows varied depending upon where they occurred during the play bout. The duration of play bows was more variable during the middle of ongoing play sequences than at the beginning of a play bout. Apparently, it is important to unambiguously signal the intent to play before play begins, but more flexibility in the play signal occurs once play is under way.

Further evidence for the flexibility of play signals comes from a comparative analysis of play faces. Pellis and Pellis (1996) found that play faces are highly variable across species and concluded that most play signals are not universal. But exactly how much variability exists in species-specific play signals? Humans are certainly capable of recognizing play behaviors in other species, and there is evidence that nonhuman species play with members of species other than their own (Brown, 1994; Mitchell and Thompson, 1990). Thus, even though play signals may not be identical across species (or even within a play context, as in the case of dog play bows), there must be something similar in the play behaviors of diverse individuals that allows participants (and observers) to recognize that the play context is in force. Griebel (personal communication) pointed out that the fact that relatively few discrete play signals have been discovered in other species (including our own) is significant, considering how many different types of play exist. She suggests that the "intent to play" may be coded in rhythm and time patterns that are significantly different from the "real thing," and as a result discrete play signals would be redundant. The lack of a one-to-one correspondence between discrete play signals and the communication of an intent to play is problematic if one assumes that animals have inflexible communication systems. However, if animal communication systems allow for at least some flexibility, then play signals may be part of a multifaceted communicative attempt to signal play rather than the sole component of such attempts.

To sum up, in order to engage in social play, animals must signal playful intent, and a variety of play signals are used for this purpose. However, too little is known about the flexibility of play signals and the manner in which they are acquired in order to truly understand the significance of such signals for the ontogeny and phylogeny of flexible communication signals. For example, do animals play *with* play signals before using them to signal play? We have seen that play signals are common during social play but rare during solitary play. When play signals do appear during solitary play, it is possible that the animal is playing with the signal rather than attempting to signal a play bout, but we lack sufficient information to test this hypothesis.

There are also unanswered questions about the use of play signals after they have been acquired. For example, we know little about how animals communicate "play themes," that is, what and how they are going to play. The earlier example of a dolphin positioning itself to be pushed is one example of a play theme being communicated, but most other examples involve play fights. The reason for this may be that play fights are easy to observe compared to other forms of social play, but it is also possible that it is more important to signal playful intent for a mock fight in order to avoid injury. Learning more about the signaling of play themes among play partners would help us to determine the role of play signals in the ontogeny and evolution of flexible communication systems.

Language Play

We have seen that some animals use play signals to communicate about play events and have suggested that it is important to determine the extent to which they play with these signals. This suggestion is based on the fact that language play is common among human children and is an important aspect of language development (Dowker, 1989; Kuczaj, 1982, 1983, 1998; Nelson, 1989; Weir, 1962). Human children use language while they play, and they certainly use language to communicate the intent to play and play themes. However, human children also play *with* language. Such play may involve the manipulation of sounds and linguistic forms, including the production of puns or verbal riddles in older children.

Kuczaj (1998) hypothesized that the more flexibility evident in a communication system, the more likely that learners of the system would play with its components. As others have noted, flexible communication systems involve more than signal production. In addition to production, users must understand when it is appropriate to use the signals in one's repertoire and how to respond appropriately to the signals of others (Marshall, Wrangham, and Arcadi, 1999; McCowan and Reiss, 1997; Seyfarth and Cheney, 1997). Play may facilitate the acquisition of signal production, signal usage, and signal comprehension, but most research in this area has focused on vocal production.

The early vocal behavior of human infants has been summarized by Oller (1980, 2000). For the purpose of this chapter, the emergence of babbling is the most important aspect of this behavior. All normal human infants babble, which was long ago recognized as a form of sound play and practice (Lewis, 1936; Jakobson, 1940; Lenneberg, 1967). Babbling has been argued to be a critical aspect of the development of a functional phonological system (Ferguson and Macken, 1983; Lieberman, 1984). For example, multisyllabic canonical babbling typically appears between six and ten months in human infants and involves different types of sound play (repetitions such as *ba ba ba* and modifications such as *ba ba di de de ga*; see Koopmans-van Beinum and van der Stelt, 1986; Oller, 1980; Stark, 1980). This type of babbling is usually accompanied by variations in voice quality, pitch, loudness, pausing, and syllable rate (Kent, Mitchell, and Sancier, 1991) and likely helps children master productive control of sounds (Lieberman, 1984).

Research involving hundreds of normally developing infants around the world has indicated that production of canonical babbling is universal—it appears no normally developing infant fails to do it (Holmgren et al., 1986; Koopmans-van Beinum and van der Stelt, 1986; Murai, 1963; Nakazima, 1962; Oller and Eilers, 1982; Preston, Yeni-Komshian, and Stark, 1967; Vihman, 1992). Even children who will not learn a spoken language babble, although the onset of vocal babbling is delayed in deaf children (Oller and Eilers, 1988). Deaf children "babble" with gestures as well as with sounds, and this manual babbling differs from the manual activity produced by hearing infants (Petitto and Marentette, 1991). There is little evidence that other species that communicate with gestures (e.g., chimpanzees) babble manually. Some apes that that were taught to use signs to communicate with humans may have engaged in forms of manual babbling (Gardner and Gardner, 1980; Patterson, 1980), but such behavior has not been found in the spontaneous behavior of individuals learning their natural communication system.

Although manual babbling has not been documented in nonhuman species, a sort of "babbling" with sounds has been reported in birds, cotton-top tamarins, and pygmy marmosets (Marler, 1970, 1984; Elowson, Snowdon, and Lazaro-Perea, 1998; Nottebohm, 1970; Snowdon, 1990, 1997; Snowdon, French, and Cleveland, 1986; Snowdon and Elowson, 2001). The similarities and differences between babbling by human infants and that produced by other species have been described by Doupe and Kuhl (1999), Elowson, Snowdon, and Lazaro-Perea (1998), and Snowdon (1997).

Although human and animal babbling are similar in that both involve playing with sounds, there are important species differences (Oller and Griebel, 2005; Snowdon, Elowson, and Roush, 1997). For example, pygmy marmoset babbling is limited in that they do not link sounds with new functions during babbling or even use babbling to play with individual sounds. Instead, their babbling involves the repetition of alternating sequences of particular sound types, most of which can be identified clearly as pertaining to the adult repertoire of signals. The babbling reported in bird song learning may also differ from human babbling in several ways (Snowdon, Elowson, and Roush, 1997), including age of onset, gender specificity, and range of sounds produced. The extent of these differences is difficult to determine given how much remains to be learned about the ontogeny of bird song. There are approximately 4,000 songbird species, and more than 300 species each for parrots and hummingbirds, but little is known about vocal development in most of these species. Nonetheless, human babbling and the subsong "babbling" of birds may be similar in terms of the use of sounds that do not become part of the adult repertoire. Human babbling incorporates a wide range of sounds, some of which will not be used in the language(s) the child eventually learns. Similarly, bird subsong is often poorly organized, and thus the sounds produced may not occur in the adult repertoire (Marler, 1970, 1984).

Sound Play Other Than Babbling

In addition to the early sound play that characterizes the earliest phases of language development, children continue to play with sounds as their language skills improve (Dowker, 1989; Ferguson and Macken, 1983; Jespersen, 1922; Kuczaj, 1982, 1983; Weir, 1962). Playing with the sounds of words was termed "expressive sound play" by Ferguson and Macken (1983), is most common between two and five years of age, and is often accompanied by laughter (Keenan, 1974). The following example from a two-year-old child illustrates the character of this form of sound play (see Kuczaj, 1983, for a more detailed discussion of this phenomenon):

Nice doggy

Noce doggy (laughs)

Nose piggy (laughs)

Dose diggy (laughs)

Dose diggy (laughs)

Dose diggy, dose piggy, dose piggy (laughs)

Children play with the sounds of words, as well as the sounds and order of sounds in songs, chants, and rhymes. They seem to particularly enjoy producing nonsensical rhyming patterns. Preschool children repeat words and sounds that they find amusing (and sometimes continue to do so long after their audience has tired of the game). They may also try to talk "funny" by altering their voice or accent. There is no evidence for these sorts of sound play in nonhuman animals nor for the joy that accompanies such play in young children.

Children also play with the structure of the utterances they produce (Kuczaj, 1982, 1983; Weir, 1962). This form of language play takes a variety of forms, including build-ups, breakdowns, completions, and substitutions. Build-ups are sequences of utterances that become increasingly complex, as in the following example: "Balloon. Yellow balloon. Look at yellow balloons." Breakdowns are sequences of utterances that become less complex, as in the following: "Ball down. Ball. Down." Completions are utterances in which a pause interrupts the utterance, but the utterance is nonetheless finished: "And put it (pause) up there." Substitutions are sequences of utterances in which a word or part of a word is replaced: "What color dog? What color sky? What color frog?" Build-ups, breakdowns, completions, and substitutions occur both in social contexts and while the child is alone.

The nature of language play changes during the course of development (Kuczaj, 1983, 1998). Two-year-old children are more likely to play with sounds than with the structural aspects of language. Five-year-old children are as likely to play with structural forms as they are to play with sounds. The reason for this change is likely similar to that of a killer whale making its gull baiting behavior more difficult—change is necessary to keep the play activity interesting. Infants and toddlers are exploring the sound systems of their language, and so much of their language play is spent manipulating the characteristics of this system. In Piagetian terms, sound play provides young children with ample opportunities to produce moderately discrepant events that involve sound. Older children, although still perfecting their articulation skills, are faced with the daunting task of learning the syntactic characteristics of their language. Consequently, play with morphological and syntactic structure becomes more challenging than play with sounds for older children, and the opportunities for moderately discrepant events involving such structures is greater.

Recall that build-ups, breakdowns, completions, and substitutions are only considered to be forms of play when they occur in a noncommunicative context. There are two general noncommunicative contexts in which language play occurs: private speech and social-context speech. Private speech is that which is produced while the child is alone, whereas social-context speech is that which is produced in the presence of others (but appears to have no communicative intent). The following is an example of a monologue produced during private speech (from Weir, 1962, p. 184):

One two three four

One two

One two three four

One two three

Anthony counting

Good boy you

One two three

The following is an example of a monologue produced in the presence of others (recorded by the first author):

Truck mine.

Truck.

Truck.

Mine.

Mine.

That mine truck.

That mine truck.

That mine truck too.

Mine trooock (vowel drawn out).

Mine trooock.

Miiine trooock (both vowels drawn out).

Miiine trooock.

Mooo (vowel drawn out).

Harrison play mine trooock.

Mommy!

Where Harrison go?

Where Harrison go?

(child leaves to find mother)

Children appear to use monologues produced in private speech and in social-context speech to help them master the intricacies of the language they are learning (Kuczaj, 1982, 1983; Weir, 1962). They are most likely to play with and practice forms they are in the process of learning, but they also play with aspects of language that are already established parts of their linguistic repertoire. Regardless of whether language play involves known linguistic units or ones that are being acquired, both repetition and modification are likely to be incorporated.

Some members of other species also use monologues to perfect elements of communication systems they are learning. Pepperberg, Brese, and Harris (1991) reported that "Alex," an African Grey parrot being taught to produce and comprehend English words, produced words he was in the process of learning during monologues. His pronunciation of these words improved during these monologues, suggesting that he was practicing the pronunciation of these less familiar words. However, the overall frequency of these sorts of words during Alex's monologues was relatively low (~6%). He was most likely to produce known English words and sounds, but he also produced parrot calls and whistles, cage noises, ringing telephones, and other environmental sounds he had encountered. Alex, then, appears to have played with a variety of sounds during his monologues and to have used some such play to consolidate aspects of new sounds he was learning.

Alex's example might reflect his experience with humans, perhaps even the intense training to which he was exposed. However, similar forms of sound play during monologues have been reported for birds with much less experience with humans. For example, ravens also appear to practice sounds during monologues:

> Young ravens . . . engage in long monologues, unlike anything heard in adults. (They) produce a variety of sound types (low gurgling sounds, barely audible chortles, squeaks, quacks, loud yelling, trills and sounds that resemble water running over pebbles in a swift stream), some of which resemble calls typical of adults.
>
> These monologues are often accompanied by continually changing gestures and feather postures, as if the birds were play-acting. . . . For example, loud trills and rasping quorks are accompanied by "macho displays"—ear and throat feathers erect and flashing nictitating membranes. In adults, these displays are mating and assertive displays directed at mates or rivals. In young birds (during the first few months out of the nest), these monologues occur out of context, often either when the bird is alone or not directed at any other individual and not eliciting any obvious reaction from others. They do not appear serious since the bird may intermittently perch, stretch, pick at twigs, etc., and then abruptly change to a different "tune" (Heinrich and Smolker, 1998, p. 40).

How Important Is Social Interaction?

Although human children and young animals engage in solitary play and social play, social play seems particularly important for social animals. For example, play with peers increases the complexity of play behavior and facilitates the imitation of novel behaviors (Kuczaj et al., 2006; Miklosi, 1999; Reynolds, 1976; Visalberghi and Fragaszy, 1990). However, neither the extent to which animal social play with peers incorporates play with "language" nor the extent to which peer interaction affects the complexity of "language" play is clear. Miller (2000) has suggested a verbal courtship theory in which he proposes that sexual selection favored verbal ability and, consequently, social play with language. If this is so, then individuals who engaged in early forms of social play with language may have enhanced their verbal abilities and so been more likely to attract and keep a mate. Over time, the need to compete for mates via verbal courtship resulted in increasingly complex forms of verbal skills, the acquisition of which was facilitated by both solitary and social play with language.

In humans, the earliest forms of social play are most likely to involve the infant and the mother. The significance of the mother in social play is evident in other species as well (Chirighin, 1987; Cockcroft and Ross, 1987; Hoff, Nadler, and Maple, 1981; Mann and Smuts, 1999). In the case of the spotted-tail quoll, the mother produces a call that precedes social play with its offspring (Lissowsky, 1996). Of course, not all mother–offspring relationships are the same, and the quality of the bond between the mother and infant affects both social play and play in general. Human and monkey infants that are insecurely attached to their mothers play less frequently and produce less complex play (Harlow and Harlow, 1962; Higley, 1985). However, the extent to which early social play between mother and offspring incorporates "language" play seems to vary across species and across individual mother–offspring pairs within a species. As a result, little is known regarding the effect of mother–offspring interaction on "language" play. What little is known suggests that social interaction with the mother may facilitate the ontogeny of acoustic signals. For example, Goldstein, King, and West (2003) reported that seven- to eight-month-old human infants used "more mature" sounds when mothers were interacting with them.

Social interactions likely influence the behavior of even younger infants. During social interactions, human mothers and two-month-old infants both produce a range of sounds with both positive and negative expressions (Oller and Griebel, 2005). Slightly older infants (three- to six-month-olds) typically begin to explore sounds in precanonical babbling, oftentimes while alone. Is there a relationship between these two events? Do mothers who produce more sound play with their two-month-old infants have infants who subsequently babble more frequently or complexly? Or is some minimal amount of social interaction sufficient to produce "normal" babbling? Some amount of social interaction involving sound play may be necessary for normal babbling to occur, but exactly what type and how much is necessary is not clear. Hearing parents of deaf infants do interact vocally with their infants, but their infants show late onset of canonical babbling (Oller and Eilers, 1988), suggesting that the ability to interact vocally may influence the onset and perhaps the course of verbal sound play. In humans, social interaction typically results in later solitary play with sounds. Social interaction may also have been crucial for the monologues that the parrot Alex produced as he attempted to master the sounds of new words he had heard.

For birds, there are important species differences in the effect of social interaction on song quality. Some birds (faculative learners) can learn song from listening to recordings, but others (obligate learners) require social interaction (Baptista and Gaunt, 1997; Pepperberg, Naughton, and Banta, 1998). Although faculative learners can acquire songs by listening to recordings, they do better if they are able to interact socially with the bird producing the model song. There are numerous examples of the importance of social factors for avian sound learning (e.g., see Hausberger et al., this volume). For instance, African Grey parrots (*Psittacus erithacus*) are more likely to learn referential labels from live models than from videotaped models (Pepperberg et al., 1998). The types of models provided to juvenile male cowbirds also influenced their songs (West, King, and Freeberg, 1997). Some juveniles were housed with female cowbirds, which do not sing. Males in this condition apparently learned to modify their songs based on social cues provided by the females. However, their sound repertoire developed more slowly than did that of juveniles housed with canaries. Moreover, the juveniles housed with canaries produced vocalizations similar to those produced by canaries, demonstrating the sound learning flexibility of these birds. West, King, and White (2003) noted that cowbirds must learn how, when, and where to sing and argued that social interaction is critical in the acquisition of this information.

Social interaction also facilitates the vocal development and vocal production of pygmy marmosets (Elowson and Snowdon, 1994; Snowdon, Elowson, and Roush, 1997; Elowson, Snowdon, and Lazaro-Persa, 1998), cotton-top tamarins (Roush and Snowdon, 1999), marmosets (Rukstalis, Fite, and French, 2003), and Campbell's monkeys (Lemasson, Hausberger, and Zuberbühler, 2005). Effects of cross-fostering experiments are inconclusive regarding the importance of social models and vocal productive flexibility in nonhuman primates. Masataka and Fujita (1989) compared the food calls of two rhesus (*Macaca mulatto*) and one Japanese monkey (*Macaca fuscata*). These animals were cross-fostered with a mother of the opposite species. The calls made by the infants were more similar to the calls made by members of the species with which they were raised than by biological conspecifics, suggesting a capacity for vocal learning. However, Owren et al. (1993) reported that cross-fostered monkeys demonstrated limited vocal adaptability in production compared to comprehension. The extent to which social interaction can affect the vocal repertoire of nonhuman primates is still to be determined.

Future Directions (with Special Attention to Cetaceans)

There is clearly much left to learn about the role of play in the evolution and ontogeny of flexible communication systems. For example, it has been hypothesized that the more a signaling system allows for flexibility, the more likely play with the system will be observed in the young of the species (Kuczaj, 1998). In order to test this prediction, we need to learn more about the flexibility of nonhuman communication systems. Human language is unique in terms of its flexibility and productivity, but other species evidence varying degrees of flexible communication. Nonhuman primates appear to be more flexible at responding to vocalizations than at producing them (Seyfarth and Cheney, 1997; Zuberbühler, 2005). However, there is evidence for vocal flexibility among nonhuman primates (Elowson, Snowdon, and Lazaro-Perea, 1998; Snowdon, this volume; Snowdon, Elowson, and Roush, 1997). Marshall et al. (1999) reported changes in the vocalizations of captive chimpanzees following the introduction of an individual that produced a unique call termed the "Bronx cheer pant hoot." After his introduction to the group, six other males started to produce this vocalization as well, presumably as a consequence of hearing this novel vocalization. Wild chimpanzees produce different types of barks depending on the context (Crockford and Boesch, 2003) and different types of screams depending on their social role during a conflict (Slocombe and Zuberbühler, 2005).

Species that are capable of vocal learning should be more likely to engage in sound play (Kuczaj, 1998). In fact, we suspect that the capacities for vocal learning, vocal complexity, and sound play coevolved. In part, the evolution of these capacities may have been influenced by social group size. For example, Carolina chickadees produce more complex calls when they are in large groups than when they are in small groups (Freeberg, 2006), suggesting that increased group size may require more complex calls in order to effectively communicate. If similar pressures were involved in the evolution of human verbal complexity, increases in human group size may have driven the need for more complex forms of communication. Individuals that were capable of modifying their communications may have thus had a selective advantage, as would those who used play contexts to facilitate their communicative flexibility.

Vocal learning can affect the development of communication systems by influencing the production of calls, call usage, and the comprehension of these vocalizations (Marshall, Wrangham, and Arcadi, 1999). Janik and Slater (2000) suggested that the phrase "vocal learning" should be applied only to cases in which the acoustic parameters of calls are altered by social experience. Although this definition encompasses many aspects of vocal learning, it is too narrow in that it ignores vocal learning through individual practice or play.

Avian songs often undergo developmental changes before achieving a "functional" or adult sound (Baptista and Gaunt, 1997; Marler, 1970). Some species are open-ended learners and so can acquire new songs throughout much of their life span (e.g., Brittan-Powell, Dooling, and Farabaugh, 1997; Heaton and Brauth, 1999). Other birds are age-limited learners and so can acquire new songs only during specific developmental periods. If sound play is related to productive flexibility, the sound play of open-ended learners should be more complex and long lasting than that of birds that only learn songs during a sensitive period.

There is mounting evidence that vocal learning occurs during the ontogeny of communication systems for species other than humans and birds (Snowdon and Hausberger, 1997), albeit to a much lesser extent. Such modifications may be important in vervet monkey acquisition of alarm calls (Seyfarth and Cheney, 1997), Diana monkey long-distance calls (Zuberbühler, Noe, and Seyfarth, 1997), dolphin calf acquisition of whistles (McCowan and Reiss, 1997; Sayigh, Tyack, Wells, and Scott, 1990; Tyack and Sayigh, 1997), and killer whale calf acquisition of social group dialects (Ford, 1989).

Most of what is known about vocal ontogeny in dolphins involves whistles (Fripp et al., 2005; McCowan and Reiss, 1995; Reiss and McCowan, 1993). McCowan and Reiss (1995) found that with increasing age, the number of shared whistles decreased, but the number of unique individual whistles increased. This suggests that maturing dolphins were better able to produce unique whistles, which in turn implies an ability to remember familiar whistles and some sort of preference for novel whistles. Although this pattern fits well with what we know about dolphin play, it is unclear how this developmental trend benefits communication. However, it does demonstrate the possibility of communicative flexibility in this species.

Fripp et al. (2005) reported that dolphin calves in the second year of life produced signature whistles similar to those of adult dolphins in their community, suggesting that calves may model their signature whistles on those they hear. However, the adults that produced the whistles imitated by the calves were adults that the calves rarely interacted with. Fripp et al. speculated that the calves adopt this strategy to better distinguish their whistles from those of animals with which they have frequent contact. Although this is possible, it assumes that signature whistles are prevalent among dolphins (see McCowan and Reiss, 1997; Tyack and Sayigh, 1997, for arguments for and against this hypothesis). And the preference for novel whistles as models may reflect an overall preference for novelty rather than a desire to better distinguish oneself from others. This preference for novelty has been demonstrated in play behavior (Kuczaj et al., 2006) and the change of humpback whale songs from one season to the next (Noad et al., 2000).

In addition, dolphins have been found to mimic and play with computer-generated sounds introduced into their environment (McCowan and Reiss, 1995, 1997). However, the role of babbling in the spontaneous development of a dolphin's acoustic repertoire is unclear. In fact, we know little about the role of babbling in cetacean species that might be expected to demonstrate sound play during ontogeny. For example, killer whales possess dialects based on social group membership (Ford, 1989). However, the manner in which killer whale calves acquire these dialects is unknown, as is the extent to which calves play with the sounds of the dialect they are in the process of acquiring.

Species that are capable of vocal imitation should also be more likely to engage in sound play than other species (Kuczaj, 1998). Although rare among nonhuman species, vocal imitation has been demonstrated in songbirds (Brown and Farabaugh, 1997; Catchpole and Slater, 1995), dolphins (Caldwell and Caldwell, 1972; Reiss and McCowan, 1993; Richards, Wolz, and Herman, 1984; Tyack and Sayigh, 1997), harbor seals (Ralls, Fiorelli, and Gish, 1985), belugas (Eaton, 1979), and killer whales (Bowles, Young, and Asper, 1988). We have observed a young captive rough-toothed dolphin (*Steno bredanensis*) playing with a variety of sounds it heard. One of the most striking examples involved the sound of a power lawn mower. The dolphin imitated this sound quite well, but also modified its imitations to produce a variety of different related sounds.

Conclusions

We suspect that future studies of the acquisition of animal communicative systems will reveal a variety of forms of "language" play, and that novelty will be particularly important in determining the aspects of the communication system with which an animal plays. Although human language play involves the manipulation of sounds, the "language" play of other species will incorporate signals from the modality used in their communication system. For example, young Adelie penguins play with the visual displays that characterize communication among members of the species (Ainley, 1972).

The first author once asked his oldest son (who was about three and a half years old at the time) if it was "hard to learn to talk." The son replied that it was "kinda" hard because there was so much to learn. Nonetheless, he took great delight in playing with his language and engaging his parents in word games (and to this day has not hesitated to produce any pun that crosses his mind). Language play in humans may have evolved to ease the burden of acquiring the most sophisticated communication system on the planet, and similar forms of play might also have emerged as part of the evolutionary process that resulted in flexible communication systems in other species.

References

Ainley DG. 1972. Flocking in Adelie penguins. *Ibis* 114: 388–90.

Baldwin JD, Baldwin JI. 1974. Exploration and social play in squirrel monkeys (*Saimiri*). *American Zoologist* 14: 303–15.

Baptista LF, Gaunt SLL. 1997. Social interaction and vocal development in birds. In *Social Influences on Vocal Development*, ed. C Snowdon, M Hausberger, pp. 23–40. New York: Cambridge University Press.

Bateson G. 1955. A theory of play and fantasy. *Psychiatric Research Reports* 2: 39–51.

Bateson G. 1972. *Steps to an Ecology of Mind.* New York: Ballantine Books.

Bekoff M. 1975. The communication of play intention: Are play signals functional? *Semiotica* 15: 231–9.

Bekoff M. 1977. Social communication in canids: Evidence for the evolution of a stereotyped mammalian display. *Science* 197: 1097–9.

Bekoff M. 1984. Social play behavior. *Bioscience* 34: 228–33.

Bekoff M. 1995. Play signals as punctuation: The structure of social play in canids. *Behaviour* 132: 419–29.

Bekoff M, Allen C. 1998. Intentional communication and social play: How and why animals negotiate and agree to play. In *Animal Play: Evolutionary, Comparative, and Ecological Perspectives*, ed. M Bekoff, JA Byers, pp. 97–114. Cambridge, England: Cambridge University Press.

Bekoff M, Byers JA. 1981. A critical reanalysis of the ontogeny of mammalian social and locomotor play: An ethological hornet's nest. In *Behavioral Development: The Bielefeld Interdisciplinary Project*, ed. K Immelmann, GW Barlow, L Petrinovich, pp. 296–337. New York: Cambridge University Press.

Biben M, Symmes D. 1986. Play vocalizations of squirrel monkeys (*Saimire sciureus*). *Folia Primatologica* 46: 173–82.

Blomqvist C, Mello I, Amundin M. 2005. An acoustic play-fight signal in bottlenose dolphins (*Tursiops truncatus*) in human care. *Aquatic Mammals* 31: 187–94.

Bogin B. 1990. The evolution of human childhood. *Bioscience* 40: 16–25.

Bowles AE, Young WG, Asper ED. 1988. Ontogeny of stereotyped calling of a killer whale calf, *Orcinus orca*, during her first year. *Rit Fiskideildar* 11: 251–75.

Brittan-Powell EF, Dooling RJ, Farabaugh SM. 1997. Vocal development in budgerigars (*Melopsittacus undulatus*): Contact calls. *Journal of Comparative Physiology* 111: 226–41.

Brown ED, Farabaugh SM. 1997. What birds with complex social relationships can tell us about vocal learning: Vocal sharing in avian groups. In *Social Influences on Vocal Development*, ed. CT Snowdon, M Hausberger, pp. 98–127. New York: Cambridge University Press.

Brown SL. 1994. Animals at play. *National Geographic* 186: 2–35.

Burghardt GM. 2005. *The Genesis of Animal Play*. Cambridge, MA: MIT Press.

Byers JA. 1984. Play in ungulates. In *Play in Animals and Humans*, ed. PK Smith, pp. 43–65. Oxford: Basil Blackwell.

Caldwell MC, Caldwell DK. 1972. Vocal mimicry in the whistle mode in the Atlantic bottlenosed dolphin. *Cetology* 9: 1–8.

Catchpole CK, Slater PJB. 1995. *Bird Song: Biological Themes and Variations*. New York: Cambridge University Press.

Chance P. 1979. *Learning Through Play*. New York: Gardner Press.

Chirighin L. 1987. Mother–calf spatial relationships and calf development in the captive bottlenose dolphin (*Tursiops truncatus*). *Aquatic Mammals* 13: 5–15.

Cockcroft VG, Ross GJB. 1990. Observations on the early development of captive bottlenose dolphin calf. In *The Bottlenose Dolphin*, ed. S Leatherwood, RR Reeves, pp. 461–85. New York: Academic Press.

Colvin J, Tissier G. 1985. Affiliation and reciprocity in sibling and peer relationships among free-ranging among free-ranging immature rhesus monkeys. *Animal Behaviour* 33: 959–77.

Creels S, Sands JL. 2003. Is social stress a consequence of subordination or a cost of dominance? In *Animal Social Complexity*, ed. FBM de Waal, P Tyack, pp. 153–79. Cambridge, MA: Harvard University Press.

Crockford C, Boesch C. 2003. Context-specific calls in wild chimpanzees, *Pan troglodytes* verus: Analysis of barks. *Animal Behaviour* 66: 15–125.

Doupe AJ, Kuhl PK. 1999. Birdsong and human speech: Common themes and mechanisms. *Annual Review of Neuroscience* 22: 567–631.

Dowker A. 1989. Rhyme and alliteration in poems elicited from young children. *Journal of Child Language* 16: 181–202.

Drea CM, Hawk JE, Glickman SE. 1996. Aggression decreases as play emerges in infant spotted hyaenas: Preparation for joining the clan, *Animal Behaviour* 51: 1323–36.

Eaton RL. 1979. A beluga imitates human speech. *Carnivore* 2: 22–3.

Elowson AM, Snowdon CT. 1994. Pygmy marmosets, *cebuella pygmaea*, modify vocal structure in response to changed social environment. *Animal Behaviour* 47: 1267–77.

Elowson AM, Snowdon CT, Lazaro-Perea C. 1998. Infant "babbling" in a nonhuman primate: Complex vocal sequences with repeated call types. *Behaviour* 135: 643–4.

Fagen R. 1981. *Animal Play Behaviour*. New York: Oxford University Press.

Fairbanks LA. 1993. Juvenile vervet monkeys: Establishing relationships and practicing skills for the future. In *Juvenile Primates: Life History, Development and Behavior*, ed. ME Pereira, LA Fairbanks, pp. 211–27. New York: Oxford University Press.

Feddersen-Petersen D. 1991. The ontogeny of social play and agonistic behaviour in selected canid species. *Bonn. Zool. Beitr.* 42: 97–114.

Ferguson C, Macken M. 1983. *The Role of Play in Phonological Development, Vol. 4*. Hillsdale, NJ: Lawrence Erlbaum Associates.

Ford JKB. 1989. Acoustic behaviour of resident killer whales (*Orcinus orca*) off Vancouver Island, British Columbia. *Canadian Journal of Zoology* 67: 727–45.

Freeberg TM. 2006. Social complexity can drive vocal complexity: Group size influences vocal information in Carolina Chickadees. *Psychological Science* 17: 557–61.

Fripp D, Owen C, Quintana-Rizzo E, Shapiro A, Buckstaff K, Jankowski K, Wells R, Tyack P. 2005. Bottlenose dolphin (*Tursiops truncatus*) calves appear to model their signature whistles on the signature whistles of community members. *Animal Cognition* 8: 17–26.

Gard GC, Meier GW. 1977. Social and contextual factors of play behavior in sub-adult rhesus monkeys. *Primates* 18: 367–77.

Gardner RA, Gardner BT. 1980. Comparative psychology and language acquisition. In *Speaking of Apes: A Critical Anthology of Two-Way Communication with Man*, ed. TA Sebok, JU Sebok, pp. 287–329. New York: Plenum Press.

Garvey C, Berndt R. 1977. Organization of pretend play. *Catalogue of Selected Documents in Psychology* 7: Manuscript 1589.

Goedeking P, Immelmann K. 1986. Vocal cues in cotton-top tamarin play vocalizations. *Ethology* 73: 219–24.

Goldstein MH, King AP, West MJ. 2003. Social interaction shapes babbling: Testing parallels between birdsong and speech. *Proceedings of the National Academy of Sciences, USA* 100: 8030–5.

Hailman JP. 1977. *Optical Signals: Animal Communication and Light*. Bloomington: Indiana University Press.

Hailman JP, Dzelzkalns JJI. 1974. Mallard tail-wagging: Punctuation for animal communication? *American Naturalist* 108: 236–8.

Harlow H, Harlow M. 1962. Social deprivation in monkeys. *Scientific American* 207: 136–146.

Heaton JT, Brauth SE. 1999. Effects of deafening on the development of nestling and juvenile vocalizations in budgerigars (*Melopsittacus undulatus*). *Journal of Comparative Physiology* 113: 314–20.

Heinrich B, Smolker R. 1998. Play in common ravens (*Corvus corax*). In *Animal Play: Evolutionary, Comparative, and Ecological Perspectives*, ed. M Bekoff, JA Byers, pp. 27–44. Cambridge, England: Cambridge University Press.

Higley JD. 1985. Continuity of social separation behaviors from infancy to adolescence in rhesus monkeys (*Macaca mulatta*). Unpublished doctoral dissertation, University of Wisconsin, Madison.

Hoff MP, Nadler RD, Maple TL. 1981. The development of infant play in a captive group of lowland gorillas. *American Journal of Primatology* 1: 65–72.

Holmgren K, Lindblom B, Aurelius G, Jalling B, Zetterstrom R. 1986. On the phonetics of infant vocalization. In *Precursors of Early Speech*, ed. B Lindblom, R Zetterstrom, pp. 51–3. New York: Stockton Press.

Hrdy SB. 1999. *Mother Nature*. New York: Ballantine.

Jakobson R. 1940. Kindersprache, aphasie und allgemeine Lautgesetze. In *Selected Writings*, ed. R Jakobson, pp 328–401. The Hague: Mouton.

Janik VM, Slater JB. 2000. The different roles of social learning in vocal communication. *Animal Behaviour* 60: 1–11.

Jespersen O. 1922. *Language: Its Nature, Development and Origin*. London: Allen and Unwin.

Keenan E. 1974. Conversational competence in children. *Journal of Child Language* 1: 163–83.

Kent RD, Mitchell PR, Sancier M. 1991. Evidence and role of rhythmic organization in early vocal development in human infants. In *The Development of Timing Control and Temporal Organization in Coordinated Action*, ed. J Fagard, PH Wolff, pp 135–149. Oxford: Elsevier Science.

Koopmans-van Beinum FJ, van der Stelt JM. 1986. Early stages in the development of speech movements. In *Precursors of Early Speech*, ed. B Lindblom, R Zetterstrom, pp. 37–50. New York: Stockton Press.

Kuczaj SA II. 1982. Language play and language acquisition. *Advances in Child Development and Behavior* 17: 197–232.

Kuczaj SA II. 1983. "I mell a kunk!" Evidence that children have more complex representations of word pronunciations which they simplify. *Journal of Psycholinguistic Research* 12: 69–73.

Kuczaj SA II. 1998. Is an evolutionary theory of language play possible? *Current Psychology of Cognition* 17: 135–54.

Kuczaj SA II, Lacinak CT, Garver A, Scarpuzzi M. 1998. Can animals enrich their own environment? In *Proceedings of the Third International Conference on Environmental Enrichment*, ed. VJ Hare, KE Worley, pp. 168–70. San Diego, CA: The Shape of Enrichment, Inc..

Kuczaj SA II, Makecha RN, Trone M, Paulos RD, Ramos JA. 2006. The role of peers in cultural transmission and cultural innovation: Evidence from dolphin calves. *International Journal of Comparative Psychology* 19: 223–240.

Kuczaj SA II, Trone M. 2001. Why do dolphins and whales make their play more difficult? *Genetic Epistemologist* 29: 57.

Lancy DF. 1980. Play in species adaptation. *Annual Review of Anthropology* 9: 471–95.

Lemasson A, Hausberger M, Zuberbühler K. 2005. Socially meaningful vocal plasticity in adult Campbell's monkeys (*Cercopithecus campbelli*). *Journal of Comparative Psychology* 119: 220–9.

Lenneberg EH. 1967. *Biological Foundations of Language*. New York: Wiley.

Lewis MM. 1936. *Infant Speech: A Study of the Beginnings of Language*. New York: Harcourt, Brace.

Lieberman P. 1984. *The Biology and Evolution of Language*. Cambridge, MA: Harvard University Press.

Lissowsky M. 1996. The occurrence of play behavior in marsupials. In *Comparison of Marsupial and Placental Behaviour*, ed. DB Croft, U Ganslosser, pp. 187–207. Furth, Germany: Filander Verlag.

Loeven J. 1993. The ontogeny of social play in timber wolves, *Canis lupus*. Master's Thesis, Dalhousie University, Halifax, Nova Scotia.

Mann J, Smuts, B. 1999. Behavioral development in wild bottlenose dolphin newborns (*Tursiops sp.*). *Behaviour* 136: 529–66.

Marler P. 1970. Birdsong and human speech: Could there be parallels? *American Scientist* 58: 669–74.

Marler P. 1984. Song learning: Innate species differences in the learning process. In *The Biology of Learning*, ed. P Marler, HS Terrace, pp. 289–309. New York: Springer-Verlag.

Marshall AJ, Wrangham RW, Arcadi AC. 1999. Does learning affect the structure of vocalization in chimpanzees? *Animal Behaviour* 58: 825–30.

Masataka N, Fujita K. 1989. Vocal learning of Japanese and rhesus monkeys. *Behaviour* 109: 191–9.

Matsusaka T. 2004. When does play panting occur during social play in wild chimpanzees? *Primates* 45: 221–9.

McCowan B, Reiss D. 1995. Whistle contour development in captive-born infant bottlenose dolphins (*Tursiops truncatus*): Role of learning. *Journal of Comparative Physiology* 109: 242–60.

McCowan B, Reiss D. 1997. Vocal learning in captive bottlenose dolphins: A comparison with humans and non-human animals. In *Social Influences on Vocal Development*, ed. CT Snowdon, M Hausberger, pp. 178–207. Cambridge, England: Cambridge University Press.

Miller G. 2000. *The Mating Mind.* New York: Anchor Books.

Miklosi A. 1999. The ethological analysis of imitation. *Biological Reviews* 74: 347–74.

Mitchell RW, Thompson NS. 1990. The effects of familiarity on dog–human play. *Anthrozoös* 4: 24–43.

Murai J. 1963. The sounds of infants: Their phonemicization and symbolization. *Studia Phonologica* 3: 17–34.

Nakazima S. 1962. A comparative study of the speech development of Japanese and American English in childhood. *Studia Phonologica* 2: 27–39.

Nelson K. 1989. *Narratives from the Crib.* Cambridge, MA: Harvard University Press.

Noad MJ, Cato DH, Jenner MN, Jenner KCS. 2000. Cultural revolution in whale songs. *Nature* 408: 537.

Nottebohm F. 1970. The ontogeny of bird song. *Science* 167: 950–6.

Oller DK. 1980. The emergence of the sounds of speech in infancy. In *Child Phonology*, ed. G Yeni-Komshian, J Kavanaugh, C Ferguson, pp. 93–112. New York: Academic Press.

Oller DK. 2000. *The Emergence of the Speech Capacity*. Mahwah, NJ: Lawrence Erlbaum Associates.

Oller DK. 2004. Underpinnings for a theory of communicative evolution. In *Evolution of Communication Systems: A Comparative Approach*, ed. DK Oller, U Griebel, pp. 49–65. Cambridge, MA: MIT Press.

Oller DK, Eilers RE. 1982. Similarity of babbling in Spanish- and English-learning babies. *Journal of Child Language* 9: 565–78.

Oller DK, Eilers RE. 1988. The role of audition in infant babbling. *Child Development* 59: 441–9.

Oller DK, Griebel U. 2005. Contextual freedom in human infant vocalization and the evolution of language. In *Evolutionary Perspectives on Human Development*, ed. R Burgess, K MacDonald, pp 135–166. Thousand Oaks, CA: Sage Publications.

Owren MJ, Dieter JA, Seyfarth RM, Cheney DL. 1993. Vocalizations of rhesus (*Macaca mulatta*) and Japanese (*M. fuscata*) macaques cross-fostered between species show evidence of only limited modification. *Developmental Psychobiology* 26: 389–406.

Patterson FG. 1980. Innovative uses of language by a gorilla: A case study. In *Children's Language, Vol, 2*, ed. KE Nelson, pp. 497–561. New York: Gardner Press.

Pellegrini AD, Bjorklund DF. 2004. The ontogeny and phylogeny of children's object and fantasy play. *Human Nature* 15: 23–43.

Pellis SM, Pellis VC. 1996. On knowing it's only play: The role of play signals in play fighting. *Aggression and Violent Behavior* 1: 249–68.

Pepperberg IM, Brese KJ, Harris BJ. 1991. Solitary sound play during acquisition of English vocalizations by an African Grey parrot (*Psittacus erithacus*): Possible parallels with children's monologue speech. *Applied Psycholinguistics* 12: 151–77.

Pepperberg IM, Naughton JR, Banta PA. 1998. Allospecific vocal learning by Grey parrots (*Psittacus erithacus*): A failure of videotaped instruction under certain conditions. *Behavioural Processes* 42: 139–58.

Petitto LA, Marentette P. 1991. Babbling in the manual mode: Evidence for the ontogeny of language. *Science* 251: 1483–96.

Piaget J. 1951. *Play, Dreams and Imitation in Childhood.* New York: Norton.

Piaget J. 1952. *The Origins of Intelligence in Children.* New York: Norton.

Preston MS, Yeni-Komshian G, Stark RE. 1967. Voicing in initial stop consonants produced by children in the prelinguistic period from different language communities. *Johns Hopkins University School of Medicine, Annual Report of the Neurocommunications Laboratory* 2: 305–23.

Ralls K, Fiorelli P, Gish S. 1985. Vocalizations and vocal mimicry in captive harbor seals, *Phoca vitulina. Canadian Journal of Zoology* 63: 1050–6.

Rasa OAE. 1984. A motivational analysis of object play in juvenile dwarf mongooses (*Helogale undulate rufula*). *Animal Behaviour* 32: 579–89.

Reiss D, McCowan B. 1993. Spontaneous vocal mimicry and production by bottlenose dolphins (*Tursiops truncatus*): Evidence for vocal learning. *Journal of Comparative Psychology* 107: 301–12.

Reynolds PC. 1976. Play, language, and human evolution. In *Play*, ed. J Bruner, A Jolly, K Sylva, pp. 621–37. New York: Basic Books.

Richards DG, Wolz JP, Herman LM. 1984. Vocal mimicry of computer-generated sounds and vocal labeling of objects by a bottlenosed dolphin, *Tursiops truncatus. Journal of Comparative Psychology* 1: 10–28.

Roush RS, Snowdon CT. 1999. The effects of social status on food-associated calling behaviour in captive cotton-top tamarins. *Animal Behaviour* 58: 1299–1305.

Rukstalis M, Fite JE, French JA. 2003. Social change affects vocal structure in a callitrichid primate (*Callithrix kuhlii*). *Ethology* 109: 327–40.

Sade DS. 1973. An ethogram for rhesus monkeys: I. Antithetical contrasts in posture and movement. *American Journal of Physical Anthropology* 38: 537–42.

Sayigh LS, Tyack PL, Wells RS, Scott MD. 1990. Signature whistles of free-ranging bottlenose dolphins *Tursiops truncatus*: Stability and mother–offspring comparisons. *Behavioral Ecology and Sociobiology* 26: 247–60.

Seyfarth RM, Cheney DL. 1997. Some general features of vocal development in nonhuman primates. In *Social Influences on Vocal Development*, ed. C Snowdon, M Hausberger, pp. 249–73. New York: Cambridge University Press.

Sharpe LL. 2005. Play does not enhance social cohesion in a cooperative mammal. *Animal Behaviour* 70: 551–8.

Slocombe K, Zuberbühler K. 2005. Agonistic screams in wild chimpanzees vary as a function of social role. *Journal of Comparative Psychology* 119: 67–77.

Snowdon CT. 1990. Language capacities of nonhuman animals. *Yearbook of Physical Anthropology* 33: 215–43.

Snowdon CT. 1997. Affiliative processes and vocal development. In *The Integrative Neurobiology of Affiliation. Annals of the New York Academy of Sciences, Vol. 807*, ed. CS Carter, II Lederhendler, pp. 340–51. New York: New York Academy of Sciences.

Snowdon CT, Elowson AM. 2001. 'Babbling' in pygmy marmosets: Development after infancy. *Behaviour* 138: 1235–48.

Snowdon CT, Elowson AM, Roush RS. 1997. Social influences on vocal development in New World primates. In *Social Influences on Vocal Development*, ed. CT Snowdon, M Hausberger, pp. 234–48. Cambridge, England: Cambridge University Press.

Snowdon CT, French JA, Cleveland J. 1986. Ontogeny of primate vocalizations: Models from bird song and human speech. In *Current Perspectives in Primate Social Behavior*, ed. D Taub, FA King, pp. 389–402. New York: van Nostrand-Reinholt.

Snowdon CT, Hausberger M. 1997. *Social Influences on Vocal Development*. Cambridge, England: Cambridge University Press.

Špinka M, Newberry RC, Bekoff M. 2001. Mammalian play: Training for the unexpected. *The Quarterly Review of Biology* 76: 141–68.

Stark RE. 1980. Stages of speech development in the first year of life. In *Child Phonology, Vol. 1*, ed. GY Komshian, J Kavanagh, C Ferguson, pp. 73–90. New York: Academic Press.

Sutton-Smith B. 1980. Conclusion: The persuasive rhetorics of play. In *The Future of Play Theory: A Multidisciplinary Inquiry into the Contributions of Brian Sutton-Smith*, ed. AD Pellegrini, pp. 275–95. Albany, NY: State University of New York Press.

Tyack PL, Sayigh LS. 1997. Vocal learning in cetaceans. In *Social Influences on Vocal Development*, ed. CT Snowdon, M Hausberger, pp. 208–33. Cambridge, England: Cambridge University Press.

van Hoof JARAM. 1967. The facial displays of the catarrhine monkeys and apes. In *Primate Ethology*, ed. D Morris, pp. 7–68. London: Weidenfeld & Nicholson.

van Hoof JARAM, Preuschoft S. 2003. Laughter and smiling: The intertwining of nature and culture. In *Animal Social Complexity*, ed. FBM de Waal, P Tyack, P, pp. 153–79 Cambridge, MA: Harvard University Press.

Vihman MM. 1992. Early syllables and the construction of phonology. In *Phonological Development: Models, Research, Implications*, ed. C Ferguson, L Menn, C Stoel-Gammon, pp. 393–422. Parkton: York Press, Inc..

Visalberghi E, Fragaszy D. 1990. Do monkeys ape? In *"Language" and Intelligence in Monkeys and Apes*, ed. S Parker, K Gibson, pp. 247–73. Cambridge, England: Cambridge University Press.

Weir R. 1962. *Language in the Crib*. The Hague: Mouton.

West MJ, King AP, Freeberg TM. 1997. Building a social agenda for the study of bird song. In *Social Influences on Vocal Development*, ed. CT Snowdon, ME Hausberger, pp. 41–56. New York: Cambridge University Press.

West MJ, King AP, White DJ. 2003. The case for developmental ecology. *Animal Behaviour* 66: 617–22.

Zuberbühler K. 2005. Alarm calls: Evolutionary and cognitive mechanisms. In *Encyclopedia of Language and Lingusitics, Vol. 2*, ed. K Brown, pp. 143–155. Oxford: Elsevier.

Zuberbühler K, Noe R, Seyfarth RM. 1997. Diana monkey long distance calls: Messages for conspecifics and predators. *Animal Behaviour* 53: 589.

V MODELING OF THE EMERGENCE OF COMPLEXITY AND FLEXIBILITY IN COMMUNICATION

The final chapters of the volume bring to bear mathematical modeling and simulation upon the matter of communicative flexibility. In the first chapter of part V, McCowan, Doyle, Kaufman, Hanser, and Burgess present an information theory approach, deriving entropic measures to compare adult and infant bottlenose dolphin whistle communications. Results illustrate greater repetitiveness in the adult whistles. The authors also illustrate a hyperspace analog model of dolphin communications, which is a mathematically based assessment of possible functional roles played or "meanings" transmitted by individual whistle types in dolphins. The tools illustrated by McCowan and colleagues suggest methods by which the systematic nature of flexible communication can be assessed by co-occurrence and sequencing patterns of vocalizations, even in the absence of direct access to information about the functions the vocalizations serve.

In the second chapter of part V, Lachlan addresses bird song, reviewing a broad literature to which he has been a contributor and illustrating a variety of properties of learning in birds. This literature assesses both naturalistically collected observations and mathematical simulations. Lachlan considers versatility of sounds that can be learned, learning accuracy, and functional flexibility of vocalizations in a variety of species. By also taking into consideration the environmental conditions that appear to have led to song learning (clearly one of the most important factors constituting contextual flexibility), he arrives at intriguing empirically based conclusions about the origins of learning in birds.

The final chapter of the volume presents a perspective by Westermann, utilizing a self-organizing map approach to simulate learning of the human vocal system through babbling. The approach incorporates the idea that infants' awareness of their own exploratory vocalizations as well as awareness of ambient speech can warp the perception/production space of vocalizations. The simulation yields an appropriate "learned" vowel system without relying on structured interactions, on explicit imitation, nor on naming games that have been used to model evolution of lexicons in prior work. The modeling also undercuts the mirror-neuron idea, because it illustrates how a simple learning system can self-organize to yield properties promoted by mirror-neuron advocates—no long-term evolution of specialized neurons appears, in fact, to be necessary.

All the chapters in part V add to growing excitement in work on the evolution of communication systems. They illustrate how mathematical tools can be applied to model and simulate the growth of communicative power and flexibility both to characterize progressions of communicative flexibility seen in ethological observation and to simulate the circumstances under which such flexibility can both develop and evolve. By combining ethological observation and increasingly sophisticated simulations and modeling, research of the future seems bound to unveil a host of secrets about how communicative systems can evolve to achieve features of flexibility that are required for language.

13 Detection and Estimation of Complexity and Contextual Flexibility in Nonhuman Animal Communication

Brenda McCowan, Laurance Doyle, Allison B. Kaufman, Sean Hanser, and Curt Burgess

Introduction

The nature and complexity of nonhuman animal communication systems in comparison to human languages is not well understood, partly as a result of methodological limitations but also because of differences in perspective among scientists on how nonhuman animal communication systems function and are structured. For example, robust evidence is lacking for two language-like characteristics thought to underlie communication complexity—syntax and symbolism—among systems of animal communication. Two alternatives could explain this lack of evidence: (1) Nonhuman animals do not possess these human-specific adaptations in communicative capacity, or (2) effective quantitative methods for deciphering the nature and complexity of communication systems have not been fully developed or applied to the communication systems of nonhuman animals. If the latter is the case, direct evidence for truly referential or symbolic communication (Deacon, 1997) will be unlikely to surface in nonhuman animal systems until a broader approach is taken on strategies for measuring communication structure. For example, the very methods by which we study communication systems in animals almost necessarily excludes discovery of a truly symbolic system because we tend to study the production of a signal in relationship to the specific context in which it occurs. Yet in symbolic systems, communication signals can refer to objects and events that are not physically present in the immediate environment, and thus this method of "one-to-one mapping of signal to context" diminishes our ability to measure the degree to which nonhuman animals exhibit contextual flexibility and complexity in their communication systems. Many nonhuman animal communication systems are, in fact, comprised of signals that are not strongly connected to a specific context, and it may be here where we can begin to develop and examine questions about complexity and contextual flexibility in nonhuman animals.

Recent research by mathematical modelers of human language indicates that a necessary condition for symbolic communication to evolve is the development of a nested or "syntactical" communicative structure (e.g., conditional-probabilistic dependencies of communication units on each other; Cancho and Sole, 2003). It has been argued that, after a certain point, lexicon expansion alone cannot keep pace with the amount of information requiring transmission as message complexity increases (Cancho and Sole, 2003). Therefore, one approach for detecting and estimating the nature and complexity of nonhuman animal communication systems is to examine the communication systems of nonhuman animals precisely as that, "systems," by categorizing signals into types based upon signal structure, instead of context, and then calculating both the degree and composition of structural nesting as well as the co-occurrence frequency of these signals with respect to context. Two quantitative measures will be discussed in this chapter that address these issues. First, we apply what might be called an "auto-correlation" application of information theory, which can quantify the overall degree of "rule" complexity or structural nesting *within* a given communication system by using signal occurrence probabilities and their higher order dependences on each other. Second, we will examine the Hyperspace Analog to Language (HAL) model, which makes an internal examination of sequential structure by focusing on the co-occurrence of signals within a communication system to estimate contextual flexibility and the underlying cognitive mapping of a communication system. We will provide the conceptual framework behind each method as well as preliminary data on their utility as estimators of communication complexity and flexibility.

Information Theory: The "Auto-Correlation" Approach

Information theory was first introduced by Claude Shannon in the late 1940s (Shannon and Weaver, 1949) to quantify the amount of information being sent through telephone lines. It has since found wide application in fields as diverse as quantifying information flow in fiber optic computer lines to quantifying information contained in biochemical reactions involving DNA. In this work we discuss the quantification of the *internal* structural complexity of animal communication systems using examples from our work with bottlenose dolphins, *Tursiops truncatus* (McCowan et al., 1999, 2002). This application can be contrasted with the usual application of information theory to quantify information transmitted *between* individual vocalizers (which can be thought of as more of a "cross-correlation" measurement), where two messages can be compared for overlapping signal content (i.e., a measure of communicative fidelity). In our approach, the *internal structure* of a communication system is evaluated, and this can be quantified by calculating information entropies (defined below). The highest information "entropic order" that a communication system exhibits will represent that communication system's highest level of constraining "rules"—in other words, a measure of that communication system's signal dependencies on each other (conditional probabilities, somewhat like syntactical rules).

Measuring the Internal Complexity of a Communication System

In this discussion, "entropy" will be defined as the mathematical quantification of information-theoretic measures. Before measurement, the entropy can be thought of as "uncertainty," while after a measurement or specification, it may be thought of as "information." For example, if one is guessing a completely unknown word, one has a higher likelihood (less uncertainty) in English of picking a word ending with the letter "e" rather than one starting with the letter "q." However, if the word is known to start with the letter "q," one possesses more information about the word than if one knew it ended with the letter "e." Thus, the largest uncertainty is turned into the most information upon measurement or specification.

In the original demonstration of the utility of information theory Shannon and Weaver (1949) generated random English letters with no constraints on their frequencies of occurrence to calculate zero-order (approximation to the information) entropy, which resulted in essentially no information about English except the number (in bits) of the different types of English letters. The zero-order entropy is therefore as follows:

$$H_0 = \log_2 N, \tag{13.1}$$

where N is the total number of different signal types (in English the number of letters plus the "space" as the 27th unit) in the communication system. The base-2 logarithm is also used so that the information entropy measure will be in common units of binary units ("bits"). The zero-order entropy of English letters is therefore $H_0 = \log_2 27 = 4.75$ bits. Note again that we are not comparing communication systems with each other in these formulations but rather quantifying the information content within a given communication system—that is, we are using auto-correlation (Chatfield and Lemon, 1970) rather than cross-correlation (see, e.g., Quastler, 1958; Pierce, 1980).

The first-order entropy of a given communication system can be calculated by weighting the frequency of occurrence of each signal (letter, word, phoneme, whistle, chemical, or whatever the signaling units are). The equation for the first-order entropy is thus the following:

$$H(i) = H_1 = -\sum_{i}^{N} p(i) \log_2 p(i), \tag{13.2}$$

where $p(i)$ represents the probability of occurrence of each signal (i). Therefore, one must have a sufficiently large sample to allow the frequency of occurrence to approximate the real probability distribution of a given communication system's units. Since probabilities are always less than or equal to zero, the negative sign cnsures that the entropic value will be positive. For English letters, the space character occurs about 18.2% of the time, the letter "e" occurs about 10.7% of the time—being the most frequent actual letter—and so on until the letter "q" occurs about 0.1% of the time, being the least frequent (along with "x" and "z"). The first-order entropy for English letters is therefore as follows:

$$H_1 = -0.182 \log_2(0.182) - 0.107 \log_2(0.107) \cdots - 0.001 \log_2(0.001) = 4.03 \text{ bits}.$$

This value is less than H_0 because there are now increased constraints on what can occur (e.g., a "q" cannot happen very often), so the information entropy does not contain as many possibilities, or bits (it is no longer just a uniform random distribution of letters).

It should be noted that the linguistic relationship known as Zipf's Law is essentially a fitted plot of the logarithm-base-10 of the frequency of occurrence probabilities that make up the elements to be summed in the first-order entropy (equation 13.2 above). However, it should also be noted that each component of the first-order entropy is also weighted again by the probability of occurrence itself (see the review in McCowan et al., 2005), so that the two functions are not linearly related.

One type of second-order entropy is calculated by taking the frequency of two-signal occurrences, a calculation that is often called the "joint entropy." In English letters this would correspond to the frequency of occurrence of digrams. The joint entropy is specified by the following:

$$H(i,j) = -\sum_{i,j}^{N} p(i)p_i(j) \log_2 p(i)p_i(j), \tag{13.3}$$

where $p_i(j)$ represents the conditional probability that "event" (e.g., letter) j occurs given that "event" i has already occurred and is nonzero. The joint probability is sometimes written as $p(i,j) = p(i)p_i(j)$, and only equals $p(i)p(j)$ if the two events i and j are completely uncorrelated (i.e., are completely unrelated, independent probabilities). At the second-order entropic level, conditional probabilities are thus introduced.

In the hierarchy of our auto-correlation approach, the second-order entropy, also known as the "conditional entropy," is calculated as follows:

$$H_j(i) = H_2 = -\sum_{i,j}^{N} p(i)p_i(j) \log_2 p_i(j). \tag{13.4}$$

The second-order entropy thus incorporates the information added by the second event alone, once the first event has already been taken into account. The conditional probability, $p_j(i)$, measures the degree of dependence of a second signal on the occurrence of the signal preceding it, (a first-order Markov chain). When considering two messages (e.g., one sent and one received), for example, the second-order entropy can be calculated to ascertain the amount of cross-correlation between the two signals (i.e., the fidelity of the transmission). However, for our auto-correlation application, the second-order entropy is just a measure of constraints imposed on two-signal occurrences by restrictions on the occurrence of neighboring-signal events. The relationship between the conditional entropy and the joint entropy is as follows:

$$H(i,j) = H(i) + H_i(j). \tag{13.5}$$

The value of the second-order English word entropy is $H_2 = 3.32$ bits. From equation 13.5, the joint entropy has two degrees of freedom and so should be larger than the first-order entropy; if there is no conditional component, then the joint entropy reduces to the first-order entropy (and so on for the higher order entropies). One can then extend the notion of orders of entropy to a n-th order (see Yaglom and Yaglom, 1983; McCowan et al., 1999).

It should be noted that equation 13.4 includes "complexity" due to simple repetition of a signaling unit. This is, of course, not really indicative of true "rule" structure and might be called the "null complexity." This component can be specified easily using the following:

$$H_{2(non-rep)} = -\sum_{\substack{i,j \\ i \neq j}}^{N} p(i)p_i(j) \log_2 p_i(j), \tag{13.6}$$

where the additional limit, $i \neq j$, is included in the summation term as well, with the remaining internal conditional probabilities being a measure of the relationship between *different* signals in the communication system.

Continuing in a similar manner, the third-order entropy for English letters can be estimated by calculating the frequency of occurrence of tri-grams according to the following:

$$H_{i,j}(k) = H_3 = -\sum_{i,j,k}^{N} p(i,j,k) \log_2 p_{i,j}(k), \tag{13.7}$$

where event k occurs, given that events i and j have already occurred, and we note that $p(i,j,k) = p(i,j)p_{i,j}(k)$. The third-order entropy for English letters is $H_3 = 3.01$. In like manner the n-order entropy of a given communication system can be calculated according to the following:

$$H_n = -\sum_{i,j,k,\ldots n}^{N} p(i,j,k,\ldots,n) \log_2 p_{i,j,k,\ldots n-1}(n), \tag{13.8}$$

where n is the largest string size of grouped signals with nonzero conditional probabilities (an $n - 1$ length Markov chain) and N is the total number of signals in the data set. Equation 13.8 thus represents the nth-order entropy.

Note that the bit values of the higher order entropies continue to be smaller than the preceding entropy because, as mentioned, more constraining "rules" are being imposed on the signaling units (i.e., the letters' frequency of occurrence). For English words, higher order information entropies continue to be less than the preceding entropic-order values until n-grams strings of about $n = 40$. This means that further, higher order entropies (i.e., inclusion of higher order Markov chains) do not produce any further changes in the entropic value so that the real measure of the entropy of the communication system has now been approximated. The value of n at which $H_n \cong H_{n-1}$, then, is a measure of the maximum "rule" complexity (capacity to constrain the signaling units) of that communication system, or the place where any additional conditional entropy approaches zero.

One can regress the value of the entropy against the entropic order to obtain a best fit plot of what may be called the "entropic slope" (Chatfield and Lemon, 1970; Yaglom and Yaglom, 1983; McCowan et al., 1999). This provides an overall look at the change, with increasingly higher order information entropy, of a communication system, with the input values (conditional entropies calculated from equation 13.8) dropping to zero as one reaches the maximum internal complexity of a system. The entropic slope for letter English is -0.566, a linear best fit to the values of H_0 through H_3 plotted against each of their entropic orders (0 to 3) as mentioned above. The fit does not necessarily have to be linear, and one may also want to plot the nonrepetitious conditional entropies for a better idea of the true, underlying structure.

Application to the Bottlenose Dolphin Whistle Communication System

In the application of entropic measures to other species' communication systems, we want to identify the frequency distribution and conditional probabilities of the signals within that communication system, in order to derive the entropic orders or the constraining multigram Markov "rules" of that communication system. Thus, information theory might be used to quantify the structural complexity of animal communication systems as well as measure important changes in the communicative structure under different conditions—for example, noise in the environment, other sources of stress, correlations of social conditions with communication complexity, and so forth (see, e.g., McCowan et al., 1999, 2002).

In order to derive the orders of entropy, two important conditions must be met: (1) The signals in the communication system must be correctly classified into independent units, which can be a difficult task (see Reiss and McCowan, 1993; McCowan, 1995; Janik, 1999; McCowan et al., 1999; Herman, 2002; McCowan and Reiss, 2001), and (2) a sufficiently large sample of the communication system must be available for unbiased sampling of signals, that is, the sample must be large enough to be approximately ergodic (i.e., in this case meaning that the statistics of the sample do not undergo a discontinuity; see McCowan et al., 1999, 2002).

To provide an example, we compare the information entropy measures of two data sets—the whistle repertoire of infant bottlenose dolphins (less than twelve months old) and the whistle repertoire of adult bottlenose dolphins (more than ninety-six months old) to determine whether the complexity (i.e., the "rule" structure) of the whistle repertoire has increased from infant to adult, and, if so, by how much. Signaling data sets (of nearly the same size) of bottlenose dolphin whistle vocalizations are compared with each other (over an age range), the signals of which were each classified in an identical manner (the signal contour classification scheme of McCowan, 1995). Further precautions should be taken when comparing different species and communicative behaviors, of course—especially when comparing nonhuman communication systems (e.g., vocal or gestural signals) with the various forms of human communication systems (e.g., letters, words, phonemes). We also need to quantify the error bars of the resulting entropic values, and for this we use the selective-sampling technique commonly referred to as the "bootstrap" approach, where random data set models are generated with the statistical weights of the original data set with differences between the generated model data sets providing the standard deviation (e.g., Efron and Tibshirani, 1998).

Data were analyzed for eight adults (greater than ninety-six months old, $N = 527$, the total number of sample signals) and four infant (less than twelve months old, $N = 365$) captive bottlenose dolphins. Sequences of whistles were analyzed by individual vocalizers and not as interactions between individuals. Signals were classified according to the *K*-means cluster analysis based on a 60-point contour sampling of the signal spectrograms as noted in McCowan et al. (1999) and McCowan et al. (1999, 2002). A limited amount of data constrained our analysis, in this case, to no higher than the second-order entropy. Applying equations 13.1, 13.2, and 13.4, and calculating the entropic slope, we obtained the values listed in table 13.1. The statistical significance of the entropic values, as derived from 1,000 bootstrap tests for each data set, are also given in table 13.1. Between infant bottlenose dolphins (less than twelve months) and adult bottlenose dolphins (more than ninety-six months) the significance of the difference can be determined using a two-sample heteroscedastic t test (Law and Kelton, 2000) at a confidence value of more than 99.9% ($p < .001$).

Table 13.1 presents the difference in the zero-order entropy, H_0, from infant to adult bottlenose dolphins, which was significant at the 99.9% level (a difference of

Table 13.1
Bootstrap results for entropic parameters of bottlenose dolphin whistle vocalization

	Adult	Infant	Difference (bits)
Zero-Order Entropy			
Mean	4.592	4.134	0.46
Variance	0.013	0.012	
t-Statistic	92.4		
First-Order Entropy			
Mean	2.440	2.349	0.09
Variance	0.020	0.020	
t-Statistic	16.5		
Second-Order Entropy			
Mean	1.295	1.797	–0.50
Variance	0.020	0.006	
t-Statistic	−124.0		
Entropic Slope			
Mean	−1.653	−1.175	0.48 bits/order
Variance	0.006	0.005	
t-Statistic	−141.7		

Two-sample unequal variance t-test. Sample size = 364. Number of iterations = 1000, 99.9% t-critical: one-tail = 3.094, two tail = 3.295.

$\sigma = 3.2$). The adult H_0 is higher than the infant H_0, which simply means that the number of whistle signal types (i.e., whistle "vocabulary") increased significantly, from eighteen to twenty-four whistle types (i.e., by a bit more than 30%) with age (we excluded whistle types that occurred less than 0.1% of the time for both adult and infant data sets).

The change in the first-order entropy, H_1, was not found to be significantly different between infants and adults (a difference is $\sigma = 0.6$). This indicates that the whistle signals that were made by both infant and adult bottlenose dolphins had a similar overall frequency-of-occurrence distribution and did not significantly differ in this way. We note that the Zipf slopes of these two distributions, however—as was found in McCowan et al. (2002)—were significantly different. These can differ because the Zipf diagram weights all frequencies of occurrence equally in a linear fit to a $\log_{10}$—$\log_{10}$ scale plot, while the calculation of the first-order entropy weights each $\log_2$ value of the information by its frequency of occurrence, $p(i)$ again, as can be seen in equation 13.2. Thus, the addition of many low-frequency-of-occurrence signals in the infant data set—which later have a higher frequency of occurrence in the adult data set—would lower the Zipf slope (i.e., bring it closer to horizontal) while making essentially no difference in the first-order entropy, because very low frequencies of occurrence do not contribute much in the calculation of the first-order entropy but are weighted equally in a Zipf plot. The Monte Carlo "bootstrap" spread in the Zipf slope might subsequently also be expected to be broader for the infant data set than in the adult data set, as is shown in McCowan et al. (2005).

A shown in table 13.1, the second-order entropy, H_2, for bottlenose dolphin whistle vocalizations changes significantly with age from infants to adults (a difference of $\sigma = 4.4$), the second-order entropy for adults being significantly lower than that for infants. This indicates that there are more constraints (i.e., "rules") on the distribution of the digram (two whistles at a time) structure in the adult bottlenose dolphin whistle vocalization communication system than in the infant vocalization communications. This would indicate that conditional probabilistic "nesting" of signals—that is, structural "rules" (including repetition, the "null rule")—develops with age. As noted, such nesting "rules" could have important adaptive value because of increased possibilities for message error recovery.

Figure 13.1 illustrates the concept of the entropic slope, showing H_{slope} (error bars are about the size of the symbols) for both the adult and infant bottlenose dolphin whistle data sets, and figure 13.2 shows the data resulting from the bootstrap experiments calculating the value of H_{slope} for these two populations. As shown in table 13.1 and figure 13.2, the entropic slopes for the infant compared to the adult bottlenose dolphin whistle data sets are significantly different ($\sigma = 6.4$). This perhaps best illustrates that the overall change in the "structure" of whistle vocalizations—from infant to adult—of the bottlenose dolphin repertoire, as quantified by information theoretic measures, includes a statistically significant increase in the "rule" structure (i.e., conditional probabilistic component of the signal frequency of occurrence structure). The fact that the entropic slope is steeper means that higher order entropy (i.e., that takes into account n-gram or higher Markov-chain structure) is increasingly required to obtain the true entropy of the bottlenose dolphin whistle data set with increasing age. Thus, additional constraints or rules are being imposed on possible vocalization production as infants mature in this species (not, at least entropically, dissimilar to the addition of grammatical rules during human language acquisition, e.g.).

We also examined how much of this structural increase with age may be due to repetition, which gives additional second-order entropic structure due to an increased probabilistic interdependence not between signals of different types but only between signals of the same type. Using equation 13.6 to determine the amount of structure due to repetition, we obtain the nonrepetitious results listed in table 13.1. The basic number of signal types does not change when examining only the nonrepetitious signals and thus does not change in either the infant or adult signal sets. The isolation of nonrepetitive signals also did not change the values for the first-order entropies. The most significant difference between the values calculated using the whole data set versus those calculated using the nonrepetitious component of the data is in the second-order entropy value (and reflected in the change in the entropic slope). This

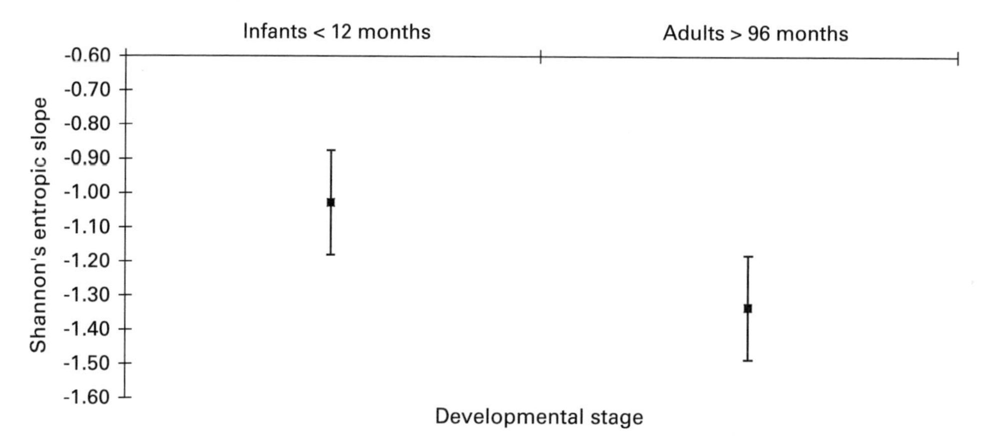

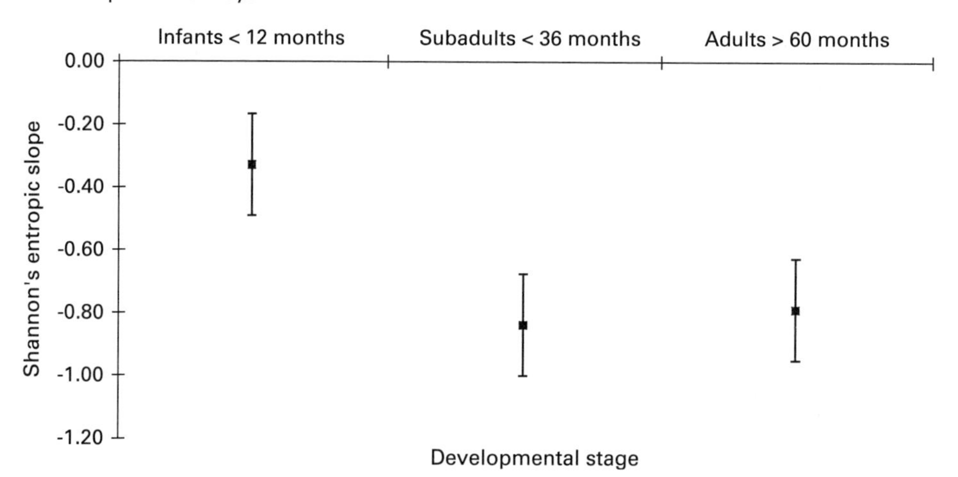

Figure 13.1
Entropic slopes (a measure of organization complexity) and their standard errors for developing infant and adult (A) bottlenose dolphin whistle repertoires and (B) squirrel monkey chuck call repertoires. Steepening of slope indicates increased communicative structure within the communication system. Reprinted from McCowan et al. (2002).

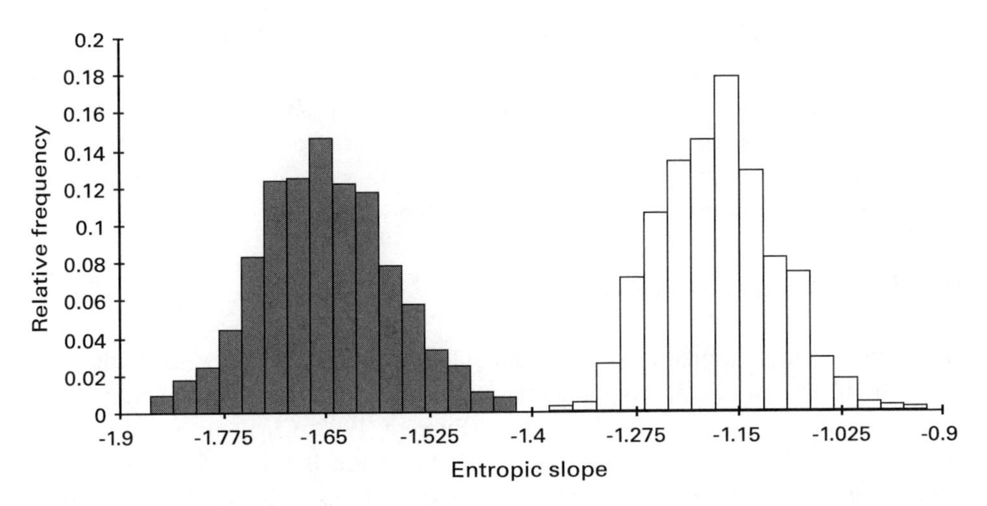

Figure 13.2
Distribution of entropic slope values generated from Monte Carlo (i.e., "bootstrap") simulations on adult (left, shaded) and infant (right, white) bottlenose dolphins. One thousand iterations of entropic slope calculations were conducted on the probability structure of the frequency of use of whistle types for adult and infant dolphins, respectively, and the results were compared using a heteroscedastic *t* test (see Law and Kelton, 2000). For adults (left distribution) the mean and standard deviation values were -1.65 ± 0.08, and for infants (right distribution) the values were -1.18 ± 0.07. The entropic slope differed between infant and adult bottlenose dolphins by more than six standard deviations.

indicates that repetition plays a larger role in the whistle repertoire of the adult as compared with the infant bottlenose dolphin. Repetitiveness is a sign of control in emerging motoric systems—as new elements come to be commanded, their newly appearing control is often manifested in stereotyped repetition (rhythmic hand banging, kicking, and reduplicated babbling are all examples of such repetitive actions that appear as children gain command over hand and arm movement, leg movement, and vocalization). Perhaps the greater repetitiveness of the adult dolphin vocalizations is an indicator of a higher level of control over the system's elements (Thelen, 1981).

Conclusion for Information Theory Section

Information theory can provide a powerful set of tools for the objective examination of communication complexity development within a species and may also have potential application across species because it provides dimensionless quantitative measures (e.g., entropic orders) that allow the comparison of the structural and organizational complexity of multimodal communication systems of different species. However, two important criteria must be met for information theory to be successfully used: (1) A sufficiently large sample of the communication system must be available for unbiased sampling of signals, and (2) the signal elements must be categorized into relevant signal types.

In addition, while information theory provides an overall estimation of the amount of complexity or "nesting rules" in a communication system, it does not directly examine the internal structure of these underlying nesting rules in relationship to contextual or cognitive maps. We will therefore turn to our second estimation tool, the HAL model, which provides a complementary set of tools for deciphering the internal structural mapping of signal sequences.

Hyperspace Analog to Language Approach

A variety of computational models of human language have been developed that examine both the form-based components (sound and visual form) and conceptually based components (semantics) of language (Burgess, 1998; Christenson, Allen, and Seidenberg, 1998; Gallistel and Gelman, 2000; Lund and Burgess, 1996). While human language is more complex than animal vocalizations, it is possible that cross-disciplinary research involving models intended for human language research may shed light on some of the questions raised by animal communication researchers. Sophisticated statistical analyses of human language have generated considerable insight into the acquisition of language (Li, Burgess, and Lund, 2000b) (Li, Burgess, and Lund, 2000a), the syntactic processing of language (Burgess and Lund, 1997), and the conceptual development of language (Burgess and Lund, 2000). Research discussed earlier in this chapter (McCowan, Hanser, and Doyle, 1999, 2002) has shown that the distribution of whistles in the repertoire of the bottlenose dolphin has a pattern with certain characteristics which are similar to human language Higher order statistical structural analysis of whistle sequences may provide a complementary approach in estimating the amount of information in the whistle stream, supplementing that provided by the information theory approach (see above).

In order to attempt to develop a conceptual model of dolphin communication using higher order statistical modeling methodologies employed in human language research, it is important that there be statistical structure in the whistle stream. It seems clear that this is the case (McCowan, 1995; McCowan and Reiss, 1995). Examining conceptual constructs in a high-dimension model requires the building of a *concept space*. The concept space used in this example is two-dimensional, although theoretically there can be any number of dimensions. The two dimensions in a concept space could be, for example, social behaviors and predation behaviors. These dimensions are gradations of intensity. For example, a particular behavior such as alarm calling would be high on the social dimension and low on the predation dimension (since not becoming prey is the opposite of finding it). This behavior would correspond to a point in the concept space (see figure 13.3). The concept space, for all intents and purposes, represents a map of behaviors and their relationships. The research described here is both an attempt to place behaviors in the concept space according

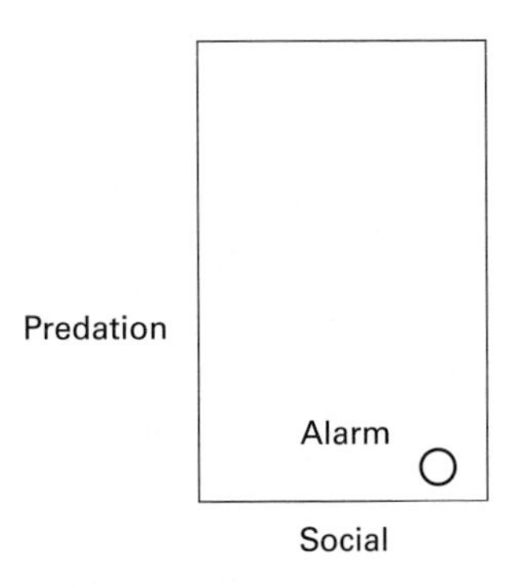

Figure 13.3
Possible dimensions along a concept space.

to their corresponding vocalizations and an example of two possible dimensions of the concept space based on the spatial relationships between the behaviors and the statistical properties of the whistle sequences. To do this, a human language model using higher order statistical processing is used to see if conceptual (semantic) structure can be bootstrapped from the statistical properties of the whistles. In human language this is analogous to determining what a word means by an analysis of the context in which it is used (see Burgess and Lund, 2000).

A Computational Approach to Concept Acquisition

The HAL model is a high-dimensional model that provides a method for the contextual analysis of language (Lund and Burgess, 1996). HAL uses word order in a sentence to compute co-occurrence values between words in a particular body of text (a *corpus*). Weighted co-occurrence values are calculated by use of a sliding window (typically ten words long), which assigns a co-occurrence value to each pair of words in the window based on the number of intervening words. A matrix is created by encoding these values. Co-occurrences between a particular word and those that precede it are encoded in rows, while those that follow it are encoded in the columns of the matrix. Once a matrix is formed, each word can be represented by a vector comprised of its row co-occurrence values followed by its column co-occurrence values (see table 13.2). Figure 13.4 is a visual representation of the vectors of four words taken from a HAL model trained on 320 million words of Usenet text (darker cells reflect larger weighted co-occurrence values). The gray-scaled representations show how *dolphin* and *fish* are more similar than *fish* and *car*. Of course, a dolphin is not a fish; however, they do share considerable contextual information in the statistical language stream (i.e., a paragraph about where fish live would look very similar to a paragraph about where dolphins live). Theses vectors can also be visualized by using multidimensional scaling (MDS), as shown in figure 13.5. Words are placed close

Table 13.2
Example of co-occurrences between a particular word and those that precede it. Example sentence: Studying animal vocalizations is fun

	Studying	animal	vocalizations	is	fun
Studying	0	0	0	0	0
animal	4	0	0	0	0
vocalizations	3	4	0	0	0
is	2	3	4	0	0
fun	1	2	3	4	0

Vector for "animal" 4 0 0 0 0 0 0 4 3 2

dolphin
fish
car
truck

dolphin	14.31	67.82	83.61	69.90	4.66	2.05	14.06
fish	9.49	51.99	59.07	48.20	4.54	2.18	4.08
car	6.54	40.72	21.98	40.87	3.54	1.01	1.85
truck	8.92	49.77	29.28	36.58	3.50	1.68	1.78

Figure 13.4
Gray-scaled and numerical representation of Hyperspace Analog to Language (HAL) vectors for dolphin, fish, car, and truck. Small numerical values and darker color represent smaller HAL co-occurrence values. This example was created from a matrix built from a Usenet corpus.

together in MDS as a function of the similarity of vectors. It is also important to note that in MDS, multidimensional space is reduced to two-dimensional space. There is considerable reduction in spatial resolution in this transformation for data visualization. For example, when distance is computed in the high-dimensional space, *radio* is found to be closer to *television* than *boy*, but the collapsing of the space to the two dimensions obscures this. To determine the spatial relations definitively one needs to conduct an analysis of variance in the high-dimensional space comparing the intragroup distances to the extragroup distances.

The HAL model of dolphin semantics that will be discussed here used as its input corpus the whistle sequences from eight adult bottlenose dolphins described in McCowan and Reiss (1995a, 1995b) and above. Whistle sequences were obtained from vocalizing individuals and not as vocal interactions between dolphins. These data provide the HAL model with a segmented stream of whistles whose sequence was maintained. Using the whistles in this form makes the assumption that a whistle roughly corresponds with a word in human language. While we have no way of knowing with any certainty of the validity of this assumption, it seemed superficially plausible and provides a starting point for this research. Within this dolphin popula-

Figure 13.5
Multidimensional scaling solution illustrating the concept space for dolphin, fish, car, and truck. This example was created from a matrix built from a Usenet corpus.

tion, twenty-seven distinct whistle types were found, representing a total of 434 whistle tokens (McCowan and Reiss, 1995a, 1995b). As a result, the contextual vector for each whistle in the lexicon consists of twenty-seven row elements followed by twenty-seven column elements, one element for each co-occurrence value between that particular whistle type and the others in the set.

The Mapping Problem

The HAL model has been used to simulate a wide variety of memory and language phenomena and has been used with adult language (Burgess and Lund, 2000), children's language (Li, Burgess, and Lund, 2000a, 2000b), and autistic language and has been robust across different languages (Conley, Burgess, and Glosser, 2001; Nicholson, Buchanan, Marshall, and Catchpole, 2007). However, there are two very major challenges in the application of the HAL model to dolphin communication. McCowan et al. (1999, 2002) have shown that there is statistical regularity in the dolphin communication stream. It is not clear if these statistical patterns provide for both the categorization of the whistles and the creation of a concept space (or map). The second challenge is determining what the concepts are. Whistles have to be mapped to concepts, but this mapping is far from a one-to-one correspondence of signal (or whistle) to context (or concept). The goal of using the HAL model is to use the spatial coordinates of each whistle to locate behaviors on a behavioral map. One could then examine the placement of behaviors along each dimension in order to derive the higher level concepts they might represent. This challenge becomes much greater than if we were working with human language, for which we already know which sounds or letters are associated with each word. The mapping process described above has to (1) translate the whistle sequences into a spatial map that would provide a first estimation of the conceptual dimensions, (2) relate the whistles patterns to behaviors, and, then, (3) from the organization of behaviors derive the conceptual structure.

Methods and Results

Translating the Whistle Sequences into a Spatial Map

This step is accomplished by submitting the dolphin whistle sequences to the HAL model to develop the co-occurrence vectors that are associated with the whistles. Each whistle was uniquely identified with a number. This unique number was substituted for every instance of the particular whistle in the vocalization stream, so that the input to HAL consisted of a stream of numbers that corresponded to the whistle sequence. One whistle within the corpus, labeled as whistle type 2, occurred far more frequently than any of the other whistles. It is suspected that this whistle is a contact call; as a result, analyses were done both with and without this whistle type. The vector representations were submitted to a MDS procedure—an example output can be seen in figure 13.6. The pattern of results is clearly nonrandom and supports previous

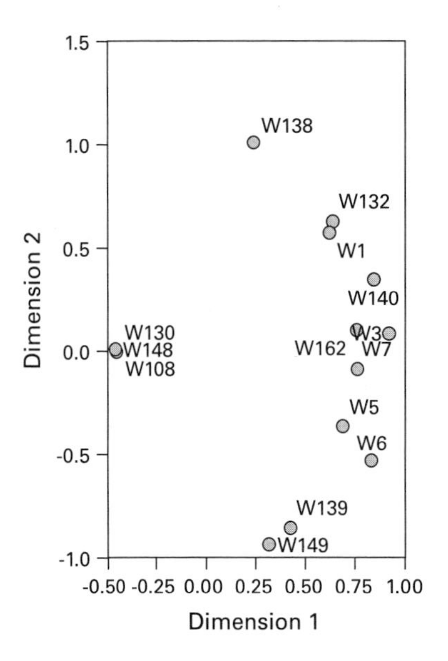

Figure 13.6
Multidimensional scaling of Hyperspace Analog to Language vectors from the dolphin corpus; window size 3, without whistle type 2, 5 element vectors. Note that not all points are visible.

findings that there exists a statistical regularity in the vocalizations of bottlenose dolphins (McCowan, Hanser, and Doyle, 1999). The HAL model would not generate whistle vectors that would produce clustering if the vocalizations were random (Lund and Burgess, 1996). Not only do the vocalizations in this corpus provide reliable clustering, there appears to be a two-dimensional pattern of clustering; this will be addressed in the next section. Additionally, in order to address the reliability of the clustering patterns, a split-half reliability procedure was conducted. The original bottlenose dolphin vocalization corpus was split, and new matrices were built for each half. The two new matrices showed very similar clustering patterns. Out of twenty-six whistle types (analyses conducted without whistle type 2, proposed to be a contact call), only four to five whistle types were misclassified when a comparison of clustering was made between the full matrices and the partial matrices (this includes examination of clustering on both dimensions 1 and 2). While the reliability is encouraging, this result also speaks to the robustness of the procedure given that the whistle corpus is not large to begin with.

Relating Whistle Frequencies to Behavior Patterns

Statistical regularities in human language exist, and the HAL model has been used to reflect the conceptual relationships that emerge from those regularities. Something similar appears with the dolphin whistle sequences, and subsequently the problem becomes one of basic interpretation. With human language, one already knows what each of the lexicon items means and can use this knowledge to interpret clustering in a MDS. There is no such advantage when analyzing animal vocalizations.

The technique used here was to examine the two dimensions of the MDS separately. On dimension 1 there are clearly two different clusters of whistles (see figure 13.7A). To place a behavior on the map, we examined the whistles that occurred with it and placed it according to the relative number of whistles from each side of the dimension. For example, all of the whistles that occurred with the behaviors *aggression* and *social play* were in the cluster on the right side of dimension 1, so these behaviors fit on the right side on this dimension. The behaviors *discipline* and *object play*, which have a high proportion of "left-side" whistles are placed on the left side of the dimension, although close to the dividing line because they are not entirely composed of "left-side" whistles (see figure 13.7B). Note that the placement of a behavior depends on proportion, so a "left-side" behavior contains a higher proportion of "left-side" whistles than other behaviors that may contain little or even no "left-side" whistles. Therefore, it is possible that a "left-side" behavior may actually contain more "right-side" whistles than "left-side" whistles.

Emergent Conceptual Spaces

By plotting behaviors in a conceptual space using the method described above, we may be able to hypothesize about what each of the dimensions means. For example,

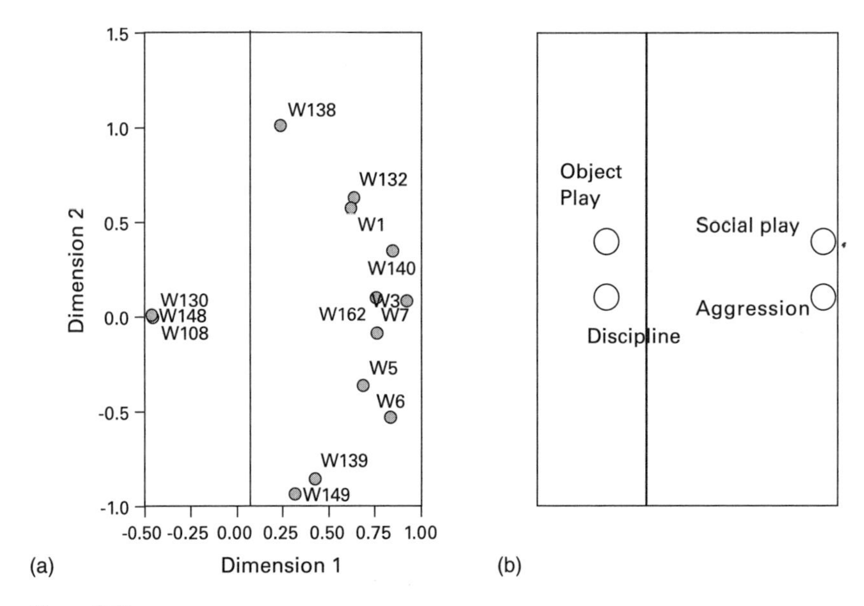

Figure 13.7
In (A), multidimensional scaling of Hyperspace Analog to Language vectors is shown; window size 3, without whistle type 2, 5 element vectors. Dividing line between clusters on dimension 1 shown. Note that not all points are visible. (B) shows placement of behaviors on dimension 1 of the concept space. Not all behaviors are shown.

dimension 1 might show a measure of cognitive complexity in behavior, while dimension 2 might be a measure of activity complexity. On dimension 1, the behaviors *aggression*, *social play*, *social approach*, and *hydrophone* fall to the right of the spectrum (all their whistles belong to the right cluster on the MDS). The behaviors *approach related*, *depart related*, *swim*, and *swim together* also fall on the right side of the spectrum, although a little more toward the center as they contain several left-side whistles. The behaviors *discipline*, *affiliation*, *social swimming*, *object play*, and *vocalization toward orcas* fall toward the left on dimension 1, as they contain proportionally more left-side whistles than other behaviors. It is possible that this grouping on dimension 1 may represent behaviors that decrease in cognitive requirements along a left-to-right continuum. For example, *discipline* may require more cognitive thought than *aggression*; *object play* (with human-made objects) may be a more advanced concept than *social play*, and *affiliation* may require an awareness of social status not required for basic social contact. Additionally, we see that four of the right-side behaviors may be a manifestation of a specific behavioral sequence that is often observed: (1) Mother and calf are swimming together (*together*), (2) calf goes to investigate something (*depart related*), (3) mother follows calf (*swimming*), (4) mother

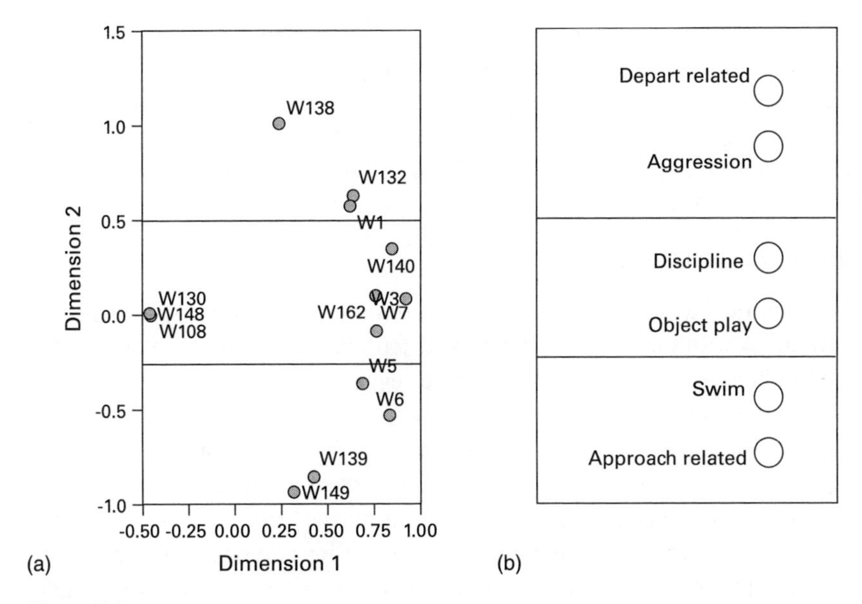

Figure 13.8
In (A), multidimensional scaling of Hyperspace Analog to Language vectors is shown; window size 3, without whistle type 2, 5 element vectors. Dividing lines between clusters on dimension 2 shown. Note that not all points are visible. (B) shows placement of behaviors on dimension 2 of the concept space. Not all behaviors are shown.

approaches calf (*approach related*), (5) the two return to swimming together (*together*).

Turning our attention to dimension 2 (see figure 13.8A, 13.8B), the behaviors *affiliation, social play, social approach, approach related, social swimming*, solo *swimming*, and mother and infant swimming *together* have the highest proportion of whistles that fall toward the bottom of dimension 2 (note—again, this does not mean they have a majority of whistles in the bottom grouping; only that they have a higher proportion of these than the other behaviors). We also see that the behaviors *hydrophone, object play, discipline, sex*, and *vocalization toward orcas* are matched almost entirely to whistles in the middle of dimension 2 and that the behaviors *depart related* and *aggression* are matched to whistles in the top portion of this dimension. Continuing with our theoretical example, this dimension could represent a continuum of levels of activity, with the behaviors at the bottom of the dimension (*affiliation, social play, social approach, approach related, social swimming, swimming*, and *together*) being basic swimming behaviors, and the behaviors just above those on the dimension (*hydrophone, object play, discipline, sex*, and *vocalization toward orcas*) seeming much more active. Social play is the only behavior in the first cluster that is not a "basic swimming" behavior; however, a case could be made that this behavior consists of large amounts of swimming (i.e., chases), which would place it with other swimming behaviors. The behaviors *depart* and *aggression* occur at the top of the dimension. The placement of *aggression* here certainly seems plausible, as it requires a high activity/energy level. The placement of *depart* is the only unexplained behavior in this particular hypothetical schema.

Theoretical Potential and Challenges

Hypothesizing about the conceptual structure of dolphin (or another species') communication requires a number of inferences and tentative theoretical linkages. One problem is how the stream of whistles should be segmented. How does a researcher determine where one whistle ends and the next begins? Many systems for segmentation have been developed, but their confirmation is extremely problematic (Janik, 1999; McCowan, 1995). Additionally, there is the inherent problem that human observers may not perceive as important (or for that matter perceive at all) the characteristics an animal is using for segmentation. To further compound the problem, there may be morphological structure internal to the whistle (or other call types) that has not been examined. A variety of other problems develop when humans have to make inferences:

- How should behaviors be categorized? Should social behaviors between related animals such as mothers and infants be separate from social behaviors between unrelated animals?
- What temporal structure should be used for evaluating context? Should whistles be paired with behaviors that occur concurrently, before, or after?
- How are decisions about the lexicon made? Should researchers include vocalizations such as contact calls and whistles from immature animals in the analysis?
- What decisions should be made about the lexical elements themselves? Should researchers look only at whistles and their sequences, or should the pulses the whistles are composed of be examined as well?

These initial results with the HAL model and whistle sequences are encouraging, but human judgments were an important aspect of interpreting them, and therefore caution should be exercised. Additionally, behavioral and cognitive evidence are required to test any potential hypothesis derived from a computational model.

There is another type of modeling that may provide additional insights. Simple recurrent network (SRN) models use context to make predictions—based on words (whistles)—of what behavior will occur next. SRNs and the HAL model have different architectures; however, both capitalize on context and SRN output has been used to validate HAL output (Burgess and Lund, 2000). Thus, the SRN could provide both a validation of the preliminary results reported here and further insights into the relationship between whistles and behaviors.

Human language and animal communication have been traditionally seen as categorically different. Research with learning mechanisms such as the HAL model has demonstrated that a broad range of human language capacity can be accounted for by simple inductive learning (Burgess and Lund, 2000). A process such as inductive learning and the representations formed by it suggest that the underlying component for a language may be cognitive rather than language specific. This idea has gained increasing support among linguists over the last decade (Burgess and Lund, 1997; Gallistel and Gelman, 2000; Landauer and Dumais, 1997). The promise of such a view is that language can be seen as starting with a set of basic cognitive processes rather than a complex linguistic system. An analogous conclusion has been reached concerning marine mammal communication by Herman and Uyeyama (1999). Dolphins communicate in a cognitively and socially complex environment (Herman, 1991; Morisaka and Connor, 2007), and if a cognitive architecture underlies the communication process, the high-dimensional modeling approach is a likely candidate for deciphering the complexity in the communication stream.

Acknowledgments

We would like to thank Jon Jenkins, Neal Heather, and Mary Lou Felch for their help with this work. We would also like to thank the participants of the Contextual Flexibility Workshop at the Konrad Lorenz Institute for their helpful input on these issues.

References

Buchanan L, Kiss I, Burgess C. 2000. Word and nonword reading in a deep dyslexic: Phonological information enhances performance. *Brain and Cognition* 43: 65–8.

Burgess, C. 1998. From simple associations to the building blocks of language: Modeling meaning in memory with the HAL model. *Behavior Research Methods, Instruments, Computers* 30: 188–98.

Burgess C, Lund K. 1997. Modeling parsing constraints with high-dimensional context space. *Language and Cognitive Processes* 12: 177–210.

Burgess C, Lund K. 2000. The dynamics of meaning in memory. In *Conceptual and Representational Change in Humans and Machines*, ed. AB Markman, pp. 117–56. Mahwah, NJ: Lawrence Erlbaum Associates.

Cancho RF, Solé RV. 2003. Least effort and the origins of scaling in human language. *Publications of the American Academy of Sciences* 100: 788–91.

Chatfield C, Lemon R. 1970. Analyzing sequences of behavioral events. *Journal of Theoretical Biology* 29: 427–45.

Christenson MH, Allen J, Seidenberg MS. 1998. Learning to segment speech using multiple cues: A connectionist model. *Language and Cognitive Processes* 13: 221–68.

Conley P, Burgess C, Glosser G. 2001. Age and Alzheimer's: A computational model of changes in representation. *Brain and Cognition* 46: 86–90.

Connor RC, Mann J, Tyack PL, Whitehead H. 1998. Social evolution in toothed whales. *TREE* 13: 228–32.

Deacon TD. 1997. *The Symbolic Species*. New York: Norton.

Efron B, Tibshirani RJ. 1998: *An Introduction to the Bootstrap*. Boca Raton, FL: CRC Press.

Elman JL. 1990. Finding structure in time. *Cognitive Science* 14: 179–211.

Elman JL. 1993. Learning and development in neural networks—The importance of starting small. *Cognition* 48: 71–99.

Elman JL. 1995. Language as a dynamical system. In *Mind as Motion: Explorations in the Dynamics of Cognition*, ed. TV Gelder, pp. 195–223. Cambridge, MA: MIT Press.

Herman LM. 1991. What the dolphin knows, or might know, in its natural world. In *Dolphin Societies: Discoveries and Puzzles*, ed. KS Norris, pp. 349–64. Los Angeles: University of California Press.

Herman LM. 2002. Vocal, social, and self-imitation by bottlenosed dolphins. In *Imitation in Animals and Artifacts*, ed. K Dautenhahn, CL Nehaniv, pp. 63–108. Cambridge, MA: MIT Press.

Herman LM, Uyeyama RK. 1999. The dolphin's grammatical competency: Comments on Kako. *Animal Learning and Behavior* 27: 18–23.

Janik VM. 1999. Pitfalls in the categorization of behaviour: A comparison of dolphin whistle classification methods. *Animal Behaviour* 57: 133–43.

Landauer TK, Dumais ST. 1997. A solution to Plato's problem: The latent semantic analysis theory of acquisition, induction, and representation of knowledge. *Psychological Review* 104: 211–40.

Law AM, Kelton WD 2000. *Simulation Modeling and Analysis*. Boston, MA: McGraw-Hill.

Li P, Burgess C, Lund K. 2000. The acquisition of word meaning through global lexical co-occurrences. In *The Proceedings of the Thirtieth Annual Child Language Research Forum*, ed. EV Clark, pp. 167–78. Stanford, CA: Center for the Study of Language and Information.

Lund KC, Burgess C. 1996. Producing high-dimensional semantic spaces from lexical co-occurrence. *Behavior Research Methods, Instruments, Computers* 28: 203–8.

McCowan B. 1995. A new quantitative technique for categorizing whistles using simulated signals and whistles from captive bottlenose dolphins (*Delphinidae*, *Tursiops truncatus*). *Ethology* 100: 177–93.

McCowan B, Doyle LR, Jenkins JM, Hanser SF. 2005. The appropriate use of Zipf's Law in animal communication studies. *Animal Behaviour* 69: F1–F7.

McCowan B, Hanser SF, Doyle LR. 1999. Quantitative tools for comparing animal communication systems: Information theory applied to bottlenose dolphin repertoires. *Animal Behaviour* 57: 409–19.

McCowan B, Hanser SF, Doyle LR. 2002. Using information theory to assess the diversity, complexity, and development of communicative repertoires. *Journal of Comparative Psychology* 116: 166–72.

McCowan B, Reiss D. 1995a. Quantitative comparison of whistle repertoires from captive adult bottlenose dolphins (*Delphindae Tursiops truncatus*): A re-evaluation of the signature whistle hypothesis. *Ethology* 100: 193–209.

McCowan B, Reiss D. 1995b. Whistle contour development in captive-born infant bottlenose dolphins (*Tursiops truncatus*): Role of learning. *Journal of Comparative Psychology* 109: 242–60.

McCowan B, Reiss D. 2001. The fallacy of 'signature whistles' in bottlenose dolphins: A comparative perspective of 'signature information' in animal vocalizations. *Animal Behaviour* 62: 1151–62.

Pierce JR. 1980. *An Introduction to Information Theory: Symbols, Signals, and Noise*. Toronto, Canada: Dover Publications.

Quastler H. 1958. A primer on information theory. In *Symposium on Information Theory in Biology*, ed. HP Yockey, RL Platzman, H Quastler, pp. 3–49. New York, NY: Pergamon Press.

Reiss D, McCowan B. 1993. Spontaneous vocal mimicry and production by bottlenose dolphins (*Tursiops truncatus*): Evidence for vocal learning. *Journal of Comparative Psychology* 107: 301–12.

Shannon CE, Weaver W 1949. *The Mathematical Theory of Communication.* Urbana, IL: University of Illinois Press.

Thelen, E. 1981. Rhythmical behavior in infancy: An ethological perspective. *Developmental Psychology* 17: 237–57.

Yaglom AM, Yaglom IM. 1983. *Probability and Information*. Dordrecht: D. Reidel Publishing Company.

14 The Evolution of Flexibility in Bird Song

Robert F. Lachlan

Introduction

A recent widely reported story (e.g., CNN, 2006) tells how a pet African grey parrot, Ziggy, mimicked the speech of one of its owners. This became news because the speech that Ziggy chose to mimic was a telephone conversation between the girlfriend of his owner and a man with whom she was having an affair. When his owner heard Ziggy's rendition of "I love you, Gary," in an apparently identifiable impersonation of his girlfriend, it led to the end of the relationships: first between him and his girlfriend, but rather sadly, also between him and Ziggy, since the parrot didn't cease its taunts. A quick search on the internet revealed a report, perhaps apocryphal but somewhat believable, of an almost identical story involving a mynah. Unlike the African grey parrot, the mynah is an oscine, or songbird, and a relative of another famous mimic, the European starling (featured in numerous local news reports because of their mastery of mobile phone ring tones). Clearly, some bird species possess remarkable flexibility in their vocal behavior. In fact, the accumulated experimental and field evidence suggests that all or most parrots and songbirds learn their vocalizations from others, as do at least some hummingbirds (as far as I am aware, though, there are no stories of hummingbirds ending human relationships or mimicking phones). In this chapter, I will focus on communicative flexibility in songbirds primarily because they are by far the most studied of these taxa.

Many of the examples in this chapter will involve the chaffinch (*Fringilla coelebs*), a Eurasian species whose ability to learn songs was one of the first to be investigated (Marler, 1952; Thorpe, 1958). Chaffinches are not spectacular song learners compared to mynahs or starlings, with their seemingly effortless ability to imitate any sound in their environment. Instead, chaffinches are perhaps more representative of an "average" song learner: Each male chaffinch learns around two or three fairly stereotyped song types (see figure 14.1). Each song begins with a section of repeated, tonal syllables and ends with a more broadband, unrepeated "flourish." However, as shown in figure 14.1, the chaffinches learn the structure of each element within

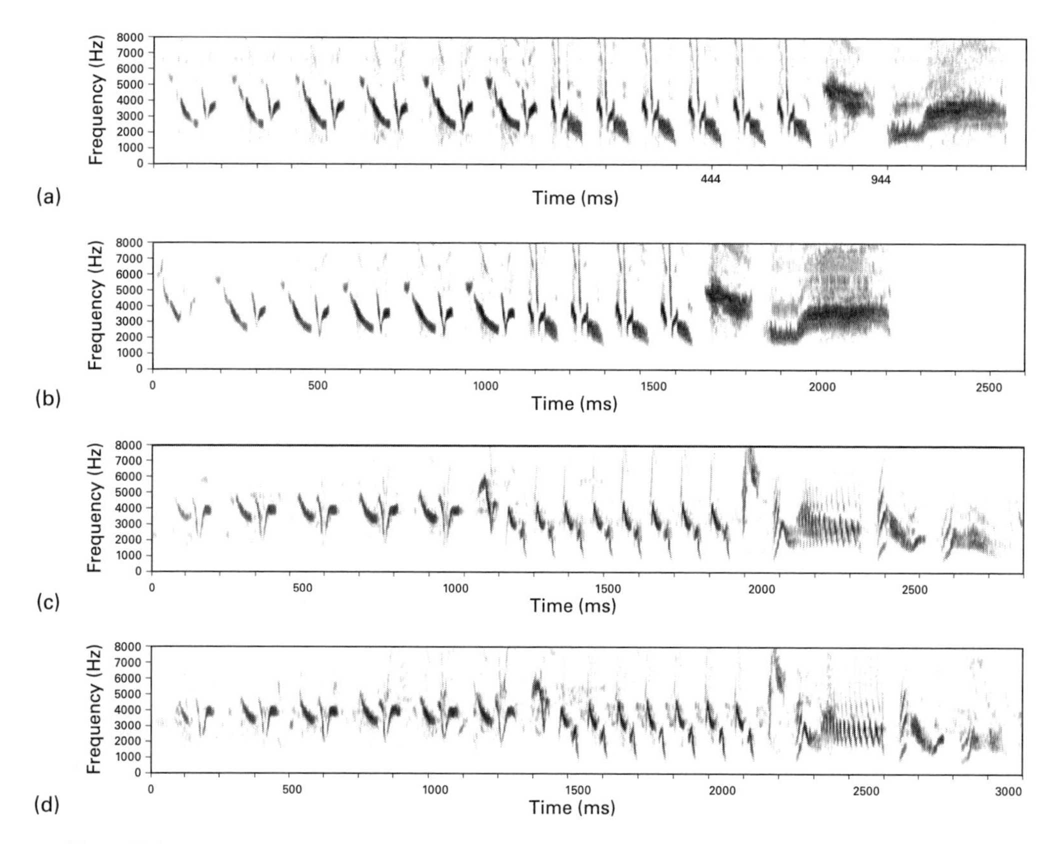

Figure 14.1
Song learning in chaffinches: (A), (B) represent a tutor (A)–tutee (B) pair in the laboratory (Riebel and Slater, 1999); (C), (D) are songs of a neighboring pair of chaffinches recorded on Pico, Azores.

the song and the sequence and hierarchical arrangement of elements remarkably accurately both in the field (figure 14.1C, 14.1D) and in the laboratory (figure 14.1A, 14.1B; Thorpe, 1958; Riebel and Slater, 1999). In contrast, chaffinches raised in acoustic isolation in the laboratory produce songs that, while recognizable as those of a chaffinch, are clearly aberrant (Thorpe, 1958).

The unusual flexibility in the signal structure of bird song has been intensively studied over the last fifty years. These studies demonstrate clearly that songbirds normally socially learn their songs, and a consequence of this seems to be a remarkable level of variability within populations (see Catchpole and Slater, 1995, and Marler and Slabbekoorn, 2004 for reviews). In the case of chaffinches, the number of song types appears to rise approximately in direct proportion to the number of individuals sampled (Lachlan, unpublished data). However, there is no consensus or even strong evidence about why song learning evolved, or why it is maintained.

An obvious hypothesis is that different song types communicate different messages. In human language, the thousands of different words we know are each associated with different meanings, while combinatorial semantics provides us with unlimited communicative scope. In the terms introduced earlier in this volume (Griebel and Oller, this volume), language has a high degree of signal and functional flexibility. The variability of learned bird song suggests a high level of signal flexibility, but what about functional flexibility?

Functions of Bird Song

To answer this, we have to identify the communicative functions of song. Who are birds singing to, and what are their songs communicating? Ethologists have provided two fairly well established answers to these questions. (A third, less common function of song is as a nonsexual signal involved in affiliative social interactions. This function appears to be restricted to group-living species and is also discussed by Hausberger et al., this volume).

Songs may be addressed to other territory holders, where they are used in a context of territorial defense. This can be deduced from two lines of evidence. First, playing back song inside or near a bird's territory elicits a vocal or aggressive reaction (see McGregor, 1992, for a review). Second, if a male is removed from a territory, playback of song inside the territory significantly increases the length of time until the territory is taken over by another male (Krebs, 1977; Nowicki et al., 1998).

The other general function of song is to attract females as mates. Again, this can be deduced from the sexual displays females produce upon being played back song (King and West, 1977), from the fact that song appears to be a positive stimulus in operant conditioning tasks for female songbirds (e.g., Riebel, 2000), and from the finding that playback of male song in the wild attracts female birds (Eriksson and Wallin, 1986; Johnson and Searcy, 1996). It appears that bird songs are therefore rather typical advertising signals, similar in function to the croaks of frogs or the chirps of crickets, and there is considerable evidence that they are adapted for this function. Songs are often loud and conspicuous. Moreover, in the same way that peacocks' tails appear to be visual extravagances that have evolved to demonstrate their owners' fitness, songs of some species appear to be exaggerated through their complexity, with females preferring males with the largest repertoire sizes (e.g., Buchanan and Catchpole, 1997).

The signal flexibility of bird song may not, therefore, be reflected in comparable flexibility in their functions, but nevertheless, in many bird species there appears to be a dual function of song (repelling males and attracting females), allowing a limited type of "many-to-many" mapping of signals to functions. Playback experiments on chaffinches (Riebel, 1998; Leitão and Riebel, 2003) have found that while male and females both respond most to complete songs, males respond most to songs with a relatively long trill (the first part of the song), while females respond most to songs with a relatively long flourish (the final part of the song). However, data such as these still have not provided evidence of functional flexibility in bird song, since it appears that all chaffinches treat songs with relatively long trills or flourishes in the same way. In fact, it seems almost impossible that the dual functions of bird song could be responsible for the signal flexibility of bird song.

On the other hand, a clue is provided by the very accuracy of song learning. As figure 14.1 illustrates, birds like chaffinches appear to imitate songs extremely accurately. This is unlike, for example, the variability in young babies' babbling (see Oller and Griebel, this volume), another example where the variability in the signal appears to exceed the complexity of its functions. If alternative song types do not communicate different messages, why do birds learn them so precisely? In the next three sections, I will review attempts to measure the accuracy of song learning and theories for why birds learn accurately.

Song Diversity and the Accuracy of Learning

If we are to understand why songs are learned accurately, we must first have some idea about how accurately they are learned. Unfortunately, song learning is a difficult trait to pin down in the field: Normally there is considerable delay between when the memorization of a song occurs and when it is first produced. This means that it has proven difficult to track down pairs of "tutors" and "tutees." An alternative is to attempt to interpret patterns of song diversity within populations. There have been a large number of studies describing "microgeographic" variation (variation between neighboring individuals rather than between neighboring populations) in bird song. These studies typically involve classifying songs into discrete types simply through visual inspection of sound spectrographs by trained observers. Such classifications therefore tend to restrict themselves to the binary decision of whether song types are the same or not rather than to how different the song types are.

Using this approach, a wide variety of patterns of variation have been described (Mundinger, 1982; Handley and Nelson, 2005). For example, sedentary populations of white-crowned sparrows (Baptista, 1977) have a regimented pattern of geographic variation where all individuals within a certain area share the same song type, and there are sharp boundaries between what have been called song "dialects." Populations with dialects represent one end of a spectrum. Great tits (*Parus major*) are somewhat less extreme (McGregor and Krebs, 1982): They share songs preferentially with neighbors, but not strongly enough for distinct dialects to appear. Chaffinches, however, are near the other end of the spectrum. Despite their ability to learn songs accurately, chaffinches do not show any tendency to share song types with neighbors more than males slightly further away (Slater et al., 1980; Lachlan and Slater, 2003).

For most species, the diversity of song types found in a local area is vastly greater than might be expected for genetic traits. This indicates that the accuracy of song learning is much lower than that of genetic transmission, not an especially surprising conclusion. On the other hand, in species such as white-crowned sparrows, where hundreds of individuals may share the same dialect, it would seem likely that learning must be extremely accurate. But is it really the case that white-crowned sparrows are more accurate song learners than chaffinches?

Since song is socially learned from other individuals, patterns of microgeographic variation are the outcome of cultural evolution (Cavalli-Sforza and Feldman, 1981; Boyd and Richerson, 1985). There are a number of parameters that are likely to affect this pattern of variation:

1. *Rate of cultural mutation* For variability to exist, there must be a source of novelty at an individual level. It is not generally established whether novel songs are beneficial or not, whether they are a result of "innovation" or are merely "mistakes in learning." Therefore, I prefer the neutral term "cultural mutation," an analogy of genetic mutation, which describes only the fact that a novel form has been produced.

2. *Dispersal* Songs may be dispersed by the movement of individuals after they have learned their repertoire but also as a result of songs being broadcast through the atmosphere.

3. *Repertoire size* Increasing the number of songs each individual sings will influence how many song types a young bird will hear. Notably, it appears that species with large repertoire sizes rarely possess dialects with sharp boundaries.

4. *Song preferences* Various suggestions have been made about selection pressures imposed by the learning preferences of individuals. However, two that seem most general are a preference for population-typical songs (i.e., a factor that imposes some sort of filter on the type of song that may be learned; Marler, 1997) and a preference for more common song types (e.g., a conformist bias; Beecher and Brenowitz, 2005). The latter is particularly likely to affect the pattern of geographic variation in song.

5. *Demography* The population size, density, age, and geographic structure will determine how many different song types an individual will hear and are therefore all likely to influence the pattern of geographic variation in song (Williams and Slater, 1990; Lachlan and Slater, 2003; Laiolo and Tella, 2005; Gammon et al., 2005).

Various studies have combined some subset of these parameters into cultural evolutionary models of bird song. One aim of such models is to take different values for these parameters and observe the patterns of geographical variation that result (e.g.,

Goodfellow and Slater, 1986; Williams and Slater, 1990). For example, mosaic patterns or dialects tend to arise when individuals have low repertoire sizes and require either a conformist bias (see below) or a fragmented habitat landscape.

A more challenging use for cultural evolutionary models of bird song is to estimate the parameters of individual song-learning behavior using observed patterns of geographic variation obtained from field data. The fact that there are so many parameters that influence geographic variation makes it rather difficult to ascribe a difference in song variation to one particular parameter. For example, several field studies have found that migratory populations have greater microgeographic song diversity than nonmigratory populations of the same species (e.g., Ewert and Kroodsma, 1994). This may be due to higher mutation rates, but it could equally well be due to increased song dispersal distances or increased mortality rates in the migratory populations. To distinguish between these alternatives, it is necessary to measure as many of these parameters as possible.

Several cultural evolutionary models have used such an approach (Burnell, 1998; Slater et al., 1980; Lynch and Baker, 1993; Lynch, 1996; Payne, 1996; Lachlan and Slater, 2003). Some of these models are not spatially explicit and therefore do not distinguish between immigration and cultural mutation as factors that lead to new songs appearing in a population. As a result, Slater et al. (1980), using a statistical approach adopted from population genetics, estimated that cultural mutation and immigration together occurred at a rate of around 0.1 per transmission, an estimate that matched an analysis of historical patterns of song change (Ince et al., 1980). A more recent study modeled more explicitly the process of song dispersal using spatial simulations in order to distinguish between immigration and mutation (Lachlan and Slater, 2003). This process found that much of previous estimate was indeed due to immigration, and the cultural mutation rate was estimated at only around 0.01 per transmission (i.e., around one in one hundred songs in the population have been learned inaccurately). This estimate remained approximately constant across several different field sites. It therefore appears that Scottish chaffinches learn songs with a very high degree of accuracy. Interestingly, Lachlan and Slater's (2003) analysis also showed that chaffinches avoid learning from neighbors: The simulations with the best fit to the field data involved individuals learning their songs from four territories away from the territory they themselves occupied. Since this distance coincides with the natal dispersal distance of British chaffinches (Paradis et al., 1998), the simplest hypothesis is that many chaffinches learn their songs before they disperse from their father's territory (see also Slater et al., 2000).

One consequence of cultural evolution is that patterns of geographic variation in signals are partly at the mercy of ecological and demographic variation. Small, less dense, or recently established populations are predicted to have reduced cultural diversity. This raises an interesting question (Kroodsma, 1996): If song sharing between individuals plays any role in communication, then shouldn't individuals attempt to regulate the pattern of geographical variation to optimize whatever role it plays in communication? No evidence of such adaptation has been found so far. In fact, the cultural mutation rates in small isolated island chaffinch populations with high song-sharing levels between neighbors are estimated to be no different from those in more typical mainland populations (Lachlan and Slater, 2003), suggesting that chaffinches, at least, have no strategy for adapting their song-learning behavior to regulate geographic variation in song.

Why Do Birds Learn Song Accurately?

Not enough studies have been carried out to make accurate comparative assessments of the song-learning behavior of dialect and nondialect species (although see Handley and Nelson, 2005, for a first attempt). However, it is clear that even some nondialect species like chaffinches learn their songs with remarkable precision. The development of these precise imitations is a time-consuming process: Young birds typically pass through a phase of several weeks during which their songs develop into their final form. What is the reason for birds' investment in accurate learning? There are two general hypotheses.

The first hypothesis is that birds benefit from accurately learning a song, but that it doesn't matter which song in the population they pick. If you remember, chaffinches experimentally raised in acoustic isolation produce songs that are not quite normal for their species, a phenomenon that has been found in most other bird species that have been tested. Playback studies to song sparrows (Searcy et al., 1985) and starlings (Chaiken et al., 1997) have found that birds discriminate against such isolate songs. Gammon and colleagues (Gammon et al., 2005) discovered a natural experiment that may have mimicked the isolation-rearing design. Black-capped chickadees (*Parus atricapilla*) are very common across much of their range and sing an extremely stereotypical "fee-bee" song. Gammon et al. focused on one of the margins of their range, in the uplands of Colorado where the birds were only found in isolated clumps of a few territories each. Clearly, birds in these clumps receive less experience of chickadee song during development than do birds in densely populated regions, and Gammon et al. (2005) found that these few birds in the peripheral populations sang a much greater range of song types than all the birds in the rest of the species' range. It is possible that black-capped chickadees and other birds are simply less likely to produce entirely species-typical songs without accurate learning, or appropriate social feedback during song development (see also Hausberger et al., this volume). If this is the case, then birds may learn songs accurately simply to improve their chances of developing a species-typical song that is easily recognized by conspecifics.

There are some problems with the hypothesis that accurate song learning is simply an insurance mechanism to ensure birds develop species-specific song. Song learning often goes beyond what would be necessary to ensure recognition. In the case of British chaffinch populations, it appears that they accurately learn not only the elements that make up the song but also the sequence of those elements and their hierarchical organization into "syllables" (several elements in a row that are repeated a number of times within the song). However, in the Azores, it appears that while elements and syllables are accurately learned, whole song types are almost never learned intact (Lachlan, Jonker, and Koese, unpublished data). Instead, syllables from different songs are recombined to produce novel sequences. Given the fact that we could detect nearly no structural differences between Azorean and British chaffinch song, it seems very unlikely that the reason British birds learn syllable sequences so precisely is simply to ensure that they produce a more typical chaffinch song type.

The second general hypothesis that might explain why birds learn songs accurately is that there is some communicative function served by singing one particular song type compared to another. In particular, research has focused on the idea that there is a benefit to singers conforming to local song traditions, a hypothesis that entails a type of contextual flexibility. Accurate song learning may be seen as a necessary but not sufficient condition for this hypothesis, which has been explored in considerable detail in field experiments (reviewed in Beecher and Brenowitz, 2005). Detailed studies of western populations of song sparrows (*Melospiza melodia*), which show high levels of song sharing between neighbors, have found a correlation between the number of song types a male shares with his neighbors and his ability to hold on to his territory (Beecher et al., 2000; Wilson et al., 2000). Of course, cause and effect are not differentiated in a correlation, and one alternative explanation for this correlation is that weaker birds may be forced to disperse further to find an empty territory and are consequently also less likely to share songs with neighbors. This latter hypothesis has some empirical support from experiments conducted on montane white-crowned sparrow populations (McDougall-Shackleton et al., 2002). However, other evidence supports the role of song sharing as an act of communication: Neighbouring pairs of song sparrows that shared few song types also engaged in more aggressive interactions throughout a breeding season (Wilson and Vehrencamp, 2001), and the very pattern of communicative interactions in song sparrows points to a communicative role for song sparrows. Song sparrows have an individual repertoire of around eight or nine song types and often attempt to match the song type used by their rival during territorial interactions. This behavior, called "matched countersinging," has been documented in several other species (Beecher and Brenowitz, 2005).

There is also some evidence that females prefer males that share songs with others in their region. Initial studies focused on white-crowned sparrows and produced mixed results (Baker and Cunningham, 1985; Chilton et al., 1990). Interest in this field has recently been revived, however, by clearer experimental demonstration that females develop song preferences in a similar way to that in which males learn songs (Riebel, 2000, 2003). We should distinguish production and perception learning here: Female white-crowned sparrows typically do not sing themselves (as is typical for most temperate female songbirds). Similarly, while male song sparrows learn to produce a repertoire of eight or nine song types, they also appear to be able to learn to recognize at least up to thirty songs of their neighbors that they do not produce themselves (Stoddard et al., 1992).

In summary, there is growing empirical evidence that some species of songbirds react differently when singers produce an accurate imitation of a song type that they themselves have learned. Given that this difference in reaction appears to increase the fitness of those that conform to local song traditions, we might call this "conformity-enforcing behavior" (Lachlan et al., 2004).

Why Might Birds Enforce Conformity?

As elegant as these empirical demonstrations may be that birds react differently when singers produce an accurate imitation of a song type that they themselves have learned, they do not provide any answer to this question: Why do receivers pay any attention to conformity? Given the fact that birds may learn to recognize many more song types than they themselves produce, a corollary question is this: Why should it make a difference whether singers produce songs that receivers themselves produce? Several hypotheses for these questions have been proposed, taking inspiration from different theories in modern communication theory.

The first group of theories is rooted in the field of honest signaling (Zahavi, 1975; Grafen, 1990). One version has been provided by Nowicki and colleagues (Nowicki et al., 1998; Nowicki et al., 2002a, b; Nowicki and Searcy, 2004). The "developmental stress" hypothesis points out that song learning often takes place during restricted sensitive phases of development and that these often correspond to periods of developmental stress. If more stressed birds are unable to learn songs as accurately (because they have less time available to practice or less brain development), this may provide an honest signal to receivers to distinguish between signalers on the basis of their early condition. While most attention has been paid to the role of female receivers, there is no reason why the developmental stress hypothesis might not also apply to male receivers. There has been some initial empirical support for the developmental stress hypothesis: Swamp sparrows that have an experimentally induced impoverished development learn songs less accurately than those that have a normal upbringing (Nowicki et al., 2002a). At the same time, congeneric female song sparrows appear to prefer accurately learned songs over less accurately learned ones

(Nowicki et al., 2002b). Together, these results provide an incomplete but tantalizing glimpse into one potential benefit of conformity.

Another way in which accurate learning confers a badge of honesty to the singer is by facilitating the "dear enemy" effect (Fisher, 1954; see also Beecher and Brenowitz, 2005). This hypothesis states that accurate learning with neighbors requires familiarity and time and that conformity therefore honestly signals that you are an established, less threatening, neighbor rather than a potential newly arrived usurper. After all, even if a song may be learned after hearing it only a very few times, it takes a significant amount of time to practice the song well enough to produce it accurately. Thus, honesty is assured. However, many young song sparrow (and other species) males spend part of their youth as "floaters," lacking a territory. They pass more or less quietly through the territories of other males and thus have ample opportunity to memorize and practice elsewhere the songs they will need to fit in, possibly removing the value of conformity as a signal of residence.

Two further hypotheses involve conformity being an example of sensory exploitation (Cheng and Durand, 2004) or a conventional signal (Vehrencamp, 2001). Singing a song type that another bird has learned may trigger a different neural sensory response (due to feedback loops the bird uses to monitor its own singing behavior) and consequently evoke a different behavioral response from singing an unshared song type. Alternatively, song sharing might be seen as a conventional signal that directs an individual's attention toward the singer. In both cases, song sharing is viewed as simply a way to direct a signal toward a specific receiver: The shared song indicates to whom it is addressed. The precise benefits of being able to control a rival male's attention have not been specified, but it seems plausible that it could increase the efficiency of signaling. If instead, the receiver is a female, it is easier to see how exploiting a bias she may have for songs she has learned would be beneficial for the signaler. However, this does not explain why females would have a preference for learned songs in the first place.

The final hypothesis that I will discuss here suggests that conformity might be an outcome of a particular game that territorial songbirds may play: how to allocate aggressive interactions between their neighbors (Lachlan et al., 2004). One simple strategy would be to choose a neighbor at random to attack. Alternatively, birds might selectively pick on neighbors that they did not share song types with. Lachlan et al. (2004) constructed a simulation model to test the relative success of these two strategies (labeled "tolerator" and "conformity enforcer," respectively). Under a wide range of conditions, "conformity enforcer" turned out to be the more successful strategy. This result depended upon one key parameter. This was the relationship between the number of fights an individual had previously engaged in and his chances of success in his next fight. "Conformity enforcer" was more successful than "tolerator" if this relationship was negative (the more previous fights you have fought, the lower your chances in the future). This assumption would make sense if birds became exhausted or overwhelmed by multiple attacks from their neighbors.

The reason for this result is quite simple: Birds used conformity as a cue to coordinate their behavior and attack nonconformers. If their opponents were likely to be exhausted from multiple attacks, then this coordination was a successful strategy. Lachlan et al. (2004) only considered these two strategies, so it is possible that other strategies could be equally or even more successful (if it is easy for birds to directly assess which of their neighbors is involved in the most number of fights, and choose him, e.g.). However, the model does make sense of the observation that neighboring pairs of song sparrows that did not share many song types engaged in more aggressive interactions than those that shared more song types (Wilson and Vehrencamp, 2001).

Despite evidence favoring these hypotheses, there are still species whose accuracy of song learning defies easy explanation, including the chaffinch. As mentioned above, the cultural mutation rate of Scottish chaffinch songs has been estimated as 0.01 (Lachlan and Slater, 2003). Yet in this study, it appears that chaffinches avoided sharing songs with neighbors and learned their songs before dispersal. For whose benefit is this feat of accurate reproduction carried out? It is possible that receivers, whether male or female, learn to recognize the large number of alternative song types necessary to guarantee familiarity with a male's song, but this seems to place an unusually large burden on the receiver of the signal.

In the last two sections, I have reviewed data that suggest that, at least in some species, songbirds use the great variation in their songs to communicate subtly different messages about the identity of the singer. This seems a clear example of functional flexibility in communication: Not only are there multiple messages to communicate but the mappings between signal and function are also flexible. A song that signals membership of a local song tradition in one location will often not communicate that message elsewhere. However, has the great signal flexibility of bird song evolved in concert with this functional flexibility, or is this functional flexibility a system of communication that simply took advantage of preexisting signal variability?

Versatility of Bird Song

The most obvious consequence of learning seems to be that, compared to most animal signals, bird songs are incredibly variable, and this variability itself varies greatly between different species. Variability in songs runs across several dimensions. For example, species vary greatly in the amount of variation found within each singing male. Some species, such as white-crowned sparrows, typically possess only one song type each. In contrast, a male nightingale will sing around 180 different song types. There is also variation in the number of different sounds produced by the population as a whole. In some species, it appears that only a finite, and sometimes small, number of sounds can be produced (Marler and Sherman, 1985; Baker and Boylan, 1995). Chaffinches, among other species, seem to be able to produce an incredibly large number of different songs. Even though chaffinch song is one of the easiest songs to recognize and apparently obeys very strict organizational rules, there seem to be hundreds of different types of elements within a population, arranged into an effectively infinite variety of song types (note that despite this potential, each individual chaffinch male only sings a repertoire of typically two or three song types, with typically fewer than fifty elements).

Bird species also vary in their versatility (*sensu* Slater and Lachlan, 2003). While some species' songs are extremely stereotyped (e.g., the simple fee-bee song of black-capped chickadees), other species may produce a song repertoire containing a wide variety of different sounds. At the extreme, birds such as starlings produce a huge variety of sounds, even mimicking mechanical sounds or the songs of other species.

The general picture of bird song presented in this chapter is one of a relatively straightforward advertising signal that just happens to be learned, and the cultural evolution that results from learning leads to variability in song structure within populations. In this view, alternative song types arise simply as a consequence of inaccurate learning. In line with this view, Lynch (1996) compared distributions of song type frequencies from sixteen populations from nine species with predictions made from a neutral cultural evolutionary model (i.e., one in which directional selection was absent) and found a close fit. This indicates that alternative song types are selectively neutral with respect to each other (or that if selective advantage occurs, it is extremely local in nature, as suggested in the last two sections).

However, if it really does not matter which song type a bird learns, why are songs so variable? This problem is enhanced by the fact that there is at least one demonstrable cost to this variability: the potential of confusion between species. Although starlings and a few other species mimic other species' songs as a matter of course, this is part of their vocal display, and is unlikely to lead to confusion with their receivers, largely because the mimicked sections of the songs are carefully encapsulated within nonmimicked species-typical components. However, it is a relatively common occurrence for ornithologists to note the occasional member of normally nonmimicking species singing a song typical of another. A visual analysis by Helb et al. (1985) of spectrograms from a range of European species suggested that 1% of songs in the population may have been learned heterospecifically.

Few hypotheses have been proposed for the evolution of song versatility. Several studies have found female preferences for larger repertoire sizes, and it is conceivable that such a preference reflects a deeper sensory bias for variability (Slater et al.,

2000). On the other hand, there is no experimental evidence for female preferences for songs unlike those found in their locale, while there are several studies demonstrating female preferences for local song types (another example of conformist bias in bird song communication).

Several experimental studies have suggested that genetically transmitted perceptual predispositions underlie song learning, ensuring that birds learn species-typical songs (Marler, 1997). For example, Whaling et al. (1997) hand raised white-crowned sparrows in isolation and played them back songs of their subspecies, as well as songs from another subspecies. They found that, as adults, the males preferentially sang the exemplars of their own subspecies. Importantly, Whaling et al. (1997) also looked at the behavior of their subjects early in life, at around twenty days posthatching, which is well before they started to produce the earliest precursors of song themselves. They discovered that the fledglings called more in response to their own subspecies' song. This indicates that the source of the bias toward own subspecies' song was at a perceptual level rather than being based on any difficulty in producing the songs of the other subspecies.

If perceptual predispositions are largely responsible for restricting what birds regard as an acceptable signal, then understanding the evolution of predispositions is equivalent to understanding the evolution of that aspect of signaling flexibility itself. The evolution of predispositions has been examined in theoretical papers by Lachlan and Slater (1999) and Lachlan and Feldman (2003). These studies examined the evolutionary interaction between socially learned songs and predispositions underlying song preferences. They are therefore models of gene–culture coevolution (Feldman and Zhivotovsky, 1992). In accord with such models the dynamics of cultural evolution can, in principle, create selection pressures that sway the direction of genetical evolution as much as the other way around.

Based upon the empirical discoveries about bird song mentioned above, it was first assumed that bird song cultural evolution is fundamentally neutral: Most song types are assumed to be selectively neutral with respect to one another. Secondly, a degree of stabilizing selection on songs was caused by the presence of the predispositions, which limited the range of songs that individuals recognized. Finally, a major assumption was that the perceptual predispositions were used both when individuals were learning songs themselves and when they were attempting to recognize the songs of others.

Given this situation, Lachlan and Slater (1999) and Lachlan and Feldman (2003) examined how well mutants whose predispositions were either more or less restrictive fared in a population of nonmutants. Mutants with a less restrictive predisposition should lose out because they have a chance of singing a song that the majority of individuals fail to recognize, but on the other hand, mutants with a more restrictive predisposition should also lose out because they would fail to recognize some bona fide conspecifics attempting to signal to them (see figure 14.2). These costs appear to be symmetrical and would at first sight appear to cancel out. However, in fairly realistic spatial simulations (Lachlan and Slater, 1999), as well as more general population genetics models (Lachlan and Feldman, 2003), the less restrictive predispositions were always considerably more successful than the more restrictive predispositions. The results depended upon cultural evolution (a similar model but without song learning found no bias in favor of less restrictive predispositions; Lachlan and Servedio, 2004): If less restrictive predispositions had been rare in the population, then males possessing them would have tended to learn from the more common males around them with narrow predispositions. In doing so, they would have effectively concealed their identities and consequently removed the frequency-dependent selection acting against them.

Over time, therefore, these models would predict that perceptual predispositions restricting what individuals recognize and learn would become progressively less restrictive. Presumably this process would continue until putative costs of less restrictive predispositions (potential confusion with other species, possible physiological costs of having a larger brain to recognize this wider range of songs, or a greater cost in time required to learn songs) overwhelmed this effect.

In the previous sections, I reviewed evidence suggesting that conforming to local song traditions may itself be important in some species. Could it be the case that versatility evolved in order for individuals to be able to conform to others' songs? Maybe not: Lachlan and Feldman (2003) also examined in their theoretical study the effect of a benefit of conformity between individuals. They found that conformist biases in communication actually led to a reduction in predisposition width. The bias was attributable to the fact that individuals with a narrower predisposition tended to learn songs from each other, increasing their level of conformity with each other and putting them at an advantage over individuals with a wider predisposition.

It is unwise to draw firm conclusions from theoretical studies that have not yet been tested empirically. However, it is intriguing to consider that versatility may have evolved simply as a by-product of the interaction of genes and culture and not because of communicative functions with which versatility may be associated.

Conclusions: Origins and Maintenance of Flexibility

Bird song provides a wealth of empirical data, which has generated a diverse set of theories for evolutionary processes that might shape how song-learning behavior evolves. Can we use these microevolutionary processes to gain insights into macroevolutionary events involved in the evolution of communicative flexibility?

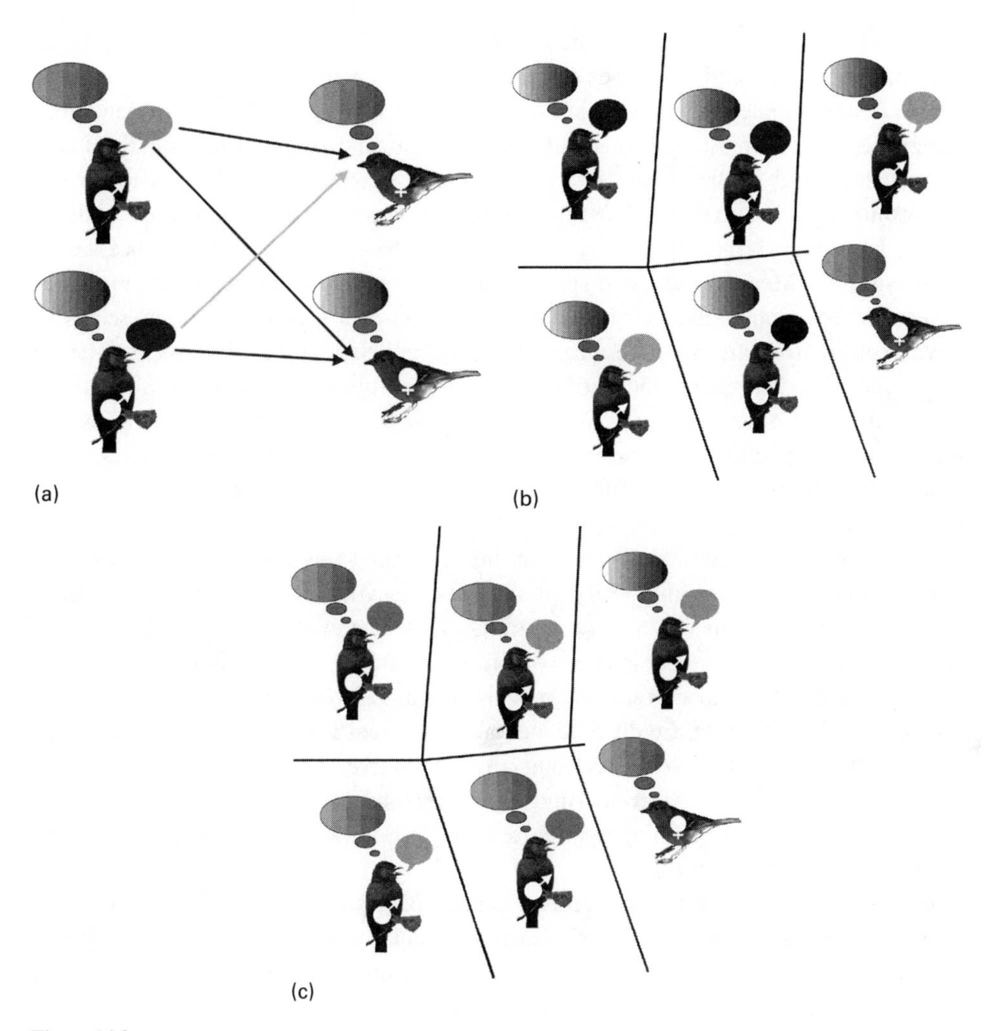

Figure 14.2
The cultural trap hypothesis. The model divides individuals between signalers (males) and receivers (females) and between those with wide perceptual predispositions and narrow predispositions (illustrated in the thought bubbles). Songs are assumed vary along a continuous gradient, here depicted as the gray scale: Wide predispositions run from black to white, while narrow predispositions only encompass an intermediate range of grays. Individual songs (in the speech bubbles) are represented as one value along this gradient. (A) depicts the compatibility of these different types of individuals. All combinations of signaler and receiver are acceptable, except when a wide-predisposition signaler (that has also learned an extreme song type) sings to a receiver with a narrow predisposition. Frequency-dependent selection then works differently on these two types of individual: (B) and (C) depict five territorial males and a female choosing between them. In (B), the narrow-predisposition allele is rare, and the female with this predisposition is faced with a limited choice, because several of the wide-predisposition males sing a song she doesn't recognize. In (C), the wide-predisposition allele is rare, but here the wide-predisposition male has learned his song from surrounding narrow-allele males and is thus sheltered from discrimination by females.

Vocal learning has only been conclusively demonstrated in three avian taxa: hummingbirds, parrots, and songbirds (oscines). Notably, within these three taxa, there are no clear examples of species that do not learn their songs (although too few hummingbirds have been studied for this to be very informative). In a few songbird species, such as grey catbirds (see Kroodsma, 1996), vocal imitation seems to be less important than in most other oscine species. In species such as the grey catbird, which have very large repertoires, songs appear to develop through processes of improvisation. Members of these species might not vocally imitate others, but from a very informal phylogenetic viewpoint, their style of vocal development seems to have evolved from an imitative ancestor (most of their relatives do appear to learn songs socially). Moreover, their songs do not resemble in any way the nonlearned vocalizations of other bird taxa: While catbird songs are very variable (demonstrating considerable versatility), and individuals possess huge song type repertoires, most nonlearning birds have particularly stereotypical songs, with very little variation within individuals.

What does this taxonomic distribution indicate about the origins of vocal learning and communicative flexibility? First, it suggests that vocal learning evolves within a nonlearning population very rarely, and possibly it has evolved on only three occasions in the 100-million-year history of birds. Whatever information we can glean about these events from current behavior is obscured by vast variations among the thousands of species that are descendents of the original vocal learners. Any hypotheses about the origin of vocal learning are speculative. The second observation is that, once it has evolved, vocal learning tends to be maintained, despite the fact that vocalizations of many learning species overlap in complexity and in function with those of nonlearning species.

In this chapter, I have focused on two aspects of flexibility in signaling: versatility (the range of signals produced) and learning accuracy. I also examined functional flexibility in bird song, in the form of song conformity between neighboring males. These three ingredients would appear to be central to the evolutionary origins of learning. From a logical standpoint, the only obvious point we can make about the order in which they evolved is that it is unlikely that imitative learning or functional flexibility could evolve without versatility.

There are several ways in which versatility might arise. First, variation in vocal signals might exist in nonlearning species as a consequence of variation in the development of the organs involved in vocal production. Second, individuality in the calls of nonlearning bird species has been found to have communicative relevance in several species (e.g., Aubin and Jouventin, 2002). One hypothesis would be that imitative learning evolved as a deceptive strategy to overcome individual recognition (so that one individual could masquerade as another). However, given the often subtle acoustic cues that are used to determine individuality, such deception might require rather good mimicry to succeed.

A larger degree of variation between individuals might arise temporarily in certain situations—for example, in a hybrid zone. An example of a hybrid zone between two nonlearning bird species has recently been investigated in Uganda, where the ranges of the doves *Streptopelia vinacea* and *S. capicola* overlap (de Kort et al., 2002). Playback experiments suggest that both species respond more strongly to conspecific coos than to heterospecific or hybrid coos. This would suggest an advantage to being able to learn the coos of either species. On the other hand, it appears that hybrid doves do not respond more strongly to either pure species or hybrid coos (den Hartog et al., 2007).

The hypothesis that a gene–culture coevolutionary relationship between songs and genetic predispositions (Lachlan and Slater, 1999) may select for versatility, as described above, may also have some relevance to this issue. In fact, Lachlan and Slater (1999) proposed the "cultural trap hypothesis" based on the idea that once versatility and learning had evolved, coevolutionary interaction would make it very difficult for it to disappear again. The very existence of learning would create a selection pressure for increased versatility, and it would appear that with increased versatility, some degree of learning may be required to develop a species-typical signal.

The models of Lachlan and Slater (1999) and Lachlan and Feldman (2003) suggest that the cultural trap might exist over a wide range of conditions. This might therefore be an important factor in the evolutionary maintenance of versatility and vocal learning. Lachlan and Feldman (2003) in addition demonstrated that this hypothesis could also explain a transition from a population where individuals only recognize one song type to a population where individuals recognize and could learn one of two song types, in other words a situation approximating the evolutionary origins of vocal learning. However, this hypothesis requires more or less simultaneous evolution of versatility and accurate learning, which seems unlikely.

The role of conformity in the evolutionary origins and maintenance of avian vocal learning is rather harder to place. It is possible that a learned preference for familiar song variants might exist before vocal production learning itself for a number of reasons (such as individual recognition), creating a conformist bias. However, would this create a selection pressure for production learning? Imitative vocal learning would result in a degree of conformity, but then again so would a reduction in versatility itself. As mentioned above, once imitative song learning evolves, selection for conformity may well lead to a reduction in versatility.

In summary, it is easy to imagine several scenarios that may have caused the evolution of vocal learning. None seem especially probable, but perhaps that is why vocal production learning has only arisen infrequently. What is much clearer, however, is that once vocal production learning had evolved, novel evolutionary mechanisms would be quickly created, such as cultural evolution of songs. The cultural trap (Lachlan and Slater, 1999) would lead to greater versatility and conformist biases, which would lead to more accurate learning. We might expect that once song learning had evolved, for whatever reasons, there would have been a rapid burst of evolution as a consequence of such mechanisms. One unfortunate ramification of this pattern of evolution may be that any phylogenetic clues about the origin of learning would have been quickly obscured.

Instead, in this chapter, what I hope I have illustrated is that within the confines of relatively simple communicative functions, bird song provides a cornucopia of different types and levels of signal structure, organization, versatility, and complexity, along with a variety of subtleties that have been added to the basic communicative functions. Bird song is a large research field, and I have really only been able to provide a taste of this diversity. Within all these examples, however, a few somewhat general principles are beginning to develop, and these may turn out to be increasingly relevant to our broader understanding of the evolutionary interactions involved in learned communication systems.

Acknowledgments

K. Riebel and S. Nowicki provided useful comments on the manuscript. I would also like to thank K. Riebel for providing sound recordings, and the Nederlandse Organizatie voor Wetenschappelijk Onderzoek (Netherlands Organization for Scientific Research) for funding.

References

Aubin T, Jouventin P. 2002. How to vocally identify kin in a crowd: The penguin model. *Advances in the Study of Behavior* 31: 243–77.

Baker MC, Boylan JT. 1995. A catalog of song syllables of indigo and lazuli buntings. *Condor* 97: 1028–40.

Baker MC, Cunningham MA. 1985. The biology of bird-song dialects. *Behavioral and Brain Sciences* 8: 85–133.

Baptista LF. 1977. Geographic variation in song and dialects of the Puget Sound white-crowned sparrow. *Condor* 79: 356–70.

Beecher MD, Brenowitz EA. 2005. Functional aspects of song learning in songbirds. *Trends in Ecology and Evolution* 20: 143–9.

Beecher MD, Campbell SE, Nordby JC. 2000. Territory tenure in song sparrows is related to song sharing with neighbors, but not to repertoire size. *Animal Behaviour* 59: 29–37.

Boyd R, Richerson PJ. 1985. *Culture and the Evolutionary Process.* Chicago: Chicago University Press.

Buchanan KL, Catchpole CK. 1997. Female choice in the sedge warbler, *Acrocephalus schoenobaenus*: Multiple cues from song and territory. *Proceedings of the Royal Society of London B* 264: 521–6.

Burnell K. 1998. Cultural variation in savannah sparrow, *Passerculus sandwichensis*, songs: An analysis using the meme concept. *Animal Behaviour* 56: 995–1003.

Catchpole CK, Slater PJB. 1995. *Bird song: Biological themes and variations.* Cambridge, England: Cambridge University Press.

Cavalli-Sforza LL, Feldman MW. 1981. *Cultural Transmission and Evolution: A Quantitative Approach.* Princeton, NJ: Princeton University Press.

Chaiken M, Gentner TQ, Hulse SH. 1997. Effects of social interaction on the development of starling song and the perception of these effects by conspecifics. *Journal of Comparative Psychology* 111: 379–92.

Cheng MF, Durand SE. 2004. Song and the limbic brain—A new function for the bird's own song. *Annals of the New York Academy of Sciences* 1016: 611–27.

Chilton G, Lein MR, Baptista LF. 1990. Mate choice by female white-crowned sparrows in a mixed dialect population. *Behavioural Ecology and Sociobiology* 27: 223–7.

CNN. January 17th 2006. Available at www.cnn.com/2006/WORLD/europe/01/17/uk.parrot/index.html.

de Kort SR, den Hartog PM, ten Cate CJ. 2002. Diverge or merge? The effect of sympatric occurrence on the territorial vocalizations of the vinaceous dove *Streptopelia vinacea* and the ring-necked dove *S. capicola. Journal of Avian Biology* 33: 150–8.

den Hartog PM, de Kort SR, ten Cate CJ. 2007. Hybrid and parental species vocalizations are effective within, but not outside, an avian hybrid zone. *Behavioural Ecology* 18: 608–614.

Eriksson D, Wallin L. 1986. Male bird song attracts females—A field experiment. *Behavioural Ecology and Sociobiology* 19: 297–9.

Ewert DN, Kroodsma DE. 1994. Song sharing and repertoires among migratory and resident rufous-sided towhees. *Condor* 96: 190–196.

Feldman MW, Zhivotovsky LA. 1992. Gene–culture coevolution: Towards a general theory of vertical transmission. *Proceedings of the National Academy of Sciences USA* 89: 11935–8.

Fisher RA. 1954. Evolution and bird sociality. In *Evolution as a Process*, ed. JS Huxley, A Hardy, E Ford, pp. 71–83. London: Allen & Unwin.

Gammon DE, Baker MC, Tipton JR. 2005. Cultural divergence within novel song in the black-capped chickadee (*Parus atricapillus*). *Auk* 122: 853–71.

Goodfellow DJ, Slater PJB. 1986. A model of bird song dialects. *Animal Behaviour* 34: 1579–80.

Grafen A. 1990. Biological signals as handicaps. *Journal of Theoretical Biology* 144: 517–46.

Helb H-W, Dowsett-Lemaire F, Bergmann H-H, Conrads K. 1985. Mixed singing in European songbirds, a review. *Zeitschift für Tierpsychologie* 69: 27–41.

Ince SA, Slater PJB, Weismann, C. 1980. Changes with time in the songs of a population of chaffinches. *Condor* 82: 285–90.

Handley HG, Nelson DA. 2005. Ecological and phylogenetic effects on song sharing in songbirds. *Ethology* 111: 221–38.

Johnson LS, Searcy WA. 1996. Female attraction to male song in house wrens (*Troglodytes aedon*). *Behaviour* 133: 357–66.

King AP, West MJ. 1977. Species identification in the North American cowbird: Appropriate responses to abnormal song. *Science* 195: 1002–4.

Krebs JR. 1977. Song and territory in the great tit *Parus major*. In *Evolutionary Ecology*, ed. B Stonehouse, C Perrins, pp. 47–62. Baltimore, MD: University Park Press.

Kroodsma DE. 1996. Ecology of passerine song development. In *Ecology and Evolution of Acoustic Communication in Birds*, ed. DE Kroodsma, EH Miller, pp. 3–19. Ithaca, NY: Cornell University Press.

Lachlan RF, Feldman MW. 2003. Evolution of cultural communication systems: The coevolution of cultural signals and genes encoding learning preferences. *Journal of Evolutionary Biology* 16: 1084–95.

Lachlan RF, Janik VM, Slater PJB. 2004. The evolution of conformity enforcing behaviour in cultural communication systems. *Animal Behaviour* 68: 561–70.

Lachlan RF, Servedio MR. 2004. Song learning accelerates allopatric speciation. *Evolution* 58: 2049–63.

Lachlan RF, Slater PJB. 1999. The maintenance of vocal learning by gene–culture interaction: The cultural trap hypothesis. *Proceedings of the Royal Society of London B* 266: 701–6.

Lachlan RF, Slater PJB. 2003. Song learning by chaffinches: How accurately and from where? A simulation analysis of patterns of geographical variation. *Animal Behaviour* 65: 957–69.

Laiolo P, Tella JL. 2005. Habitat fragmentation affects cultural transmission: Patterns of song matching in Dupont's lark. *Journal of Applied Ecology* 42: 1183–93.

Leitão AM, Riebel K. 2003. Are good ornaments bad armaments? Male chaffinch perception of songs with varying flourish length. *Animal Behaviour* 66: 161–7.

Lynch A. 1996. The population memetics of birdsong. In *Ecology and Evolution of Acoustic Communication in Birds*, ed. DE Kroodsma, EH Miller, pp. 181–97. Ithaca, NY: Cornell University Press.

Lynch A, Baker AJ. 1993. A population memetics approach to cultural evolution in chaffinch song: Meme diversity within populations. *American Naturalist* 141: 597–620.

MacDougall-Shackleton EA, Derryberry EP, Hahn TP. 2002. Nonlocal male mountain white-crowned sparrows have lower paternity and higher parasite loads than males singing local dialect. *Behavioral Ecology* 13: 682–9.

Marler P. 1952. Variation in the song of the chaffinch, *Fringilla coelebs*. *Ibis* 94: 458–72.

Marler P. 1997. Three models of song-learning: Evidence from behavior. *Journal of Neurobiology* 33: 501–16.

Marler P, Sherman V. 1985. Innate differences in singing behavior of sparrows reared in isolation from adult conspecific song. *Animal Behaviour* 33: 57–71.

Marler P, Slabbekoorn H, eds. 2004. *Nature's Music: The Science of Birdsong*. Amsterdam: Elsevier Press.

McGregor PK, ed. 1992. *Playback and Studies of Animal Communication*. New York: Plenum Press.

McGregor PK, Krebs JR. 1982. Song types in a population of great tits (*Parus major*): Their distribution, abundance and acquisition by individuals. *Behaviour* 79: 126–52.

Mundinger PC. 1982. Microgeographic and macrogeographic variation in the acquired vocalizations of birds. In *Acoustic Communication in Birds, Vol. 2: Song Learning and Its Consequences*, ed. DE Kroodsma, EH Miller, pp. 147–208. New York: Academic Press.

Nowicki S, Searcy WA. 2004. Song function and the evolution of female preferences—Why birds sing, why brains matter. *Annals of the New York Academy of Sciences* 1016: 704–23.

Nowicki S, Searcy WA, Hughes M. 1998. The territory defenses function of song in song sparrows: A test with the speaker occupation design. *Behaviour* 135: 615–28.

Nowicki S, Searcy WA, Peters S. 2002a. Brain development, song learning and mate choice in birds: A review and experimental test of the "nutritional stress hypothesis." *Journal of Comparative Physiology A* 188: 1003–14.

Nowicki S, Searcy WA, Peters S. 2002b. Quality of song learning affects female response to male bird song. *Proceedings of the Royal Society of London B* 269: 1949–54.

Payne RB. 1996. Song traditions in indigo buntings: Improvisation, dispersal and extinction in cultural evolution. In *Ecology and Evolution of Acoustic Communication in Birds*, ed. DE Kroodsma, EH Miller, pp. 198–220. Ithaca: Cornell University Press.

Paradis E, Baillie SR, Sutherland WJ, Gregory R. 1998. Patterns of natal and breeding dispersal in songbirds. *Journal of Animal Ecology* 67: 518–36.

Riebel K. 1998. Testing female chaffinches song preferences by operant conditioning. *Animal Behaviour* 56: 1443–53.

Riebel K. 2000. Early exposure to song leads to repeatable preferences for male song in female zebra finches. *Proceedings of the Royal Society of London B* 267: 2553–8.

Riebel K. 2003. The 'mute' sex revisited: Vocal production and perception learning in female songbirds. *Advances in the Study of Behavior* 33: 49–86.

Riebel K, Slater PJB. 1999. Do male chaffinches *Fringilla coelebs* copy song sequencing and bout length from their tutors? *Ibis* 141: 680–3.

Searcy WA, Marler P, Peters SS. 1985. Songs of isolation-reared sparrows function in communication, but are significantly less effective than learned songs. *Behavioural Ecology and Sociobiology* 17: 223–9.

Slater PJB, Ince SA, Colgan PW. 1980. Chaffinch song types: Their frequencies in the population and distribution between the repertoires of different individuals. *Behaviour* 75: 207–18.

Slater PJB, Lachlan RF. 2003. Is innovation in bird song adaptive? In *Animal Innovation*, ed. SM Reader, KN Laland, pp. 117–36. Oxford: Oxford University Press.

Slater PJB, Lachlan RF, Riebel K. 2000. The significance of learning in signal development: The curious case of the chaffinch. In *Animal Signals: Adaptive Significance of Signalling and Signal Design in Animal Communication*, ed. Y Espmark, T Amundsen, G Rosenqvist, pp. 401–13. Trondheim, Norway: Tapir Publishers.

Stoddard PK, Beecher MD, Loesche P, Campbell SE. 1992. Memory does not constrain individual recognition in a bird with song repertoires. *Behaviour* 122: 274–87.

Thorpe WH. 1958. The learning of song patterns by birds, with especial reference to the song of the chaffinch, *Fringilla coelebs. Ibis* 100: 535–70.

Vehrencamp SL. 2001. Is song-type matching a conventional signal of aggressive intentions? *Proceedings of the Royal Society of London B* 268: 1637–42.

Whaling CS, Soulis MM, Doupe AJ, Soha JA, Marler P. 1997. Acoustic and neural bases for innate recognition of song. *Proceedings of the National Academy of Sciences USA* 94: 2694–8.

Williams JM, Slater PJB. 1990. Modelling bird song dialects: The influence of repertoire size and numbers of neighbours. *Journal of Theoretical Biology* 145: 487–96.

Wilson PL, Towner MC, Vehrencamp SL. 2000. Survival and song-type sharing in a sedentary subspecies of the song sparrow. *Condor* 102: 355–63.

Wilson PL, Vehrencamp SL. 2001. A test of the deceptive mimicry hypothesis in song-sharing song sparrows. *Animal Behaviour* 62: 1197–1205.

Zahavi A. 1975. Mate selection: A selection for a handicap. *Journal of Theoretical Biology* 53: 205–14.

15 Development and Evolution of Speech Sound Categories: Principles and Models

Gert Westermann

Introduction

Flexible communication relies on efficient ways to encode information. In spoken language this efficiency is achieved by using a fixed repertoire of speech sounds—phonemes—that are combined in different ways to form words. Spoken communication in most languages is based on combinations of between twenty and forty different phonemes (Maddieson, 1984) instead of the far greater number of possible sounds that can in principle be produced by the human speech apparatus. These phonemic repertoires show both cultural and universal aspects: Between different languages the number of phonemes can differ significantly, from around ten to over 100 (de Boer, 2000), suggesting cultural convention as the origin of the sound systems of particular languages. On the other hand, several phonemes are shared between most languages. For example, 87% of the 451 languages in the UCLA Phonological Segment Inventory Database (Maddieson, 1984) use [a], 89% use [k], and 94% use [m] (de Boer, 2000). These universal tendencies might be explained by the constraints of the human speech apparatus that make the production of some sounds easier than others and therefore favor their use.

Two questions about the emergence of speech sound repertoires in a language are as follows: How does a society of speakers initially establish the convention of a specific repertoire? And how does a newborn member of this society adapt to the specific repertoire of her language environment? These two questions are closely related because they might share a common answer: Both language evolution and child acquisition of the sound patterns of a language should be constrained by the perceptual and articulatory systems of language users and learners. In language evolution, the forms that become established in a linguistic community might be selected for the ease and reliability with which they can be produced and perceptually discriminated. Likewise, the same perceptual and articulatory constraints will bias an infant to preferentially produce and perceive certain sounds. For example, it has been found that forms that are easiest to produce appear first during infant babbling (Locke, 1983) and that the sounds that are articulated well in babbling form the basis of an infant's first words (Vihman and Miller, 1988). If common constraints underlie language evolution and language development, this leads to a situation in which an infant has to learn a conventional linguistic repertoire that is optimally adapted to these constraints because it has emerged from constraints similar to those that act on her learning mechanisms. This coupling between evolution and development will lead to the optimal learning and cultural transmission of forms.

In this chapter I focus on the constraints and mechanisms at work in the development and evolution of phonetic repertoires to establish a stable system of discrete speech sounds that can form the basis of a flexile communicative system. The main hypothesis put forward here is that the biological constraints of the speech system together with neurobiological constraints of experience-dependent cortical development shape the speech sound repertoire against the background of an infant's exploratory vocalizations in the babbling phase. This hypothesis is implemented in a connectionist neural network model of the development of vowel sounds in infancy. The model employs a neurally plausible mechanism that leads to the emergence of discrete vowel categories, and to adaptation to an ambient language. An extension of the model is suggested that will account for the emergence of shared speech sound inventories in language evolution. Finally, I suggest a general principle of cortical development, mutually constraining topographic mappings, and discuss its relevance to the concept of mirror neurons and to an evolutionary account of how word–meaning relationships become established in a language community.

Babbling as the Foundation for Speech

At the end of their first year of life most infants begin to speak their first words. The preceding months see the laying of the groundwork for this ability. During this period, between six months and one year of age and beyond, infants show a remarkable development both in their articulation and in their perception of speech sounds.

The finding that babbling lays the foundations for phonological development is fairly recent, and earlier research denied such a link, albeit based on false assumptions. For example, Jakobson (1941) claimed that babbling and speech were independent because between the end of babbling and the beginning of speech was a silent period in which the infant did not produce any utterances. Although this view went unchallenged for a long time, it has since been shown to be false: Many studies have made clear that the transition from babbling to speech is smooth and continuous, with an infant's first words overlapping with late babbling, and with late babbling sounds used for the infant's first words (Locke, 1983). A different argument for the presumed independence of babbling and speech came from Lenneberg's (1967) claim that hearing and congenitally deaf children produced the same babbling sounds whereas their word production differed. However, more recent studies have uncovered important differences between the babbling of hearing and deaf infants. For example, whereas hearing infants begin canonical babbling (the production of syllables) between six and ten months of age, the onset of this stage in deaf infants is delayed to between eleven and twenty-five months and is characterized by a reduced set of speech-like babbling sounds (Stoel-Gammon and Otomo, 1986; Oller and Eilers, 1988).

Further evidence for the importance of babbling for speech development comes from infants who are unable to articulate normally due to tracheostomy, where a tube is placed in the windpipe through an opening in the neck to facilitate breathing. Despite normal hearing, after the end of treatment these infants can show abnormal patterns of vocal expression with reduced speech sound repertoires (Locke, 1986; Locke and Pearson, 1990; Bleile et al., 1993; Jiang and Morrison, 2003). Given this evidence, current knowledge has firmly established the importance of babbling for the normal development of speech. The babbling phase itself is characterized by changes both in an infant's articulations and in her ability to perceive speech sounds.

Articulatory Change during Babbling

An infant's articulations are constrained by the immaturity of her vocal apparatus and by her initial lack of control over this apparatus. However, as the vocal system develops and control is slowly gained, the sounds produced by the infant undergo considerable change during the first year of life, from crying to the syllables of her native language (for reviews, see Vihman, 1996; Jusczyk, 1997; Boysson-Bardies, 1999; Oller, 2000). Before six months of age, articulations consist mainly of isolated vowel sounds and sometimes of nasalized or glottal consonants and some fricatives. Around six months most infants begin to babble canonically (Oller, 1980), that is, they produce sequences of consonant–vowel syllables (e.g., "gagaga"). In the following months, articulations become more complex, and eventually the speech sounds produced by the infants correspond to those in their native language.

The precise timing of the shift from language universal babbling sounds to a repertoire that reflects the particular sounds of the native language (the so-called "babbling drift") is controversial. Several researchers have argued that adaptation to the native language occurs between nine and ten months of age by beginning to reflect the native distribution of vowels and consonant types (Boysson-Bardies et al., 1989; Boysson-Bardies and Vihman, 1991) and intonational patterns (Whalen et al., 1991). However, other studies of babbling drift have reported negative results with no discernible adaptation to the native language (see Oller and Eilers, 1998). Although clearly research methodologies in this field have to be refined, there is no doubt that at some point between ten and fifteen or so months of age infants do adapt to produce some language-specific sounds. This process is seen as an important step in the development of a phonological inventory, first words, and more complex linguistic structures (McCune and Vihman, 2002).

Perceptual Change during Babbling

Research on perceptual change during the first year of life has been more extensive than that on production, perhaps motivated by the finding that infants are capable of discriminating certain speech sounds from a very young age (e.g., Eimas et al., 1971; Moffitt, 1971; Trehub, 1973). It has also been shown repeatedly that infants can show a preference for their native language from birth (Mehler et al., 1988; Moon et al., 1993; Nazzi et al., 1998), pointing to a role for learning in utero.

An important result that has been replicated many times is that whereas young infants are capable of discriminating between many phonetic contrasts from many languages, between six and twelve months of age they gradually lose the ability to discriminate between contrasts that are not in their native language. For example, English-learning six-month-olds were shown to be able to distinguish phonetic pairs in English, Hindi, and Salish (a Canadian Indian dialect), but the ability to discriminate between pairs in Hindi and Salish had declined by twelve months (Werker and Tees, 1984). Similar results were obtained with vowel sounds (Polka and Werker, 1994). Vowel categories have further been shown to possess an internal structure in which infants develop perceptual prototypes for native phonetic categories (Kuhl, 1991; Kuhl et al., 1992). These prototypes act as "perceptual magnets," and the perceptual space around them is shrunk. As a consequence, phonetic variations close to the prototype are discriminated less well (because perceived sounds are mapped onto the prototype) than those that are very different from the prototypes. From this view, nonnative contrasts can be perceived when they map onto the prototypes of different phonetic categories but not when they map to the compressed regions around a single prototype.

Behaviorally then, by the end of the first year—before speaking her first words—much progress has been made in the infant's perceptual adaptation to the phonemes of her ambient language. Recent experiments have shown that the precise nature of perceptual adaptation is strongly linked with specific language experience. For example, speech discrimination ability in the infant is significantly correlated with the clarity of her mother's speech (Liu et al., 2003) and is highly sensitive to the statistical distribution of speech sounds in the native language (Maye et al., 2002).

Different theories have been put forward to explain the neural underpinnings of the changes in infant speech perception during the first year of life. According to the selectionist view (Mehler and Dupoux, 1994), neural structures supporting those speech sound contrasts that occur in the child's native language become stabilized whereas structures supporting distinctions that are not needed are lost. A counterargument to this view is that not all nonnative contrasts are lost in development.

For example, English-speaking infants and adults are still able to discriminate between different click sounds that are part of the Zulu language (Best, 1995). The loss of discrimination ability thus seems to occur only for nonnative sound pairs that are similar to those of the native language. Based on these results, Best (1995) developed the Perceptual Assimilation model, arguing that speech sounds are assimilated to native categories when possible. Thus, a nonnative contrast (such as Zulu clicks for English speakers) can be discriminated if the two sounds map onto two different phonetic categories in the native language. Discrimination is worst if the nonnative sounds map onto the same native category.

The Perceptual Assimilation model is compatible with the view suggested in this chapter where speech sound development relies on the experience-dependent reorganization of neural structures to accommodate native speech sounds based on their statistical properties. Further evidence for this view comes from research showing that at the same time as nonnative contrasts are lost, infants show an increase in their ability to discriminate native contrasts (Kuhl et al., 2006). A related hypothesis is the Native Language Neural Commitment hypothesis (Kuhl, 2000, 2004), which explains behavioral change in speech development with the construction of dedicated neural networks for the encoding of native speech. This neural commitment, while enhancing the processing of native speech, interferes with the processing of nonnative speech by effectively warping the perceptual space in favor of native contrasts (Iverson et al., 2003).

Linking Perception and Production

In theory the described developmental changes in speech perception and production could develop independently (Lenneberg, 1967). However, there is much evidence that a link exists between perception and production, that this link develops during the first year of life, and that a close coupling between perception and production might form the basis for the development of a speech sound repertoire in infants.

The strongest evidence for an effect of perception on the production of speech sounds comes from research with deaf infants described above: In the absence of auditory input, babbling is significantly delayed and characterized by a smaller sound repertoire than in hearing infants. Effects of an adaptive perception–production link have also been observed in adults. For example, when Japanese speakers were trained to discriminate the English /r/ and /l/ (Bradlow et al., 1997) or vowel sounds (Lambacher et al., 2005), their production of these sounds also improved. A different study has linked speakers' discriminative accuracy of vowels with their distinctness of articulation in American English speakers (Perkell et al., 2004).

Evidence for an effect of speech sound production on perception is more recent. The aim of this research has generally been to link speech sounds produced by an infant to the way in which different speech sounds are perceived. One such study investigated monolingual English and Welsh twelve-month-olds (Vihman and Nakai, 2003). In this study, two consonants in each language (English and Welsh) were identified that had roughly the same frequency in infant-directed speech but that were produced by the infants with different frequencies. The idea here was that any perceptual differences between the two consonants could only be explained by the differences in production because they were perceptually of the same frequency. When tested on both consonants, the infants did indeed show a novelty preference for the consonant that they produced less often. Another study (DePaolis et al., 2005) investigated twelve-month-olds' responses to heard speech sounds depending on whether these sounds were within or outside the infants' own productive repertoire. DePaolis et al. found that infants with small productive repertoires showed a familiarity preference to heard within-repertoire sounds. Infants with slightly larger repertoires, by contrast, showed a novelty preference to heard sounds outside their productive repertoire. These results further confirmed the role of the productive repertoire in the perception of speech sounds.

Taken together, the reviewed evidence suggests (1) that babbling together with exposure to an ambient language plays a crucial role in establishing the speech sound repertoire of the infant's native language, (2) that both perception and production converge on this native repertoire, and (3) that a perception–production link is established which aligns perceptual and articulatory abilities and eventually leads to preference for those speech sounds that then form the basis of the infant's first words. The developing perception–production link has been explained with the Articulatory Filter hypothesis (Vihman, 1993, 2002), which suggests that after the onset of canonical babbling, an articulatory filter begins to highlight those speech sounds in the environment that correspond to vocal patterns produced by the infant herself and facilitates motoric recall of these patterns. As a consequence, these patterns become particularly salient to the infant and can serve as building blocks for first words.

A Computational Model of Sensorimotor Integration in Speech Development

A recent model (Westermann and Miranda, 2004) has suggested an account of the neural mechanisms underlying the developing link between perception and production in the babbling phase and the resulting changes in both motor and auditory representations (see figure 15.1). This artificial neural network model was developed to explore in detail a biologically plausible mechanism for the formation of vowel repertoires, the development of perceptual and production prototypes, and the gradual adaptation of the articulatory and perceptual system to the infant's native language. The main idea behind the model is that during babbling, experience-dependent links are established between articulatory representations and the resulting speech sounds.

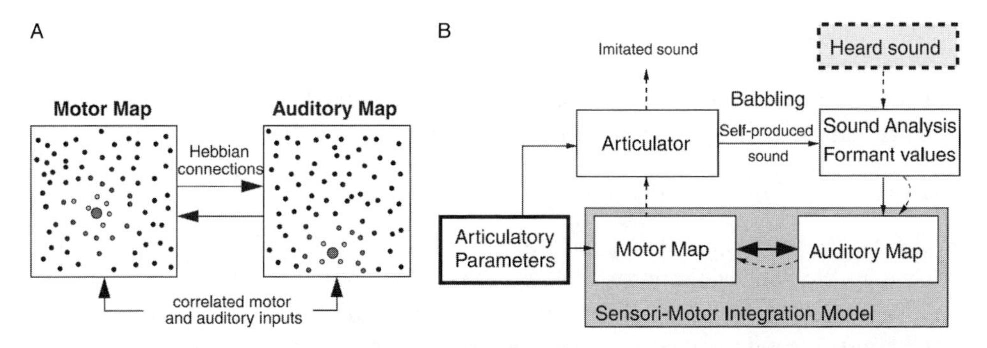

Figure 15.1
The basic architecture of the sensorimotor integration model (A) and its embedding in the experimental setup (B) to learn the coupling between articulatory parameters and self-generated sounds through babbling. Once learned, the system imitates heard sounds (dashed lines). Reprinted from Westermann G, Miranda ER, A new model of sensorimotor coupling in the development of speech, *Brain and Language* 89: 393–400, copyright (2004) with permission from Elsevier.

Motor commands and auditory signals are represented on topographic neural maps like those found in many areas of the brain. Due to the developing links between these maps, both maps become "warped" to preferentially represent sounds that can be reliably perceived and produced. Furthermore, exposure to the infant's native language leads to modification of these initial links, biasing the perception–action system to favor sounds from the native language. This adaptation to the infant's native language occurs independently of explicit imitation of heard sounds: Instead, preferential production of native speech sounds is an emergent outcome of the adaptation between perception and production.

The perception–action model consists of two cortical maps, one for motor programs and one for auditory representations of vowels (see figure 15.1A). Each map consists of 400 neurons, or units. The neurons have overlapping Gaussian (bell-shaped) activation functions and act as receptive fields for motor and auditory stimuli: A stimulus that falls into the receptive field of a unit activates this unit, and the more central to the receptive field the stimulus is, the stronger the activation of the unit. Articulatory gestures consist of a set of motor parameters that are represented by a population of active neurons on the motor map. A physical model (Praat) of the vocal system (Boersma and Weenink, 1996) is used to generate a vowel sound based on these articulatory parameters. Speech sounds in this model are restricted to vowel sounds because they can be represented as static signals. Vowel sounds are analyzed in terms of their first two formants, and this value pair activates neurons on the auditory map. The model can either "babble" by generating an articulatory gesture and "listening" to the resulting self-produced sound—articulatory parameters and resulting formant values then form the inputs to the two maps—or it can "listen" to speech sounds from the environment, in which case only the auditory map receives external input (see figure 15.1B).

The motor and auditory maps in the model are linked by developing Hebbian weights that are updated with the covariance activation rule (Sejnowski, 1977):

$$\Delta w_{ij} = (act_i - \overline{act_i})(act_j - \overline{act_j})$$

where Δw_{ij} is the weight change between a unit i on one map and a unit j on the other map, $\overline{act_i}$ and $\overline{act_j}$ are the average activations of units i and j over a certain time window, and act_i and act_j are their current activation values. The effect of this weight update is that connections between articulatory parameters and vowel sounds become strengthened if they co-occur reliably. The initial value of all weights is 0.

A motor or sound input to a map typically activates an ensemble of units, and the response to an input, that is, how the neural map "perceives" that input, is computed in a population code: The response is the vector sum of the positions (receptive field centers) of all activated units, weighted by their activation values:

$$r_x = \frac{\sum_i act_i pos_i}{\sum_i act_i},$$

where r_x is the perceived response to an input x, pos_i is the center of the receptive field for neuron i, and act_i is the unit's activation value. Such population codes have been found to encode stimulus properties in many areas of the cortex, for example, in monkey motor cortex (Georgopoulos et al., 1988).

The model develops as follows: In the babbling phase, an individual babbling event consists in choosing a random articulatory setting.[1] These motor parameters are used to simulate a vowel articulation in Praat, and the resulting sound is analyzed in terms of its first two formant values to form the input to the auditory map. Because the maps are interconnected and activation flows between the maps, the system is then allowed to settle into a stable state where unit activations no longer change. Then the weights are adapted as described above. The babbling phase consists of a large number of individual babbling events.

As a consequence of the developing connections between the maps, a map receives input not only from its own domain (articulatory or auditory) but also from the other domain. Connections between units on the maps that consistently covary are strengthened, and therefore, these units become more highly activated because they simultaneously receive external activation and activation from the other map. As a consequence, the population-coded response, which is based on the weighted activations of all units, shifts toward these more active units. This means that the responses to external stimuli on each map are shifted toward those locations in motor and perceptual space where responses reliably co-occur with responses in the other domain.

That is, the model develops preferred responses for highly correlated motor–sound pairs. Such highly correlated pairs correspond to regions in articulatory space where small changes of motor parameters lead only to small changes in the resulting sounds, that is, sounds that can be produced with high reliability and consistency. In effect, an emergent outcome of the learned integration between perception and production is that due to the nonlinear mapping between articulatory gestures and their resulting sounds, stable sounds become preferred.

Nonlinearity of the Articulation–Sound Mapping

In a first simulation, articulatory gestures were generated by the manipulation of two parameters: jaw opening and the position of the styloglossus muscle, a muscle that controls elevation of the tongue. Each of these parameters was continuously varied in eighteen steps, leading to 324 possible articulatory settings. The left panel of figure 15.2A shows this regular variation of the motor parameters on the motor map (the

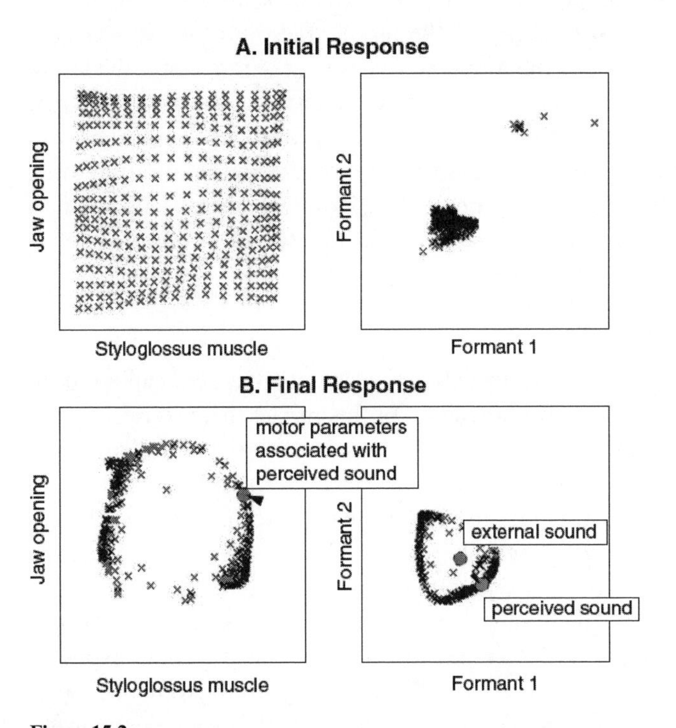

Figure 15.2
Initial (A) and final (B) responses of the sensorimotor integration model. How an external sound is perceived and imitated in terms of a prototypical sound is also shown in (B). Reprinted from Westermann G, Miranda ER, Modeling the development of mirror neurons for auditory-motor integration, *Journal of New Music Research* 31: 367–75, copyright (2002) with permission from Taylor & Francis, http://www.tandf.co.uk.

slight warping is a result of the population-decoded response with randomly placed receptive fields). The right panel of figure 15.2A shows the resulting sounds that were generated on the basis of these motor settings with Praat. It is apparent that the mapping from articulatory gesture to resulting sound is highly nonlinear.

Development of Perceptual and Motor Prototypes

The model simulated babbling by repeatedly randomly selecting one of the 324 motor parameter sets, generating the associated sound, and adapting the connections between the maps as described above. Each of these adaptations warps the representations of articulatory gestures and resulting sounds, and figure 15.2B shows the warped maps at the end of a period of babbling where both productive and perceptual prototypes developed. Instead of the full range of jaw and styloglossus positions, the model developed to prefer only certain settings. In this simulation the model in effect settled on two discrete styloglossus positions indicated by the vertical dense columns in figure 15.2B, with a larger range of jaw opening positions. It should be noted, however, that due to the simplicity of the model, these prototypes will bear little resemblance to those developed in infants who vary a large range of articulatory parameters. Heard sounds in the model were perceived in terms of the developed perceptual prototypes, which in turn activated articulatory prototypes associated with the perceived sound. This activation of the associated motor representations allowed the model to imitate heard sounds (see figure 15.2B). However, imitation was not a mere replay of the external sound but a re-creation of this sound in terms of developed perceptual and articulatory prototypes.

Adaptation to the Native Language

A second simulation examined how the articulatory–perceptual system can adapt to the sound patterns of an infant's native language. In this model there were six articulatory parameters[2] that were continuously varied as in the previous simulation, leading to a larger and more realistic repertoire of articulatory gestures and vowel sounds. Sounds were again analyzed in terms of their first two formant values. Training of the model was either by babbling only, as described above, or by randomly interspersing babbling with the formant representations of prerecorded vowel sounds from either French or German. The sounds produced by the model before any learning took place are shown in figure 15.3A. When the model babbled without exposure to sounds from an ambient language (figure 15.3B), that is, by just hearing itself, perceptual prototypes developed but naturally did not coincide with the ambient vowels. However, when ambient vowels were presented during the babbling phase, the perceptual prototypes developed to largely coincide with the external sounds (figure 15.3C and 15.3D), spreading out the initial vowel space to reflect the distribution in the native language, as described in infants (Boysson-Bardies et al., 1989). Adapta-

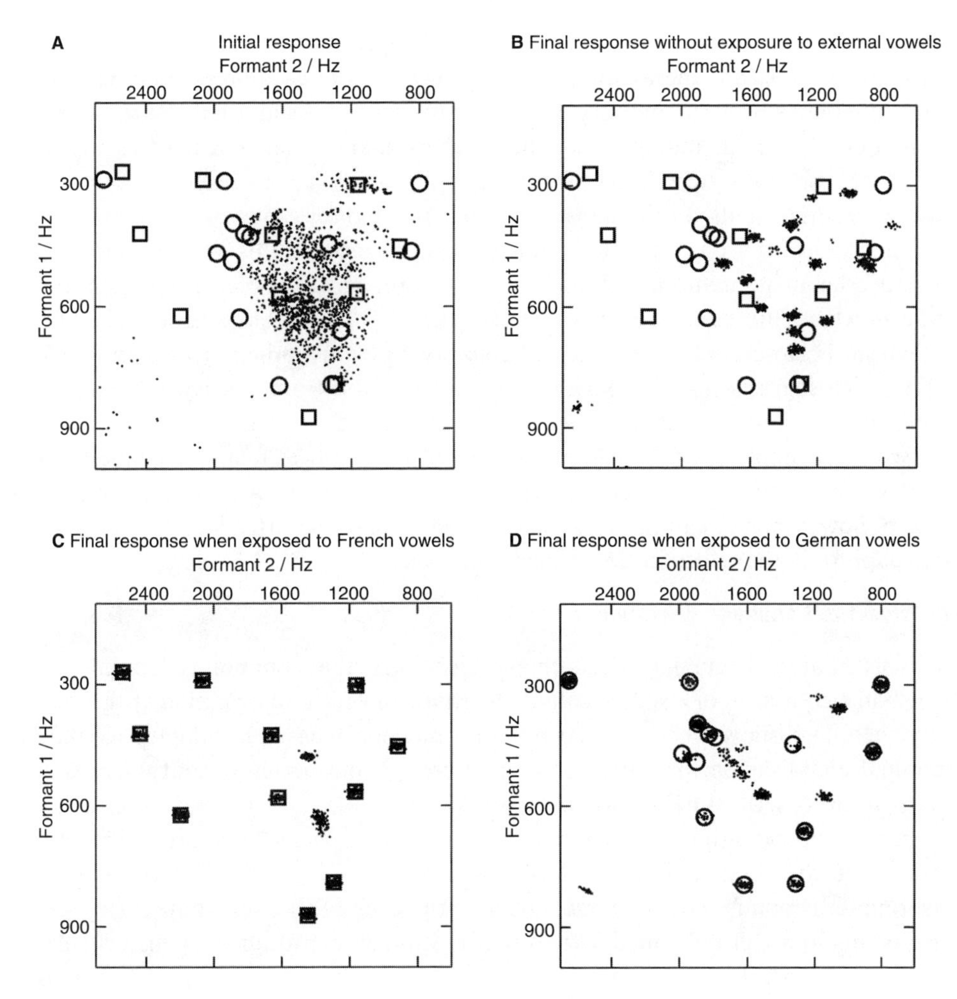

Figure 15.3
Responses of the model to self-generated sounds under different training conditions. French vowels are indicated by squares, German vowels by circles. Reprinted from Westermann G, Miranda ER, A new model of sensorimotor coupling in the development of speech, *Brain and Language* 89: 393–400, copyright (2004) with permission from Elsevier.

tion to the native language in this model does not depend on the explicit imitation of heard sounds. Instead, the adaptation mechanism works as follows: During babbling, initial links between auditory and motor neurons are established. When an external sound is heard, initially only the auditory map becomes active. However, through the previously established connections, activation also reaches the motor map. As a result, units on the motor map and the auditory map are coactive, and the connections between them will be further strengthened. In this way, sounds from the native language reinforce selectively the relevant links between perception and production and the associated sounds form the infant's babbling inventory. This mechanism is closely related to the Articulatory Filter hypothesis (Vihman, 1993, 2002), which suggests that the experience of frequently self-produced syllables sensitizes infants to similar patterns in their ambient language.

Based on a simple mechanism—the linkage of two cortical maps by connections that develop in experience-dependent ways—the model could thus give a precise account of how babbling can serve to establish perceptual and articulatory prototypes that adapt to those present in the ambient language.

The Model in Language Evolution

As discussed at the beginning of this chapter, it is likely that common constraints underlie both acquisition of a speech sound repertoire and cultural evolution of this repertoire in a linguistic society. The sensorimotor model can easily be adapted to study the initial establishment of a linguistic repertoire by considering several interacting agents. In the described model, Westermann and Miranda (2004) showed how a single such system can adjust to a (static) ambient sound system (German or French). However, in the absence of an established ambient language, different agents with sensorimotor couplings could interact and adapt to each other's repertoires. This scenario points to a cultural sound system that is strongly constrained by the articulatory and perceptual constraints of the individual agents. Assuming that regions of stability in motor and perceptual space are similar between different agents, such stable sounds would become strong candidates for the basis of a culturally shared vowel system. A study with a model similar to the one described here has found that in a "cultural-babbling" scenario stable phonetic repertories can indeed emerge (Oudeyer, 2005).

Such a model of the evolution of speech sounds differs from most other suggested models in two important ways. First, it is based on neurobiological principles and takes into account the perceptual and articulatory constraints present in an individual agent. Second, it does not rely on structured interactions between agents and explicit imitation as in "imitation games" (de Boer, 2000) for the emergence of vowel systems or complex "naming games" used to model the evolution of a shared lexicon (e.g., Steels, 2003).

The nonlinearity of the mapping between speech sound production and perception has been used by Stevens (1972, 1989) to explain the distributional characteristics of the human speech sound system. However, the model described here makes less strong claims: It merely suggests that stable sounds become preferred, but it is likely that subsequent to this initial bias other constraints will alter the sound system further. One such additional constraint is distinctiveness (see also Lindblom and Engstrand, 1989; Kingston and Diehl, 1994): Two stable sounds that are very similar would not be good candidates for a communicative repertoire because they cannot reliably be discriminated in a noisy environment.

Mutually Constraining Topographic Maps as a Universal Principle in Evolution and Development

The model described in this chapter suggests an intriguing possibility: Perhaps correlation-based couplings between topographic cortical maps that warp representations on each map are a ubiquitous principle of brain organization. Topographic feature maps have been identified in many areas of the brain, from visual and auditory to somatosensory and motor areas, and it has been hypothesized that topographic organization is a pervasive aspect of cortical organization that also applies to higher areas with more abstract representations (Simmons and Barsalou, 2003). Furthermore, the integration between different domains, both multisensory and sensorimotor, seems to be a widespread principle in brain processing. In adults this integration is strong and can be manipulated with incongruent signals (McGurk and MacDonald, 1976). On the neural level, multisensory integration has been found to lead to altered activation in the topographic maps responsible for unimodal processing (Macaluso and Driver, 2005). Some of these principles of multisensory integration have been modeled with a neural network model very similar to the one described here (Westermann, 2001). In language processing it has been found that words from different categories differentially activate cortical regions. For example, hearing action verbs such as "walking" and "talking" elicits different activations over motor cortex depending on the body part to which they refer (the legs or the face, in this instance; Pulvermüller, 2001). Outside the primary motor and sensory cortices these couplings might be mediated by intermediate areas that receive inputs from multiple domains and provide feedback to the domains of integrated signals. Other research has shown an interaction between categorization and perception (Schyns et al., 1998), with a warping of perceptual space induced by knowledge of category membership (Goldstone, 1995). In sensorimotor integration research it has been shown that one's own actions alter the perception of the simultaneous actions of others (Hamilton et al., 2004).

In the case of language evolution, a model of the evolution of a shared lexicon (Zuidema and Westermann, 2003) showed that communicative success is established faster if word and meaning space are topographically organized, and a mapping between these spaces respects the topographic relationship. This implies that similar meanings come to be expressed by similar sounding words. While such "sound symbolism" has often been eyed with suspicion since de Saussure's (1916) postulate of the "arbitrariness of the sign," several psychological studies have found evidence that language users are sensitive to such a link (e.g., Kelly, 1992). A study of English found that there is a small but significant global relationship between phonological word form and meaning (Shillcock et al., 2001). Combined with our results from modeling language evolution (Zuidema and Westermann, 2003), this suggests that when new words are created in a language, they can benefit from respecting form–meaning congruency. Observing the relationship between form and meaning facilitates interpretation of unknown words or words under noisy transmission by mapping on the form–meaning relationship of similar sounding words. However, this mapping *regularity* stands in conflict with the need for *distinctiveness* (Zuidema and Westermann, 2003): Individual forms can be identified best if they are maximally distinct from other forms. A possible resolution of this conflict would be for new words to respect form–meaning congruency, and when these forms become more established and it becomes less important to be able to infer their meaning, they could change to increase distinctiveness from other forms.

In summary, the principle of mutually constraining topographic maps offers possible explanations of phenomena in areas as diverse as speech processing, word learning, categorization, and language evolution.

Mirror Neurons

The hypothesis that interacting, mutually constraining cortical maps might be a pervasive principle of brain organization sheds a new light on the role of mirror neurons. Mirror neurons are a class of neurons, described in both monkeys and humans, that are activated when an action is performed but also when a similar action is observed in someone else (Gallese et al., 1996; Rizzolatti et al., 1996). These neurons have been celebrated as a mechanism that allows for imitation and for the understanding of others' actions and intentions (Rizzolatti and Craighero, 2004; Iacoboni et al., 2005), and they have been argued to take a central role in the evolution of language (Rizzolatti and Arbib, 1998; Arbib, 2005).

In the sensorimotor integration model described in this chapter mirror neurons emerge naturally on the basis of correlated activity between neurons in different domains. Babbling serves to establish a bidirectional link between perception and production, with the consequence that heard sounds activate not only neurons on the auditory map but also motor neurons. These motor neurons are the same that would be activated if the model itself produced the heard sound and can therefore be construed as auditory mirror neurons. Visual mirror neurons responding to mouth movements (Ferrari et al., 2003) could develop in much the same way in a model enhanced with a visual input map: When observed mouth gestures correlate with heard sounds, which in turn correlate with articulatory parameters, associative links between visual and articulatory maps can also develop (Westermann and Miranda, 2004).

These results question the special status attributed to mirror neurons (see also Hurford, 2003). In particular they suggest that mirror neurons are an emergent property of experience-dependent brain development in which neural populations encoding correlated stimuli develop interconnections that can lead to activation of neurons in one domain solely on the basis of external input to another domain. The bidirectional links that develop between the maps in the computational model predict that auditory speech input should activate motor neurons, and motor actions should generate activations of auditory neurons even in the absence of acoustic input.

Extending the model with a visual map would further lead to motor and auditory cortex activations on the basis of visual inputs alone. Empirical evidence for these predictions has recently been established in a series of studies of speech processing: Both passive listening to speech and silent lipreading have been found to activate speech motor areas (Watkins et al., 2003; Wilson et al., 2004), and both lipreading and own silent speech modulate the activation of auditory cortex (Paus et al., 1996; Numminen and Curio, 1999; Watkins et al., 2003). The fact that neurons that respond to correlated input in other domains seem so pervasive in the cortex makes it unlikely that these neurons assume a special status when they are in the motor cortices. Instead, mirror neurons are one piece of evidence of the correlation-based integration of neural populations across domains.

Conclusions

In this chapter I have outlined a view of the development of a repertoire of discrete speech sounds that is based on developing links between perception and production during the infant's babbling phase. Speech sound categories emerge from biological constraints and initially center on those sounds that form plateaus of stability in sensorimotor space. They can therefore be produced and perceived in a stable and robust manner and form the basis of a flexible system of communicative sounds. Exposure to an ambient language warps both the articulatory and auditory spaces to adapt to the speech sounds of that language in production and perception. A neural network model was described that served to suggest a neural mechanism underlying this adaptation process. This mechanism consists in the experience-dependent establishment of connections between neurons with correlated activity across domains. The system adapts to correlations present in the environment to reflect these correlations in its internal structure, and discrete prototypical sounds emerge from this adaptation process. I have argued that the same process might be at work in development and language evolution: Developmentally it explains how an individual adapts to the speech sounds of her native language; in terms of evolution it suggests how a repertoire of shared speech sounds can become established in a community of speakers sharing similar biological constraints.

An interesting hypothesis offered by Owren and Goldstein (this volume) is that babbling might be mediated subcortically as an initially automatic process that then leads to the formation of cortical representations for speech sounds. This process would mirror other cases in which subcortical areas generate behavior that trains the cortex, such as in face perception, where the subcortical visuomotor pathway biases newborn infants toward orienting to faces, enabling the development of cortical face-specific representations (Morton and Johnson, 1991).

An important question arises from the simplicity and pervasiveness of the adaptation mechanisms described in this chapter. Multisensory and sensorimotor integration as well as warping of topographic maps through altered input have also been described in nonspeaking animals. Furthermore, songbirds and other animals have been found to develop repertoires of communicative sounds that are discrete and flexible, without reaching the complexity of the human language system (see other chapters in this volume). What is needed, then, to develop a flexible repertoire of communicative sounds, and how does this relate to language? A perspective on this question is offered by Scott and Johnsrude (2003), who argue that speech and language are conceptually distinct in that the perception of speech sounds as communicative acts is mediated by phylogenetically older brain systems that are shared across different species, whereas language, characterized by syntax, generative aspects, and complex semantics, may be unique to humans. In this way the speech perception system would form a necessary but not sufficient basis for the development of language in humans.

Another aspect of language that might explain some of the differences between human and animal communication systems is the vast amount of input that human infants receive. It has been estimated that, depending on family background, children hear between three and eleven million words per year (Hart and Risley, 1995), that is, up to 2,000 words per waking hour. It is unlikely that exposure to communicative signals in young animals is similarly vast. Given that the amount of language input in children has been closely linked with their linguistic performance (Hart and Risley, 1995), it is an intriguing possibility that both in language development and evolution there might be a critical level of communicative input from adults to their offspring that is not present in nonhumans.

Notes

1. In reality, articulatory parameters would, of course, form smooth trajectories and lead to alternating vowels and consonants rather than sequences of isolated vowel sounds. The model presented here makes these simplifications to investigate its ability in principle to account for changes in perception and production during babbling.

2. These were styloglossus muscle (backward and upward tongue movement), hypoglossus muscle (backward and downward tongue movement), levator palatini muscle (raising of the velum), cricothyroid muscle (vocal chord stretching), interarytenoid muscle (vocal chord adduction), and lung pressure.

References

Arbib MA. 2005. From monkey-like action recognition to human language: An evolutionary framework for neurolinguistics. *Behavioral and Brain Sciences* 28: 105–24.

Best CT. 1995. A direct realist perspective on cross-language speech perception. In *Speech Perception and Linguistic Experience: Theoretical and Methodological Issues in Cross-Language Speech Research*, ed. W Strange, pp. 167–200. Baltimore: York Press.

Bleile KM, Stark RE, McGowan JS. 1993. Speech Development in a Child after Decannulation—Further Evidence that Babbling Facilitates Later Speech Development. *Clinical Linguistics & Phonetics* 7: 319–37.

Boersma P, Weenink D. 1996. *Praat, a System for doing phonetics by computer, Technical Report 132*: Institute of Phonetic Sciences of the University of Amsterdam.

Boysson-Bardies B de. 1999. *How Language Comes to Children*. Cambridge, MA: MIT Press.

Boysson-Bardies B de, Halle P, Sagart L, Durand, C. 1989. A cross-linguistic investigation of vowel formants in babbling. *Journal of Child Language* 16: 1–17.

Boysson-Bardies B de, Vihman, MM. 1991. Adaptation to language: evidence from babbling of infants according to target language. *Language* 67: 297–319.

Bradlow AR, Pisoni DB, Akahane-Yamada R, Tohkura Y. 1997. Training Japanese listeners to identify English /r/ and /l/: IV. Some effects of perceptual learning on speech production. *Journal of the Acoustical Society of America* 101: 2299–310.

de Boer B. 2000. Self-organization in vowel systems. *Journal of Phonetics* 27: 441–65.

DePaolis R, Vihman M, Keren-Portnoy T. 2005. Output as input: Combining experimental and observational methods, *10th IASCL*. Berlin, Germany.

de Saussure, F. 1916. *Course in General Linguistics*. La Salle, IL: Open Court.

Eimas PD, Siqueland ER, Jusczyk P, Vigorito J. 1971. Speech perception in infants. *Science* 171: 303–6.

Ferrari PF, Gallese V, Rizzolatti G, Fogassi L. 2003. Mirror neurons responding to the observation of ingestive and communicative mouth actions in the monkey ventral premotor cortex. *European Journal of Neuroscience* 17: 1703–14.

Gallese V, Fadiga L, Fogassi L, Rizzolatti G. 1996. Action recognition in the premotor cortex. *Brain* 119: 593–609.

Georgopoulos AP, Kettner RE, Schwartz AB. 1988. Primate motor cortex and free arm movements to visual targets in three-dimensional space. II: Coding of the direction of movement by a neural population. *Journal of Neuroscience* 8: 2928–37.

Goldstone RL. 1995. Effects of categorization on color perception. *Psychological Science* 6: 298–304.

Hamilton A, Wolpert D, Frith U. 2004. Your own action influences how you perceive another person's action. *Current Biology* 14: 493–8.

Hart B, Risley TR. 1995. *Meaningful Differences in the Everyday Experience of Young American Children.* Baltimore, MD: Brookes.

Hurford J. 2003. Language beyond our grasp: What mirror neurons can, and cannot, do for language evolution. In *Evolution of Communication Systems: A Comparative Approach*, ed. DK Oller, U Griebel, pp. 297–313. Cambridge, MA: MIT Press.

Iacoboni M, Molnar-Szakacs I, Gallese V, Buccino G, Mazziotta JC, Rizzolatti G. 2005. Grasping the intentions of others with one's own mirror neuron system. *Plos Biology* 3: 529–35.

Iverson P, Kuhl PK, Akahane-Yamada R, Diesch E, Tohkura Y, Kettermann A, Siebert C. 2003. A perceptual interference account of acquisition difficulties for non-native phonemes. *Cognition* 87: B47–B57.

Jakobson R. 1941. *Kindersprache, Aphasie, und allgemeine Lautgesetze*. Uppsala: Almqvist & Wiksell.

Jiang D, Morrison GAJ. 2003. The influence of long-term tracheostomy on speech and language development in children. *International Journal of Pediatric Otorhinolaryngology* 67: S217–S220.

Jusczyk PW. 1997. *The Discovery of Spoken Language*. Cambridge, MA: MIT Press.

Kelly MH. 1992. Using sound to solve syntactic problems: The role of phonology in grammatical category assignments. *Psychological Review* 99: 349–64.

Kingston J, Diehl RL. 1994. Phonetic knowledge. *Language* 70: 419–54.

Kuhl PK. 1991. Human adults and human infants show a "perceptual magnet effect" for the prototypes of speech categories, monkeys do not. *Perception and Psychophysics* 50: 93–107.

Kuhl PK. 2000. A new view of language acquisition. *Proceedings of the National Academy of Sciences, USA* 97: 11850–7.

Kuhl PK. 2004. Early language acquisition: Cracking the speech code. *Nature Reviews Neuroscience* 5: 831–43.

Kuhl PK, Stevens E, Hayashi A, Deguchi T, Kiritani S, Iverson P. 2006. Infants show a facilitation effect for native language phonetic perception between 6 and 12 months. *Developmental Science* 9: F13–F21.

Kuhl PK, Williams KA, Lacerda F, Stevens KN, Lindblom B. 1992. Linguistic experience alters phonetic perception in infants by 6 months of age. *Science* 255: 606–8.

Lambacher SG, Martens WL, Kakehi K, Marasinghe CA, Molholt G. 2005. The effects of identification training on the identification and production of American English vowels by native speakers of Japanese. *Applied Psycholinguistics* 26: 227–47.

Lenneberg E. 1967. *Biological Foundations of Language*. New York: Wiley.

Lindblom B, Engstrand O. 1989. In what sense is speech quantal? *Journal of Phonetics* 17: 107–21.

Liu H-M, Kuhl PK, Tsao F-M. 2003. An association between mothers' speech clarity and infants' speech discrimination skills. *Developmental Science* 6: F1–F10.

Locke JL. 1983. *Phonological Acquisition and Change*. New York: Academic Press.

Locke JL. 1986. Speech perception and the emergent lexicon. In *Language Acquisition*, ed. P Fletcher, M Garman, pp. 240–50: Cambridge University Press.

Locke JL, Pearson DM. 1990. Linguistic significance of babbling—Evidence from a tracheostomized infant. *Journal of Child Language* 17: 1–16.

Macaluso E, Driver J. 2005. Multisensory spatial interactions: A window onto functional integration in the human brain. *Trends in Neuroscience* 28: 264–71.

Maddieson I. 1984. *Patterns of Sounds*. Cambridge, England: Cambridge University Press.

Maye J, Werker JF, Gerken L. 2002. Infant sensitivity to distributional information can affect phonetic discrimination. *Cognition* 82: B101–B111.

McCune L, Vihman MM. 2002. Early phonetic and lexical development: A productivity approach. *Journal of Speech, Language & Hearing Research* 44: 670–84.

McGurk H, MacDonald J. 1976. Hearing lips and seeing voices. *Nature* 264: 746–8.

Mehler J, Dupoux E. 1994. *What Infants Know*. Oxford: Blackwell.

Mehler J, Jusczyk P, Lambertz G, Halsted N, Bertoncini J, Amiel-Tison C. 1988. A precursor of language acquisition in young infants. *Cognition* 29: 143–78.

Moffitt AR. 1971. Consonant cue perception by 20-week-old to 24-week-old infants. *Child Development* 42: 717.

Moon C, Cooper RP, Fifer WP. 1993. 2-day-olds prefer their native language. *Infant Behavior & Development* 16: 495–500.

Morton J, Johnson MH. 1991. Conspec and conlern—A 2-process theory of infant face recognition. *Psychological Review* 98: 164–81.

Nazzi T, Bertoncini J, Mehler J. 1998. Language discrimination by newborns: Toward an understanding of the role of rhythm. *Journal of Experimental Psychology: Human Perception and Performance* 24: 756–66.

Numminen J, Curio G. 1999. Differential effects of overt, covert and replayed speech on vowel-evoked responses of the human auditory cortex. *Neuroscience Letters* 272: 29–32.

Oller DK. 1980. The emergence of the sounds of speech in infancy. In *Child Phonology, Vol. 1: Production*, ed. GY Komshian, J Kavanagh, C Ferguson, pp. 93–112. New York: Academic Press.

Oller DK. 2000. *The Emergence of the Speech Capacity*. Mahwah, NJ: Lawrence Erlbaum Associates.

Oller DK, Eilers RE. 1988. The role of audition in infant babbling. *Child-Development* 59: 441–9.

Oller DK, Eilers RE. 1998. Interpretive and methodological difficulties in evaluating babbling drift. *Revue Parole* 7/8: 147–64.

Oudeyer P-Y. 2005. The self-organization of speech sounds. *Journal of Theoretical Biology* 233: 435–49.

Paus T, Perry DW, Zatorre RJ, Worsley KJ, Evans AC. 1996. Modulation of cerebral blood flow in the human auditory cortex during speech: Role of motor-to-sensory discharges. *European Journal of Neuroscience* 8: 2236–46.

Perkell JS, Guenther FH, Lane H, Matthies ML, Stockmann E, Tiede M, Zandipour M. 2004. The distinctness of speakers' productions of vowel contracts in related to their discrimination of the contrasts. *Journal of the Acoustical Society of America* 116: 2338–44.

Polka L, Werker JF. 1994. Developmental changes in perception of nonnative vowel contrasts. *Journal of Experimental Psychology: Human Perception and Performance* 20: 421–35.

Pulvermüller F. 2001. Brain reflections of words and their meaning. *Trends in Cognitive Sciences* 5: 517–24.

Rizzolatti G, Arbib M. 1998. Language within our grasp. *Trends in Neuroscience* 21: 188–94.

Rizzolatti G, Craighero L. 2004. The mirror-neuron system. *Annual Review of Neuroscience* 27: 169–92.

Rizzolatti G, Fadiga L, Gallesea V, Fogassi L. 1996. Premotor cortex and the recognition of motor action. *Cognitive Brain Research* 3: 131–41.

Schyns PG, Goldstone RL, Thibaut J. 1998. The development of features in object concepts. *Behavioral and Brain Sciences* 21: 1–54.

Scott SK, Johnsrude IS. 2003. The neuroanatomical and functional organization of speech perception. *Trends in Neurosciences* 26: 100–7.

Sejnowski TJ. 1977. Storing covariance with nonlinearly interacting neurons. *Journal of Mathematical Biology* 4: 303–12.

Shillcock R, Kirby S, McDonald S, Brew C. 2001. *The relationship between form and meaning in the mental lexicon.* Unpublished manuscript.

Simmons WK, Barsalou LW. 2003. The similarity-in-topography principle: Reconciling theories of conceptual deficits. *Cognitive Neuropsychology* 20: 451–86.

Steels L. 2003. Evolving grounded communication for robots. *Trends in Cognitive Sciences* 7: 308–12.

Stevens KN. 1972. The quantal nature of speech: Evidence from articulatory-acoustic data. In *Human Communication: A Unified View*, ed. JEE David, PB Denes, pp. 51–66. New York: McGraw-Hill.

Stevens KN. 1989. On the quantal nature of speech. *Journal of Phonetics* 17: 3–45.

Stoel-Gammon C, Otomo K. 1986. Babbling development of hearing-impaired and normally hearing subjects. *Journal of Speech and Hearing Disorders* 51: 33–41.

Trehub SE. 1973. Infants' sensitivity to vowel and tonal contrasts. *Developmental Psychology* 9: 91–6.

Vihman MM. 1993. Variable paths to early word production. *Journal of Phonetics* 21: 61–82.

Vihman MM. 1996. *Phonological Development*. Oxford: Blackwell.

Vihman MM. 2002. The role of mirror neurons in the ontogeny of speech. In *Mirror Neurons and the Evolution of Brain and Language*, ed. MI Staminov, V Gallese, pp. 305–14. Amsterdam: John Benjamins.

Vihman MM, Miller R. 1988. Words and babble at the threshold of language acquisition. In *The Emergent Lexicon*, ed. MD Smith, JL Locke, pp. 151–83. New York: Academic Press.

Vihman MM, Nakai S. 2003. *Experimental Evidence for an Effect of Vocal Experience on Infant Speech Perception.* Paper presented at the Proceedings of the 15th International Congress of Phonetic Sciences, Barcelona.

Watkins KE, Strafella AP, Paus T. 2003. Seeing and hearing speech excites the motor system involved in speech production. *Neuropsychologia* 41: 989–94.

Werker JF, Tees RC. 1984. Cross-language speech-perception—Evidence for perceptual reorganization during the 1st year of life. *Infant Behavior and Development* 7: 49–63.

Westermann G. 2001. A model of perceptual change by domain integration. In *Proceedings of the 23rd Annual Conference of the Cognitive Science Society*, ed. JD Moore, K Stenning, pp. 1100–5. Hillsdale, NJ: Lawrence Erlbaum Associates.

Westermann G, Miranda ER. 2004. A new model of sensorimotor coupling in the development of speech. *Brain and Language* 89: 393–400.

Whalen DH, Levitt AG, Wang Q. 1991. Intonational differences between the reduplicative babbling of French-learning and English-learning infants. *Journal of Child Language* 18: 501–16.

Wilson SM, Saygin AP, Sereno MI, Iacoboni M. 2004. Listening to speech activates motor areas involved in speech production. *Nature Neuroscience* 7: 701–2.

Zuidema W, Westermann G. 2003. Evolution of an optimal lexicon under constraints from embodiment. *Artificial Life* 9: 387–402.

Contributors

Stéphanie Barbu
UMR CNRS 6552 Ethologie-Evolution-Ecologie, Université de Rennes 1

Curt Burgess
Department of Psychology, University of California, Riverside

Josep Call
Max Planck Institute for Evolutionary Anthropology, Leipzig

Laurance Doyle
SETI Institute, Mountain View, California

Julia Fischer
Research Group Cognitive Ethology, German Primate Center, Göttingen

Michael H. Goldstein
Department of Psychology, Cornell University

Ulrike Griebel
The University of Memphis and The Konrad Lorenz Institute for Evolution and Cognition Research

Kurt Hammerschmidt
Research Group Cognitive Ethology, German Primate Center, Göttingen

Sean Hanser
Graduate Group in Ecology, University of California, Davis

Martine Hausberger
UMR CNRS 6552 Ethologie-Evolution-Ecologie, Université de Rennes 1

Laurence Henry
UMR CNRS 6552 Ethologie-Evolution-Ecologie, Université de Rennes 1

Allison B. Kaufman
Department of Neuroscience, University of California, Riverside

Stan Kuczaj
Marine Mammal Behavior and Cognition Group, Department of Psychology, University of Southern Mississippi

Robert F. Lachlan
Institute of Biology, Leiden University

Brian MacWhinney
Department of Psychology, Carnegie Mellon University

Radhika Makecha
Marine Mammal Behavior and Cognition Group, Department of Psychology, University of Southern Mississippi

Brenda McCowan
Department of Population Health & Reproduction, School of Veterinary Medicine and California National Primate Research Center, University of California, Davis

D. Kimbrough Oller
School of Audiology and Speech–Language Pathology, The University of Memphis and The Konrad Lorenz Institute for Evolution and Cognition Research

Michael J. Owren
Department of Psychology, Georgia State University

Ronald J. Schusterman
Marine Biologist–Research and Adjunct Professor of Ocean Sciences University of California, Santa Cruz

Charles T. Snowdon
Department of Psychology, University of Wisconsin, Madison

Kim Sterelny
Philosophy Program, Victoria University of Wellington and RSSS, Australian National University

Benoît Testé
Laboratoire de Psychologie Sociale, Université de Rennes 2

Gert Westermann
Department of Psychology, Oxford Brookes University

Index

Acoustic analysis
  acoustic structure, 42–44, 48, 96, 98–104, 108, 112
  formant, 105, 106, 151, 333, 334, 336
  pitch, 48, 85, 95–98, 101, 104, 105, 108, 125, 148, 149, 157, 179, 180, 182, 262
  spectrographic, 58, 63, 66, 97, 130, 287, 316
Adaptation, 15, 74, 76, 111, 113, 125, 132, 140, 193, 198, 200, 201, 203, 215, 220, 224–226, 289, 311, 328–333, 336, 338, 341, 342. *See also* Natural selection
Addington, R., 84–86, 88
Aggression, 10–12, 23, 27, 28, 31, 32, 52, 71–73, 123, 126, 133, 142, 143, 151, 200, 244, 260, 297–300, 307, 312, 314, 315
Articulation, 101, 102, 146, 147, 148, 150–152, 158, 162, 172, 173, 180–183, 203, 264, 328, 331, 334, 335. *See also* Speech
Attention
  allocation of, 133
  joint, 132, 159, 160, 233
  pointing, 13, 33, 130, 159, 202–204, 233, 244–249, 330
  shared, 132
  states of, 239
Audience effects, 17, 25, 81, 122, 239, 249
Austin, J. L., 11–13, 72
Australopithecus, 194

Babbling
  canonical, 5, 99, 103, 150, 152, 162, 169, 170, 172, 175, 176–178, 181, 183–185, 187, 262, 267, 329, 332
  infant, 26, 34, 43, 80, 81, 98, 99, 103, 132, 133, 139, 162, 169, 170–173, 175, 176–179, 181–185, 187, 189, 191, 203, 204, 262, 263, 267, 270, 279, 291, 308, 327–334, 336, 338, 340, 341, 342, 343
  in the pygmy marmoset, 99, 263
Baldwin, J. M., 199, 255, 258, 259
Baptista, L., 20, 42, 268, 269, 308
Bat, 45, 48, 66, 80, 172
  greater spear-nosed, 80
  sac-winged, 172
Bates, E., 140, 158, 161, 163, 235, 236
Beau Geste hypothesis, 21
Beecher, M. D., 309, 312, 314
Bickerton, D., 5, 160, 161, 164, 193, 216, 222, 223
Bipedalism, 139, 193, 194, 197, 209
Bird, 42, 44, 268, 269, 320, 321
  African Grey parrot, 42, 266, 268, 305
  barnacle goose, 124
  bird song, 8, 18, 19, 20, 21, 79, 113, 122, 203, 263, 279, 306, 307–310, 315–318, 320, 322
    countersinging, 21, 312
    plastic song, 131
    subsong, 64, 130, 132, 143, 263
  black-capped chickadee, 125, 269, 311, 316
  budgerigar, 43
  chaffinch, 305–312, 315, 316
  cowbird, 42, 43, 46, 49, 123, 127, 268
  dove, 184, 321
  great tit, 308
  hand-raised, 60, 131
  kingbird, 72
  lesser skylark, 125
  nightingale, 125, 315
  oscine (songbird), 4, 8, 15, 18–21, 41, 42, 64, 72, 77, 88, 112, 122, 126, 129, 130–132, 143, 170, 199, 263, 271, 305–307, 313, 314, 315, 320, 342
  parrot, 4, 41, 42, 43, 50, 112, 132, 199, 263, 266, 268, 305, 320
  passerine, 43
  pigeon, 217
  robin, 125
  song sparrow, 311–315
  starling, 20, 111, 125, 126, 129–134, 305, 311, 316
  swamp sparrow, 313
  white-crowned sparrow, 42, 308, 309, 312, 313, 315, 317
  wood duck, 184
  wren, 20
Bonding, 25, 26, 34, 35, 113, 139, 193, 196–200, 209
Bootstrap, 170, 179, 287, 288, 289, 291

Brain
  brain stem, 95, 110, 156, 171, 177, 180, 185, 200, 201
  cerebellum, 86, 109
  cerebral cortex, 173
  cingulate cortex, 86, 87, 106, 107, 109, 173, 174
  corticobulbar, 173, 175, 177–179, 186, 187
  hippocampus, 86
  hypothalamus, 86, 106, 173
  imaging, 109
  limbic cortex, 109
  medial preoptic area, 86
  mirror neuron, 328, 335, 340, 341
  motor cortex, 109, 110, 173, 177, 201, 204, 334, 339
  neocortex, 174, 200, 201
  neuron, 279, 280, 334
  nucleus accumbens, 87, 106
  parietal cortex, 197, 198
  periaqueductal gray, 106, 110, 156, 173, 200
  plasticity, 41, 44, 46, 49, 63, 65, 77, 98, 110, 112, 122, 173, 195, 196, 199, 223, 224
  putamen, 86, 109
  substantia nigra, 86

Call, J., 5, 196, 233
Calls (nonspeech communicative vocalizations). *See also* Vocalization
  alarm, 9, 10, 12, 13, 16, 18, 27, 28, 48, 49, 73–76, 81, 83, 84, 88, 100, 110, 111, 124, 171, 174, 248, 270, 292
  contact, 11, 100, 296, 297, 300
  convergence of, 100, 101
  distress, 28, 50
  duets, 124
  food call, 46, 75, 76, 78, 80, 81, 84–86, 268
  mobbing call, 83
Camouflage, 15, 17, 18, 19, 30, 31, 32
Canalization, 184
Captivity, 52, 53, 60, 76, 80, 83, 126
Caregiver, 3, 15, 26, 99, 147, 178–185
Cephalopod
  cuttlefish, 32
  squid, 31, 32
Cetacean, 5, 21, 23, 24, 50, 65, 129, 233, 268, 270
  dolphin, 22–24, 45, 47, 48, 51, 54, 56, 65, 66, 80, 112, 259, 260, 261, 270, 271, 279, 282, 286–297, 300, 301
  whale, 19–22, 35, 47, 48, 51, 66, 256, 257, 270
Cheney, D., 44, 48, 49, 65, 73, 77, 81, 93, 94, 98, 111, 121, 123, 124, 172, 174, 199, 248, 262, 269, 270
Child
  abuse, 26
  deafness, 262, 328 (*see also* Deafness)
  development, 158, 163 (*see also* Ontogeny)
  language, 140, 204 (*see also* Language acquisition)
  learning, 233 (*see also* Learning)
  rearing, 207, 253

Chomsky, N., 47, 150, 153, 202, 206
Cognition
  categorization, 157, 198, 295, 339, 340
  conceptual space, 297
  planning, 10, 46, 122, 194, 206, 218
  social, 207
Communication
  acoustic, 30, 44, 112, 257
  affective, 13, 24, 45, 46, 48, 75–77, 83, 86, 103, 105, 109, 146, 151, 152, 174, 186, 268, 269, 309
  affiliative, 11, 22, 52, 73, 80, 147, 151, 298, 299, 307
  complexity, 9, 141, 281, 286
  context
    negative, 176
    positive, 176
    social, 11, 85, 130, 133, 257, 264
  deception in, 12, 15, 17–19, 21, 26–29, 31, 32, 35, 200, 217, 219, 220–222, 224, 245, 321
  face-to-face, 24, 146, 147, 182, 196, 199, 200, 202
  flexibility in, 4, 5, 7–9, 15, 17, 19, 27, 34, 35, 71, 87, 140, 203, 257, 258, 269, 270, 279, 280, 305, 318, 320 (*see also* Flexibility)
  functions of (usage), 12, 143, 146, 153, 154, 158, 307, 312, 318, 322
  gestural, 5, 14, 33, 34, 122, 140, 182, 196, 200–203, 209, 222, 223, 233, 235–238, 239, 242–245, 247, 248, 249, 287, 333, 336
    gestural origins of language, 33
  honest, 219, 221
  nonverbal, 102–104, 109, 110, 112, 127
  pragmatics of, 72, 76, 122, 218, 222–224
  rules of, 127, 134 (*see also* Rules)
  symbolic, 5, 204, 247, 281, 340 (*see also* Symbolization)
  triadic, 12, 72
Communication system features (design features or properties)
  arbitrarity, 45–49, 54, 59, 63, 174, 220, 226, 229, 340
  conventionality or conventionalization, 162, 164, 215, 314, 328
  discreteness, 224
  efficiency, 156, 314, 327
  expressivity, 146, 152
  imitation, 16, 17, 20, 21, 23, 29, 42, 45, 112, 113, 142, 143, 153, 182, 198, 199, 202, 203, 254, 258, 266, 271, 279, 305, 308, 313, 320, 333, 336, 338, 340
  intentionality, 13, 133, 139, 144, 228
  involuntarity, 10, 41, 46, 157, 174
  productivity, 201, 204, 206, 269
  recombinability, 164
  recursion, 47, 193, 201, 202, 204–209, 222, 223
  referentiality, 8, 12–14, 42, 49, 73–75, 99, 122, 160, 161, 163, 202, 209, 224, 226–228, 244, 245, 247, 248, 268, 281
  specialization, 10

voluntarity, 7, 44, 45, 46, 48, 53, 63, 66, 101, 109, 113, 139, 152, 153, 157, 158
Communication systems
modeling of, 5, 139, 149, 279, 280, 300, 301, 335, 340
natural logic in evolution of, 139, 141, 144, 145, 148, 152, 154–156, 158–161, 164, 201
Comparative research, 7, 16, 19, 33, 35, 41, 43, 45, 47, 65, 66, 93, 112, 121, 127, 139, 183, 233, 260, 311
Condillac, E. B. de, 329, 336
Conditioning, 42–47, 49, 50, 54–56, 63, 65, 76, 79, 87, 98, 107, 112, 173, 179, 307
affective, 76
classical, 46
instrumental (operant), 46
sexual, 86
vocal, 45, 46, 49, 54, 56, 59, 60, 63, 67, 173
Connectionism, 139, 328
Connotation, 14
Context sensitivity, 123, 134. *See also* Flexibility
Cooperation, 77, 216–219, 229, 236, 246, 258
Cortisol, 87, 105
Coupling of signal and function, 10, 328, 331, 333, 337. *See also* Stereotypy; Decoupling
Courtship, 10, 11, 16, 19–21, 23, 28, 29, 31, 32, 41, 75, 126, 133, 184, 200, 267
Crying, 13, 14, 102, 104, 109, 146, 169, 175, 178, 180–182, 329. *See also* Infancy, cry
Cultural trap hypothesis, 319, 321, 322

Darwin, C., 156, 227, 228, 229
Dawkins, R., 21, 51, 215, 217
Deacon, T., 25, 196, 239, 244
Deafness, 5, 33, 35, 95, 96, 164, 172, 182, 262, 267, 328, 329, 331. *See also* Hearing impairment
Decoupling of signal and function, 13, 14, 257. *See also* Flexibility, Coupling
Dialect, 42, 50, 131, 199, 225, 270, 308–311, 330
Dog, 76, 205, 207, 208, 261, 264
Dolphin, 22–24, 45, 47, 48, 51, 54, 56, 65, 66, 80, 112, 259–261, 270, 271, 279, 282, 286–297, 300, 301. *See also* Cetacean
Donald, M., 197, 201, 202
Dunbar, R., 23, 25, 34, 196, 216, 225, 229

Elman, J. L., 5, 164, 195
Elowson, M., 26, 75, 77, 80, 81, 84, 85, 88, 94, 98, 99, 101, 170, 172, 259, 263, 268, 269
Emergence, 3, 5, 16, 19, 102, 139, 144, 145, 148, 150, 151, 153, 161, 162, 164, 169–171, 175, 176, 178, 180–184, 187, 194, 196, 197, 200, 202, 203, 206, 233, 246, 253–258, 262, 327, 328, 338
Emergentism, 206, 297, 333, 335, 341
Emotion
affect, 13, 24, 45, 46, 48, 75–77, 83, 86, 103, 105, 109, 146, 151, 152, 174, 186, 268, 269, 309
aggression, 10–12, 23, 31, 32, 71, 73, 123, 133, 142, 143, 200, 260, 297–300
emotional expression, 99, 103, 107
expression of distress, 3, 4, 11, 14, 28, 50, 64, 179
facial expression, 24, 64, 146, 147, 151, 152, 177, 181, 182, 184, 195, 196, 198–200, 202, 217, 221, 236, 238, 239, 240–242, 248, 339, 342
fussing, 147
laughter, 23, 24, 50, 104, 169, 175, 181, 263
smile, 152
states of, 103, 104, 106, 107, 176, 248
Entropy, 283–289
Epigenesis, 158
Ethology, 4, 10, 24, 41, 71, 143, 307, 347
Evolution. *See also* Natural selection
adaptation, 170, 186
cultural, 124, 195, 209, 216, 222, 228, 229, 246, 249, 309–311, 315, 316–319, 321, 322, 327, 328, 338
gene-culture, 187
language, 4, 5, 15, 19, 33, 35, 47, 121, 122, 129, 139, 140, 142, 153, 158, 164, 197, 201, 206, 215–218, 222, 226, 233, 327, 328, 340, 342

Farabaugh, S. M., 122, 124, 269, 271
Feldman, M. W., 309, 317, 318, 321
Fernald, A., 24, 179
Fischer, J., 85
Fish, 30, 32, 53–55, 57, 59, 83, 248, 256, 293–295
Fitch, W. T. S., 43, 47, 93, 97, 106, 113, 153, 186, 202, 219
Fitness signals, 15, 19–21, 26, 143, 146. *See also* Signals
Flexibility in communication and cognition
cognitive, 169, 195, 254, 255, 257, 258
contextual, 7, 11, 17, 32, 35, 45, 75–78, 81, 85–88, 122, 127–132, 139, 141–147, 149, 151, 153, 155, 157, 159, 161, 163–167, 186, 187, 248, 279, 281, 282, 301, 312
functional, 7, 9, 12, 14, 16–19, 21, 23, 27, 29, 32, 33, 139, 142–146, 151, 154, 164, 279, 307, 308, 315, 320
signal, 7, 17–19, 21, 23, 25–27, 29, 32, 34, 139, 142–146, 148, 154, 307, 308, 315
vocal, 7, 8, 15, 25, 33, 110, 139, 144, 148, 152, 156, 269
FOXP2, 113, 206
Frog, 264, 307

Gallistel, C. R., 292, 301
Geary, D., 193, 199, 209
Genetics, 41, 42, 44, 79, 80, 100, 101, 111, 158, 177, 309, 321
Gesture combinations, 235, 237, 238, 247. *See also* Communication, gestural
Ghiglione, R., 8, 123, 124, 128
Gibson, J. J., 27, 203
Goldstein, M., 43, 49, 267
Goodall, J., 200

Gould, S. J., 195
Grammar, 5, 101, 122, 140, 142, 154, 160–164, 186, 201, 203, 206, 208, 217, 221, 222, 224, 225, 226, 247, 264, 281, 292, 342
Griebel, U., 253, 261, 263, 267
Group size in human evolution, 25, 34, 35, 196, 269

Hamilton, W. D., 339
Hammerschmidt, K., 44, 172
Hausberger, M., 19, 20, 24, 56, 110, 268, 270
Hauser, M., 47, 76, 84, 93, 106, 111, 125, 129, 150, 153, 173, 174, 186, 193, 202
Hearing impairment, 102–104, 171, 175, 176, 181, 185, 187
Hebb, D. O., 334
Hockett, C., 9, 194, 197, 203, 204
Hominin (or Hominid) evolution, 4, 5, 7, 15, 16, 25, 26, 33–35, 139–141, 144, 156, 158–164, 193, 194, 196–201, 206, 217, 219, 220, 222, 229
Homo erectus, 195, 197, 201–203, 209
Homology, 121
Homoplasy, 121
Hunting, 29, 31, 74, 122, 199, 218, 257
Hurford, J., 341

Illocutionary force,11–15, 28, 72, 163, 222. *See also* Perlocutionary force; Meaning; Pragmatics of communication
Imitation, 16, 17, 20, 21, 23, 29, 42, 45, 112, 113, 142, 143, 153, 182, 198, 199, 202, 203, 254, 258, 266, 271, 279, 305, 308, 313, 320, 321, 333, 336, 338, 340
Infancy, 14, 85, 129, 144, 145, 184, 187, 194, 195, 328. *See also* Mimesis and Mimicry
- cry, 3, 103
- development, 19, 141, 164
- monkeys, 48, 94, 267
- vocalization, 16, 26, 102, 103, 139, 145, 148, 149, 152, 181, 182, 289 (*see also* Babbling)
  - coo, 73, 94, 96, 97, 98, 104, 107, 173–175, 176, 205
  - gooing, 147, 148, 150, 152, 158, 169, 176
  - growl, 49, 55, 57, 58, 61, 63, 149, 152
  - nonlinguistic, 170, 171, 175–181, 183, 185, 187
  - quasivowel, 146, 147, 157
  - speech-like, 182
  - squeal, 13, 84, 85, 105, 149, 152, 175

Information theory, 279, 282, 283, 286, 291, 292
Infraphonology, 142, 145–148, 150, 153, 155, 157, 158
Infrasemiotics, 142, 143, 145, 151, 153
Innateness, 24, 27, 32, 47, 63, 65, 71, 74, 75, 87, 93, 98, 100, 103, 109, 110, 112, 121, 154, 155, 170, 172, 175, 178, 186, 187
Innovation, 77, 309. *See also* Flexibility
Insect
- bee, 311, 316
- firefly, 4, 28, 29

Instinct, 42, 195
Intentionality in communication, 7, 44, 45–48, 53, 63, 66, 101, 107, 109, 113, 139, 152, 153, 157, 158, 170, 171, 173, 175, 177–181, 183, 185–187. *See also* Communication system features
Invertebrate, 19, 28

Jackendoff, R., 93, 164, 222, 223
Jürgens, U., 24, 95, 96, 101, 104–110, 173, 174, 177, 200

Kent, R., 158, 181, 262
Killer whale, 21–23, 25, 125, 256, 264, 270, 271, 298, 299. *See also* Cetacean
Koopmans-van Beinum, F., 102, 146, 176, 262
Kroodsma, D., 19, 310, 320
Kuczaj, S., 23, 127, 133, 135, 157, 165
Kuhl, P. K., 122, 157, 170, 179, 182, 186, 263, 330, 331

Language
- acquisition of, 227, 289
- adjectives, 247
- change in, 223
- consonants, 150, 162, 163, 169, 176, 177, 182, 201, 329, 332, 343
- gossip, 196, 216
- grammar, 217, 221, 222
- lexical characteristics of, 142, 150, 162, 163, 203–206, 208, 209, 222, 224–229, 282, 295, 297, 300, 338, 340
- nouns, 208, 247
- phonology, 43, 99, 112, 150, 169, 170, 179, 203, 204, 262, 283, 287, 327, 328, 330, 331, 338, 340
- recursion in, 47, 193, 201, 202, 204–209, 222, 223
- sentences, 14, 24, 26, 128, 205, 207–209, 225, 293, 294
- sign language, 14, 33, 34, 83, 94, 98, 128, 161, 196, 220, 226, 236, 244, 247–249, 262
- syntax, 5, 101, 142, 154, 160–164, 203, 208, 221, 222, 224–226, 247, 264, 281–283, 292, 342
- universals of language, 132, 329
- vowels, 5, 13, 50, 51, 58, 59, 149, 150, 152, 155, 169, 170, 176, 177, 181, 182, 201, 265, 279, 328–334, 336–338, 343

Larson, C., 98, 108
Larynx, 50, 52, 59, 61, 63, 66, 98, 101, 106, 108, 109, 112, 145, 148, 171, 172, 176–181, 183, 185, 187, 201
Laughter, 23, 24, 50, 104, 169, 175, 181, 263
Learning
- adult modeling, 129, 132
- contingency, 43, 45, 46, 48, 52, 65, 183
- cultural, 246
- explicit, 124
- inhibition, 47, 124

item-based, 140, 204
learnability, 222, 223
motor, 175
origin of, 322
reinforcement, 25, 42, 43, 45, 46, 48, 53, 58, 63, 65, 181
scaffolding, 144, 218
social, 65, 80, 83, 180, 181, 183, 216
song, 43, 64, 79, 112, 113, 279, 306, 308, 309, 312, 313, 315, 317, 318, 321, 322
statistical, 87, 199, 202
trial and error, 181, 254, 256
vocal, 33, 41–51, 53–57, 59, 61, 63, 65, 67, 69, 93, 100, 112, 132, 268–270, 320, 321
Lenneberg, E., 47, 262, 328, 331
Lewontin, R. C., 215, 229
Lexicon, 142, 150, 162, 163, 203–206, 208, 209, 222, 224–229, 282, 295, 297, 300, 338, 340
Lieberman, P., 5, 103, 106, 173, 177, 204, 262
Locke, J. L., 5, 26, 34, 99, 101, 122, 123, 127, 129, 133, 181, 200, 327–329
Lorenz, K., 4, 5, 10, 35, 67, 71, 74, 143, 301, 347

MacNeilage, P. F., 5, 151, 162, 183, 201
MacWhinney, B., 154, 158, 162
Mammal, 7, 19, 21, 27, 35, 41, 43–55, 57, 59, 61, 63–67, 69, 106, 109, 110, 112, 113, 122, 143, 170, 172, 187, 233, 245, 301, 347
Manual dexterity, 139, 193, 195, 209, 233
Marler, P., 20, 42, 73, 75, 81, 101, 122, 173, 239, 263, 269, 305, 306, 309, 316, 317
Maturation, 11, 17, 43, 56, 64, 65, 95, 98, 103, 111, 112, 131, 172, 177, 183, 253
Maynard Smith, J., 9, 10, 15
McCowan, B., 22, 23, 111, 258, 262, 270, 271
Meaning, 13, 14, 66, 71, 72, 75, 76, 93, 113, 144, 150, 204, 222, 223, 238, 259, 287, 328, 340. *See also* Semantics
Memes, 209
Mental retardation, 184
Mimesis and Mimicry, 16, 17, 20, 27–29, 47, 50, 111, 197, 202, 203, 209, 256, 270, 305, 316, 321. *See also* Imitation
Minimalism, 206
Mirror neuron, 328, 335, 340, 341
Modeling (computational), 5, 139, 149, 279, 280, 292, 300, 301, 335, 340. *See also* Neural networks
hyperspace analog approach, 279, 282, 292, 294, 296, 298, 299
of internal complexity in communication, 286
Motivation, 42, 75, 86, 87, 157, 180, 247, 259
Music, 335

Natural logic, 43, 78, 79, 126, 129, 139, 141, 144, 145, 148, 152, 154–156, 158, 159–161, 164, 194, 201, 215, 235, 248, 257, 327, 329, 336, 338, 341
Natural selection
adaptation, 15, 74, 76, 111, 113, 125, 132, 140, 193, 198, 200, 201, 203, 215, 220, 224–226, 289, 311, 328–333, 336, 338, 341, 342
arms race, 15, 21, 27, 199
for variability, 9, 16, 26, 34
niche construction in, 215, 216
niche creation in, 140
sexual selection, 18, 19, 21, 29, 32, 35, 267
intersexual, 19, 20, 21
intrasexual, 19, 21
survival, 4, 87, 217, 253, 254, 255
Neanderthal, 206
Neoteny, 139, 193, 195–199, 209
Neural networks (artificial neural networks), 149, 204, 328, 331, 332, 339, 341
feedback in, 42, 43, 48, 55, 57, 67, 88, 99, 101, 104, 108, 109, 133, 157, 158, 170, 179, 184, 185, 230, 311, 314, 339
recurrent networks, 300
self-organizing maps, 279
Nottebohm, F., 20, 41, 64, 263
Nowicki, S., 307, 313, 314

Oller, D. K., 80, 99, 102, 103, 169, 175, 176, 177, 181, 184, 185, 199, 201, 253, 257, 262, 263, 267, 329
Ontogeny, 41, 43, 48, 49, 63, 64, 93, 94–96, 101–103, 112, 141, 170, 175, 176, 178, 183, 184, 186, 201, 236, 243, 253–255, 257–259, 261, 263, 267, 268, 270. *See also* Child, development
Overgeneralization, 85
Owings, D. H., 73, 74
Owren, M., 10, 44, 46, 76, 94, 111, 238, 248, 268, 342

Papoušek, M., 24, 26, 178, 180, 182
Peer interaction, 267. *See also* Sociality
Pepperberg, I. M., 14, 42, 133, 266, 268
Perception, 56, 73, 123, 128, 141, 149, 170, 174, 177, 198, 203, 279, 313, 328, 331–333, 335, 338–343
speech, 177, 328, 330–332, 342
Perlocutionary force, 11, 12, 72
Perspective taking, 140, 200, 203, 206, 207, 209, 239, 243. *See also* Illocutionary force; Meaning; Pragmatics of Communication
Pharynx, 52, 59–61, 181
Phonation, 102, 109, 134, 145–153, 155–158, 176, 181, 201
Phylogeny, 47–49, 105, 109, 112, 113, 121, 122, 141, 201, 261, 320, 322, 342
Piaget, J., 237, 254–256, 264
Pinker, S., 5, 154
Pinniped, 7, 26, 41, 49, 50–54, 56, 57, 59, 60, 63, 64–66
Plastic song, 131. *See also* Bird, bird song
Plasticity, 41, 44, 46, 49, 63, 65, 77, 98, 110, 112, 122, 173, 195, 196, 199, 223, 224

Play, 5, 8, 11, 15, 20, 23, 33, 35, 45, 57, 63–65, 93, 101, 107, 111, 124, 127, 131, 133, 146, 154, 157, 170, 173, 175, 181, 182, 188, 200, 229, 233, 237, 238, 247, 253–271, 273, 275, 277, 297–299, 314. *See also* Flexibility
- with language, 262, 264–267, 271, 332
- with vocalization, 13, 132, 143, 146, 149, 152, 233

Ploog, D., 24, 106, 108, 177, 201
Pointing, 13, 33, 130, 159, 202–204, 233, 244–249, 330
Population, 20, 75, 79, 80, 98, 100, 101, 126, 195, 206, 209, 217, 219, 309, 310, 311, 316–318, 320, 321, 333, 334, 336, 347
Posture, 42, 46, 61, 127, 130, 196, 237, 238
Povinelli, D., 159, 198, 240, 241, 245
Pragmatics of communication, 72, 122, 222, 223. *See also* Illocutionary force; Perlocutionary force
Predator, 3, 10, 13, 15, 16, 21, 27–31, 72–76, 81–84, 88, 111, 171, 172, 201, 215, 220, 248, 257
Preformationism, 154
Primate
- ape, 14, 33, 34, 94, 121, 122, 161, 174, 185, 186, 194, 198, 199, 201, 233, 235–249, 251, 262
- baboon, 46, 85, 105, 111, 197, 238, 248
- bonobo, 185, 196, 197, 238, 240, 241, 243
- Campbell monkey, 124, 125
- capuchin monkey, 74, 84, 199, 245
- chimpanzee, 23–25, 74, 75, 79, 84, 93, 100, 113, 173, 185, 196, 197, 199, 205, 219, 235, 237–249, 259, 260, 262, 269
  - pant-hoot, 100
- Diana monkey, 270
- gibbon, 94, 173, 196, 197
- gorilla, 185, 196, 243
- Howler monkey, 106
- lemur, 74, 105, 245
- macaque, 46, 73, 75, 79, 81, 94, 96, 97–100, 105, 108, 111, 173, 200, 238, 242, 245, 248
- marmoset, 26, 73, 77, 79–87, 95, 98, 99, 125, 172, 263, 268
- nonhuman, 7, 8, 12, 14, 16, 33, 34, 44, 56, 65, 66, 73, 79, 81, 87, 88, 93, 98, 99, 101–104, 107, 109, 110, 112, 121, 122, 124, 129, 141, 144, 147, 148, 150, 152, 156, 157, 162, 169, 205, 236, 245, 248, 268, 269
- orangutan, 185, 196, 198, 237, 238, 240, 241, 243, 245, 249
- pygmy marmoset, 26, 73, 77, 79–87, 95, 98, 99, 125, 172, 263, 268
  - babbling, 99
- rhesus monkey, 75, 81, 94, 96–98, 107, 109, 111, 125, 173, 174, 241, 245, 248, 259, 268
- spider monkey, 75, 84
- squirrel monkey, 81, 94–96, 104, 105–109, 125, 173, 259, 290
- tamarin, 73, 75–79, 83–85, 109, 248, 259, 263, 268
- vervet monkey, 48, 73, 75, 81, 83, 110, 111, 123–125, 171, 172, 174, 226, 227, 270
  - alarm calls, 75, 83, 171

Problem solving, 158, 235, 257
Protean behavior, 31, 32, 35. *See also* Flexibility
Protolanguage, 160, 161, 197, 217, 219–223, 226, 229
Prototype, 330, 332, 336, 338
- perceptual and motor, 330, 336

Psychology, 123, 183, 222, 347, 348

Reflex, 64, 71, 81
Releasing stimulus, 71. *See also* Innateness
Repertoire of signals, 8, 10, 18, 20–22, 24, 32, 42, 50, 57, 60, 63, 72, 73, 80, 94, 99, 101, 113, 126, 150, 153, 154, 156, 172, 176, 184, 236, 237, 243, 244, 247, 255, 262, 263, 266, 268, 270, 287, 289–292, 307, 309, 310, 312, 313, 316, 320, 327–329, 331, 332, 336, 338, 339, 341, 342. *See also* Vocal
Repetition in signaling, 26, 51, 59, 80, 82, 99, 105, 112, 148–150, 238, 254, 255, 262, 263, 266, 285, 289, 291
Riebel, K., 306, 307, 313
Ritualization of signals, 4, 10, 24, 41, 128, 148, 243
Rizzolatti, G., 200, 340
Rules
- communication, 127, 134
- contract-based, 123
- of conversation, 127

Sacks, H., 127, 128
Saussure, F. de, 142, 340
Savage-Rumbaugh, S., 14, 186, 196, 244, 245
Scaffolding in learning, 139, 144, 170, 177, 183–185, 187, 218. *See also* Learning
Schusterman, R. J., 7, 26
Searcy, W. A., 307, 311, 313
Self-organization, 149, 158, 178, 204, 234, 279
Semantics, 160, 163, 164, 171, 186, 203–205, 223, 292–294, 307, 342
Semiotics, 193. *See also* Infrasemiotics
Sensorimotor integration, 333, 335, 339, 340, 342
Seyfarth, R., 44, 48, 49, 65, 73, 77, 81, 93, 94, 98, 111, 121, 123, 124, 171, 172, 174, 199, 248, 262, 269, 270
Sign language, 14, 33, 34, 244, 247, 249
Signals
- aggressive, 11, 28, 142, 151
- communicative display, 3, 4, 18, 20, 21, 31, 35, 42, 51, 53, 57, 60, 63, 72, 88, 124, 145, 147, 148, 196, 236, 244, 247, 260, 266, 271, 307, 316
- communicative value of, 163
- conventional, 314
- coupling of signal and function, 10, 328, 331, 333, 337

cues, 3, 10, 15, 18, 24, 27, 43, 46, 47, 51, 54, 56–59, 63, 64, 107, 127, 131, 173, 204, 208, 218, 238, 245, 259, 268, 315, 321
decoupling of signal from function, 13, 14, 257
fixed, 4, 10–14, 17, 18, 24, 26, 32, 143, 144, 146, 151
gradations of, 10, 11, 292
honest signaling, 15, 52, 217–219, 313
natural signs, 217, 226, 227
olfactory, 76, 86
playback of, 51, 73–75, 94, 100, 101, 105, 130, 307, 311, 321
referential, 13, 75, 160, 161, 224, 226, 227, 244, 245, 268, 281
requests, 235, 238, 241, 242, 246–249
ritualization of, 4, 10, 24, 148, 243
territorial, 19, 20, 21, 52, 53, 72, 78, 125, 199, 307, 312, 314, 319
threat, 3, 4, 11, 24, 25, 51, 52, 63, 79, 80, 82, 83, 143, 174, 217, 220, 221, 260
warning, 4, 11–14, 18, 27, 143
Skinner, B. F., 41, 46, 64
Slater, P. J. B., 45, 47–49, 65, 79, 93, 122, 131, 269, 271, 306, 309–311, 315–318, 321, 322
Smith, W. J., 71, 72, 76
Snowdon, C. T., 25, 26, 34, 56, 94, 98, 99, 101, 122, 125, 129, 147, 148, 248, 259, 263, 268, 269, 270
Sociality and social behavior
alliances, 81, 209
attachment, 200, 208
bonding, 23, 25, 26, 34, 35, 113, 139, 193, 196–200, 209, 267
caregiving, 170
coalition building, 23
cooperation, 77, 216–219, 229, 236, 246, 258
cultural transmission, 228, 328
culture, 122, 124, 128, 169, 195, 203, 209, 216, 222, 228, 229, 246, 249, 309–311, 315–319, 321, 322, 327, 328, 338
food sharing, 77, 78
grooming, 23, 24, 25, 34, 35, 196, 237
group size, 269
of animals, 17, 266
peer interaction, 267
social construction, 207
social interaction, 12, 22, 25, 32, 42, 64, 80, 127, 131, 132, 148, 169, 171, 181, 183, 216, 218, 235, 258, 267, 268, 307
social isolation, 105, 130, 133
social learning, 65, 80, 83, 180, 181, 183, 216
social organization, 122, 132, 196, 227
social regulation, 125
social status, 77, 85, 87, 128, 298
Speech
articulation, 5, 50, 66, 101, 102, 108, 145–148, 150–152, 158, 162, 172, 173, 180–183, 203, 204, 206, 264, 327–329, 331–336, 338, 341, 343
consonants, 150, 162, 163, 169, 176, 177, 182, 201, 329, 332, 343
infant-directed, 179, 332
primitive articulation in speech development, 158, 169, 176
prosody, 105
vowels, 5, 13, 50, 51, 58, 59, 149, 150, 152, 155, 169, 170, 176, 177, 181, 182, 201, 265, 279, 328–334, 336–338, 343
Stages of development, 41, 131, 155–157, 162, 178. *See also* Child
Stages of evolution, 10, 161
Stark, R. E., 26, 143, 145, 146, 262
Stereotopy of signals, 3, 4, 10, 11, 14, 16, 24–26, 28, 32, 35, 51, 57, 59, 64, 142, 143, 148, 152, 157, 248, 291, 305, 316. *See also* Flexibility
Struhsaker, T., 16, 75
Subsong, 64, 130, 132, 143, 263. *See also* Bird, bird song
Syllables in communication systems, 20, 42, 43, 99, 108, 145, 148, 150, 151, 153, 162–164, 169, 176, 201, 262, 305, 312, 329, 338
Symbolization, 94, 140, 185, 217, 220, 226–229, 247, 281, 289. *See also* Meaning

Temporal organization, 124, 134
Thelen, E., 149, 157, 291
Theory of mind, 159
Thorndike, E., 46
Tinbergen, N., 4, 10, 71, 72, 74, 143
Tomasello, M., 5, 94, 140, 158, 159, 199, 204, 237–240, 242, 243, 245, 246, 249
Topographic maps, 328, 339, 340, 342
Tracheostomy, 329
Turn-taking in communication, 8, 82, 125, 127, 128, 129, 130, 132, 134, 182, 258, 260
Tyack, P., 19, 20, 22, 48, 50–52, 60, 65, 80, 270, 271

Usage in communication, 47, 54, 65, 75, 78, 93, 98, 110, 269. *See also* Communication, functions of

Variability (vocal), 34, 35. *See also* Flexibility
Versatility, 279, 315, 316, 318, 320, 321, 322. *See also* Flexibility
Vertebrates, 28, 106
Vihman, M. M., 99, 262, 328–330, 332, 338
Vocal
accommodation, 101, 112
canonical stage of vocal development, 150, 158
category formation, 149
development, 44, 56, 57, 60, 79, 80, 93, 94, 95, 98, 122, 132, 139, 142, 145, 152–155, 170, 171, 179, 181, 183–185, 187, 257, 263, 268, 320
expansion stage of vocal development, 148, 176
folds, 95, 96, 97, 106, 108, 173
imitation, 113, 153, 164, 203, 271, 320 (*see also* Imitation)

Vocal (cont.)
interaction, 24, 26, 123–125, 129, 130, 131, 132, 134, 294
face-to-face, 24, 147, 182, 199, 200, 202
song-matching, 126, 130
production of vocalization, 7, 25, 42, 44, 45, 55, 63, 65, 66, 79, 81, 93, 94, 98, 106, 108, 110, 112, 113, 122, 132, 141, 170–175, 177–180, 185–187, 188, 262, 268, 320–322
spontaneous, 60, 145–147
raw material of vocal categories, 145–148
repertoire, 51, 56, 77, 94, 95, 104, 121, 146, 268
signals, 10, 13, 16, 44, 50–57, 78, 121, 122, 143, 151, 187, 188, 320 (*see also* Signals)
tract, 43, 97, 106, 147, 149, 150, 173, 181, 204
usage, 21, 34, 45, 49, 65, 66, 122 (*see also* Communication)
Vocalization, 7, 8, 14, 16, 22, 25, 26, 33–35, 54, 55, 58, 72, 75, 80, 82, 85, 86, 94, 98, 99, 102–04, 106–111, 134, 139, 145–149, 152, 153, 157, 164, 170, 171, 174, 177, 178, 180, 183–185, 187, 199, 200–202, 209, 233, 257, 269, 288, 289, 291, 296–299
prelinguistic, 169, 170, 175, 176, 180, 187
Voice, 262, 264. *See also* Phonation
Vygotsky, L., 195, 207

Waddington, C. H., 5, 101, 177, 183, 193, 227, 281
West, M. J., 20, 42, 43, 49, 122, 123, 127, 128, 267, 268, 307
Westermann, G., 179

Zahavi, A., 313
Zuberbühler, K., 74, 94, 111, 173, 248, 268, 269, 270